Images and Enterprise

JOHNS HOPKINS STUDIES
IN THE HISTORY OF TECHNOLOGY

THE MECHANICAL ENGINEER IN AMERICA, 1830–1910
Professional Cultures in Conflict
Monte Calvert

AMERICAN LOCOMOTIVES
An Engineering History, 1830–1880
John H. White, Jr.

ELMER SPERRY
Inventor and Engineer
Thomas Parke Hughes

PHILADELPHIA'S PHILOSOPHER MECHANICS
A History of the Franklin Institute, 1824–1865
Bruce Sinclair

IMAGES AND ENTERPRISE
Technology and the American Photographic Industry, 1839–1925
Reese V. Jenkins

Images and

Technology and the American Photographic Industry 1839 to 1925

Reese V. Jenkins

Enterprise

The Johns Hopkins University Press · Baltimore & London

Manufactured in the United States of America

The Johns Hopkins University Press, Baltimore, Maryland 21218
The Johns Hopkins University Press Ltd., London

Library of Congress Catalog Card Number 75-11348
ISBN 0-8018-1588-6

Library of Congress Cataloging in Publication data
will be found on the last printed page of this book.

To the Memory
of My Father,
John Thomas Jenkins
(1909-65)

Contents

Preface xiii

Acknowledgments xv

Abbreviations of Frequently Cited Sources xvii

Introduction/Framework of Industrial Development 2

I Daguerreotype Period 1839 to 1855

1/The Daguerreotype Period 10

II Collodion Period 1855 to 1880

2/Glass Plate and Paper Photography 36

III Gelatin Plate Period 1880 to 1895

3/The Gelatin Revolution, 1880–95 66

4/Origins of the Roll Film System, 1884–89 96

5/The Development of Celluloid Roll Film, 1887–95 122

6/Modifications of the Roll Film System, 1890–95 134

7/Responses of the Traditional Leadership to the Technical Revolution, 1880–95 160

IV Period of the Amateur Roll Film System 1895 to 1909

8/The Corporate, Industrial, and Technological Setting 172

9/Horizontal Integration: Eastman Kodak 188

10/Vertical Integration: Eastman Kodak 236

11/Horizontal and Vertical Integration: Anthony and Scovill 246

12/The Emergence of the Cinematographic Industry 258

V Period of Silent Cinematography 1909 to 1925

13/The Cinematographic Industry: Integration and Innovation 282

14/Preservation of the Corporate and Industrial Structure 300

Conclusion 340

Appendix/Technological Diffusion 347

Primary Sources of Information 353

Index 359

Tables

1.1 Profits of Scovill New York Store, 1846–51 *21*

6.1 Blair Camera Company Financial Data, 1891–97 *144*

6.2 Eastman Companies Financial Data, 1886–95 *157*

8.1 Growth of Eastman Kodak Sales Compared to Growth of U.S. Photographic Industry, All U.S. Manufacturing, and U.S. Population, 1889–1909 *178*

8.2 U.S. Exports of Photographic Goods, 1894–1912 *179*

9.1 U.S. Photographic Paper Sales, 1895–99 *193*

9.2 Profits and Acquisition Price of Companies Combined in General Aristo Company, 1898 *200*

9.3 U.S. Photographic Paper Sales, 1902–11 *205*

9.4 Sales and Market Share of Photographic Paper Manufacturers, 1908 *206*

9.5 Comparison of Artura and American Aristotype Paper Sales, 1904–9 *206*

9.6 Eastman Kodak Roll Film Camera Sales, 1892–1900 *210*

9.7 Sales by Century Camera, Folmer & Schwing, and Rochester Optical Divisions, Eastman Kodak Company, 1903–9 *217*

9.8 Professional Photographers Using Various Brands of Dry Plates, 1900–1905 *227*

9.9 Net Sales of Leading American Dry Plate Manufacturers, 1902–10 *228*

12.1 Eastman Kodak Film Sales in the United States, 1897–1909 *279*

13.1 Leading Independent Motion Picture Companies Ranked by Contractual Consumption of Film, 1911 *288*

13.2 U.S. Exports and Imports of Cinematographic Film, 1913–20 *296*

14.1 Ansco Company Net Sales, 1910–12 *331*

14.2 Payments to Goodwin Film & Camera Company for Production Rights under Goodwin Patent *334*

Preface

The history of photography and of photographic technology has been treated in articles and books since the middle of the nineteenth century in a literature that ranges from competent and scholarly to anecdotal and popular to highly inaccurate and trivial. Among the most reliable general histories are Helmut and Alison Gernsheim's *History of Photography* (1955, 1969), Beaumont Newhall's *History of Photography from 1839 to the Present Day* (1964), Josef Eder's *History of Photography* (1945), Eaton Lothrop's, *A Century of Cameras* (1973), and Robert Taft's *Photography and the American Scene* (1938). However, with the exception of Carl Ackerman's *George Eastman* (1930) and C. E. Kenneth Mees's *From Dry Plates to Ektachrome Film* (1961), which give substantial data on certain limited aspects of the Eastman Kodak Company, the history of the American photographic industry, its leading companies, and its most notable personalities have received scant attention. Historians have lavished attention upon individual firms and written industrial histories of the textile, railroad, iron and steel, petroleum, chemical, electrical, automobile, aircraft, and machine tool industries—among others—but they have neglected the photographic industry, perhaps because it was not a leading sector of the economy.

Despite its relatively peripheral position in the economy, the industry reflected in its changes during the nineteenth and early twentieth centuries many important features of the growth of American business. Specifically, because of the technical character of its products, the historical study of the industry permits an examination of the changing relationship of technology and science to business during the period in which the United States moved from a predominantly rural and agrarian society to a predominantly urban and industrial society. Moreover, it permits an exploration, at least within the American context, of the emergence of America into a position of world leadership in an industry that European firms had previously dominated and that was linked closely to the internationally dominant German optical and fine chemicals cartels. This study will focus on the transformation in the character of the American photographic industry from 1839 to 1925 and will seek to identify some of the major forces shaping its development, directing particular attention to the role played by technology in the formulation of business strategy, the organization of enterprise, and the structure of the industry.

The study begins in 1839, the year in which the first commercial photographic process was introduced in the United States. The choice of termination date represents a compromise between the historical setting, the internal logic of the framework of the study, and the availability of manuscripts and other historical data. The silent cinematographic era passed from dominance in the late 1920s; the ownership and management of Ansco changed in 1927; and the formal retirement of George Eastman from active management of Eastman Kodak came in 1925. Therefore, the comprehensive examination of the industry ends in 1925; but, some attention is given to cinematography and to Ansco in the late years of the decade.

Some aspects of the photographic industry from 1839 to 1925 have been treated in great detail, some in a cursory manner, and others not at all. The basic principle of selection and emphasis has been general economic or technological importance, compromised only by the unavailability of primary sources. The period prior to 1880 is treated in some detail—particularly the roles played by Scovill and Anthony—but the period 1880–1910 is given much greater attention because of the important series of technological, marketing, and organizational changes that occurred with the entry of the mass consumer into photography. This emphasis also reflects the richness of primary and other historical sources available for this period. The cinematographic industry is examined in a much more cursory fashion than it deserves, but to have done otherwise would have distorted the framework of the study and greatly extended its length. For similar reasons the role that psychological need and the promotion and advertising of photography played in the creation of demand in the photographic industry has received considerably less attention than might otherwise have been justified. The areas relating to photoengraving and printing have arbitrarily been excluded from the study. The primary focus after 1880 justifiably falls on the Eastman companies, which, from the middle 1890s, were economically the most important firms in the industry, not only in the United States but in the world.

Acknowledgments

I wish to acknowledge the assistance and support of the many individuals and institutions that contributed to this study. Hayden V. White initially excited my interest in history and the history of science when I was an undergraduate. Aaron J. Ihde first directed my attention to the photographic industry and generously encouraged and guided my doctoral dissertation on the interrelations of science, technology, and the photographic industry in the Atlantic community in the nineteenth century. Also at that time Rondo Cameron stimulated my interest in the economic aspects of technical and industrial development. Although this study is the outgrowth of a doctoral dissertation, it differs substantially in scope and focus from the earlier work as a consequence of the influence of Ralph Hidy, James Baughman, and Ralph Hower.

Of critical significance to this study was the generous cooperation provided by the Eastman Kodak Company, which made available the extensive collection of George Eastman correspondence and other historical materials and artifacts. I wish to acknowledge particularly the assistance of William S. Vaughn, Albert K. Chapman, Marion B. Folsom, Thomas Robertson, Delores Stover, Donald C. Ryon, and Gail Freckleton.

Several institutions have contributed substantial financial support to this study. I am particularly appreciative to the Division of Humanities of Case Western Reserve University for its sustained support and to Harvard University and the Newcomen Society of North America for the opportunity for study and research as Harvard-Newcomen Fellow at the Harvard Graduate School of

Business Administration for the year 1969–70. Significant financial support was also contributed by the National Science Foundation (GS-1726), Northern Illinois University, and the Wisconsin Alumni Research Foundation of the University of Wisconsin.

Scholars, librarians, and relatives of participants in the American photographic industry have been most generous with their time. I thank in particular Lawrence Bachmann, Thomas Barrow, George Basalla, Elsie H. Bishop, P. W. Bishop, Kenneth Carpenter, Florence Cornwall, Joel Eastman, James Forbes, Carrol L. Gensert, Thomas B. Greenslade, Stephen Goldfarb, Edward M. Hallett, Jr., Patsy R. Hatley, Virginia R. Hawley, Russell A. Hehr, Ruth W. Helmuth, Charles A. Hill, Helen Iverson, Thelma C. Jeffries, Carl B. Kaufmann, Rudolph Kingslake, James T. Lee, Jennie S. Levey, Robert Lovett, Pat McFarland, Stephen G. Mayti, Blake McKelvey, Albert W. Mentzer, Beaumont Newhall, Elmer S. Newman, Rudolph Schiller, Margaret J. Snider, Constance Stankrauff, and Raymond Stanley. I am appreciative of the personal courtesies and assistance given by Cheryl Beswick, Margaret Boulding, Douglas Corbin, Ruth L. Kohn, Foster and Queenie Mitchem, and Marc and Judy Zicari. I also wish to thank the staffs of the many libraries and research institutions I visited, most particularly those at the George Eastman House; the Gernsheim Photographic Collection, University of Texas; Baker Library, Harvard University; Butler Library, Columbia University; the National Archives; the Smithsonian Institution; the Missouri Historical Society; the Western Reserve Historical Society; and the Cleveland Public Library.

My colleagues in the Programs in History of Science and Technology and in American Studies at Case Western Reserve University, most notably Robert Schofield, Melvin Kranzberg, and Morrell Heald, have been generous in their encouragement and assistance. Winnifred Randall and David Heald ably contributed to the illustrations in the book through their art and photographic work. Virginia Benade made numerous valuable suggestions as she typed the final manuscript. A special word of gratitude is due Barbara Kraft and others of the staff of The Johns Hopkins University Press for their generous assistance and careful production of a very complex manuscript. I am most grateful to my family, who have sacrificed, understood, and encouraged. In particular, I thank my mother, Vada F. Jenkins, who assisted with reading proofs, and my wife, Alyce Mitchem Jenkins, who not only shared the vicissitudes of the years of research and writing but also contributed substantially to the work by editing and typing the manuscript.

Abbreviations of Frequently Cited Sources

Journals, Books, and Catalogues

Anthony Cat.	E. Anthony, *A Comprehensive and Systematic Catalogue* (New York: Snelling, 1854)
Anthony Photog. Bul.	*Anthony's Photographic Bulletin*
Brit. J. Photog.	*British Journal of Photography*
Dag. J.	*Daguerreian Journal: Devoted to the Daguerreian Photogenic Art, Also Embracing the Sciences, Arts, and Literature*
Humphrey J.	*Humphrey's Journal of Photography and the Allied Arts and Sciences*
Photog. A. J.	*Photographic Art Journal*
Photog. N.	*Photographic News: A Weekly Record of the Progress of Photography*
Photog. T.	*Photographic Times*
Photog. Korresp.	*Photographische Korrespondenz: Organ der Photographischen Gesellschaft in Wein*
Tissandier Hist.	Gaston Tissandier, *A History and Handbook of Photography,* trans. J. Thomson (London: Low, Marston, Searle & Rivington, 1878)

Manuscripts and Public Documents

Full bibliographic references are given in the Primary Sources of Information section.

D & B	Dun & Bradstreet Manuscript Records
SC	Scovill Manufacturing Company Records
GEC	George Eastman Business Correspondence

GECP	George Eastman Personal Correspondence
GEN	George Eastman Notebook
EkCo. v. *Blackmore*	*Eastman Kodak Company* v. *J. Edward Blackmore*
Eastman v. *Blair*	*Eastman Company* v. *Blair Camera Company*
Goodwin v. *EkCo.*	*Goodwin Film & Camera Company* v. *Eastman Kodak Company*
U.S. v. *EkCo.*	*United States* v. *Eastman Kodak Company*

Note: References in the text to the Eastman companies will employ the company name appropriate to the period of the reference:

Eastman Dry Plate Company (1 January 1881–1 October 1884)

Eastman Dry Plate & Film Company (1 October 1884–24 December 1889)

Eastman Company (24 December 1889–23 May 1892)

Eastman Kodak Company (23 May 1892–)

Images and Enterprise

Introduction/ Framework of Industrial Development

The American photographic industry had its beginnings in 1839 with the commercial introduction of the new French process, the daguerreotype. Despite the technical complexity of a process that combined knowledge, materials, and apparatus from optics and chemistry, the practice of daguerreotypy spread across the nation within a few years by means of a small group of technically oriented operators, some of whom established galleries in cities and large communities and many of whom assumed an itinerant life style, moving from village to village in the less populated regions of the new nation. From the first the daguerreotype excited popular curiosity. People were awed by the seemingly occult workings of the photographic "alchemist" who, shrouded by his black cloth and tent, mysteriously manipulated light and chemicals to produce a mirror of nature on the silver plate. Soon, however, the unknown and mysterious process, while not yet understood, became familiar and commonplace.

Photography found its niche in American life as a producer of an occasional image of a family member, a departed loved one, a close friend, or very rarely, a landscape or a building. Consequently, a small but steady demand for such images provided the base for a very small, technically oriented industry composed initially of a few craftsmen and small-scale manufacturers from other industries who produced optical and camera apparatus, chemical materials, and ancillary supplies; the intermediary photographic supply merchants; and the professional operators, who produced their own photosensitive materials, made the exposures, and ultimately produced the finished images. From this small and unassuming beginning, the photographic industry grew during the next century into a vast and complex business in which vertically integrated corporations played the commanding role in a mass consumer market and the supply merchants and professional photographers retired to a secondary position.

From the introduction of the daguerreotype to the use of panchromatic and sound-track cinematographic film, the technology of the photographic process and the production of its requisite materials underwent a series of radical changes. The most fundamental and far-reaching innovations came in a series of changes in the form of photography (from still photographs to cinematography) and in the carrier-bases for the photosensitive material. The first major change was from the daguerreotype plate, whose carrier-base consisted of a silvered copper sheet, to collodion on glass or iron plates. The second major change was from the collodion on glass plates to gelatin on glass plates. The third major change was from gelatin on glass plates to gelatin on roll film (thin strips of paper or celluloid). A fourth change involved the adoption of celluloid roll film as the base in cinematography to replace the glass disk base initially employed.

These changes were accompanied by major upheavals that altered the mode of production, the methods of distribution and marketing, and the business conceptions and assumptions of the participants in the industry. Although the growth of the photographic industry in America, as in other nations, bore the stamp of these major technological changes, both the technological changes and the business upheavals occurred within the context of a dynamic sociocultural matrix, the specifically American characteristics of which distinctively shaped and molded the industry in the United States. Thus, the character of the accompanying business innovations reflected not just the direct impact of the new technology but also the indirect effect of major technological changes that destroyed the traditional *barriers to entry*,[1] opening the industry to the winds of change. Such upheavals facilitated the inculcation of certain new and distinctly American economic, social, and cultural values then emerging outside the industry.

Recent historical studies of the American economy, of business organization, and of industrial structure emphasize the role of the market as a stimulus to organizational and structural change;[2] moreover, one study interprets the development of an industry in terms of a sequence of stages, each characterized by a distinctive theoretical market model.[3] While the growth and changing character of the market were very important in the photographic industry, a sequence of stages characterized by distinctive product technologies seems to provide a more fundamental interpretative tool for understanding the marketing changes and the history of the industry generally. Consequently, this study will employ as a framework a sequence of five stages,[4] each stage characterized by the dominant photosensitive carrier-base or form of photography in use at the time: the daguerreotype (1839–55); wet collodion (1855–80); dry gelatin on glass plates (1880–95); gelatin on celluloid roll film, in the amateur roll film system (1895–1909); and cinematographic film (1909–25), which, without supplanting the amateur roll film system, assumed the lead in relative economic importance.

During each stage the distinctive photosensitive carrier-base

[1] "Barriers to entry" are those impediments that are often employed to discourage new firms from entering an industry. Generally such barriers to entry as patents, trade secrets, trade-name identity, and economies of large-scale production are sought by established firms in order to promote and protect their position in the industry. See Willard F. Mueller, *A Primer on Monopoly and Competition* (New York: Random House, 1970), pp. 10–20; and Richard Caves, *American Industry: Structure, Conduct, Performance* (Englewood Cliffs, N.J.: Prentice-Hall, 1964), pp. 21–28.

[2] Some of the more noteworthy illustrations include Douglass C. North, *The Economic Growth of the United States, 1790–1860* (New York: Prentice-Hall, 1961; reprint ed., New York: W. W. Norton & Co., 1966); Glenn Porter and Harold C. Livesay, *Merchants and Manufacturers: Studies in the Changing Structure of Nineteenth-Century Marketing* (Baltimore: Johns Hopkins Press, 1971); and most notably, the works of Alfred D. Chandler, Jr., including: "The Beginnings of 'Big Business' in American Industry," *Business History Review* 33 (Spring 1959): 1–31; "The Coming of Big Business," in *The Comparative Approach to American History*, ed. C. Vann Woodward (New York: Basic Books, 1968), pp. 220–35; *Strategy and Structure: Chapters in the History of American Industrial Enterprise* (Cambridge: Massachusetts Institute of Technology Press, 1962); and (with Stephen Salsbury), *Pierre S. DuPont and the Making of the Modern Corporation* (New York: Harper & Row, 1971).

[3] Alfred S. Eichner, *The Emergence of Oligopoly: Sugar Refining as a Case Study* (Baltimore: Johns Hopkins Press, 1969).

[4] In the sense of a model of ideal types as defined by Max Weber, as quoted in H. Stuart Hughes, *Consciousness and Society* (New York: Alfred A. Knopf, 1958), pp. 312–13; although the framework employed here is substantially different, my formulation of it was significantly influenced by the ideas of Thomas S. Kuhn in *The Structure of Scientific Revolutions* (Chicago: University of Chicago Press, 1962).

tended to dominate the technological conceptions of photographic technicians and businessmen and to define the boundaries within which most technological change occurred. The technological conception became an important element in a distinctive business-technological mind-set consisting of conceptions, assumptions, attitudes, and values that dominated the thinking and the *modus operandi* of the participants in the industry. Moreover, the nature of the photosensitive carrier-base set boundaries to the modes of production, the extent of simplification of the processes, and the potential size of the market.

Within each stage the market structure and price behavior can be interpreted in terms of a sequence of three phases, to each of which may apply a theoretical model of market behavior based on the assumption that the position of natural equilibrium is imperfect, not perfect, competition. The sequence consisted of an initial phase of imperfect competition, a second phase of perfect competition, and a final phase of oligopolistic competition. The rationale for such a pattern assumes that for many businessmen the primary goal was profit maximization within the legal and value system of the society at the time. Therefore, their efforts were directed toward gaining a competitive advantage that would allow them to improve their profit margins. At each stage the product innovator or innovators had an initial advantage of operating within an imperfectly competitive market structure; however, unless those innovators possessed some barrier to entry, their advantageous profits attracted imitators who contributed to the second phase of near perfect competition. As this occurred, however, profits plummeted, putting pressure upon the businessmen to seek methods of reestablishing former profit levels. The success of a few businessmen in this endeavor generated the third phase of oligopolistic competition.

This general pattern within stages and within sectors of the industry during each stage did not apply to every sector at every stage because varying strategies and possibilities for barring entry to the industry abridged or prolonged a given phase at the expense of another. Typically, however, a few firms came to dominate the industry during each stage, and as they did so, the leadership of these firms sought to fortify and expand their market positions by exploiting to the maximum the protective features that could be derived from the conceptions and assumptions characteristic of that stage.

The introduction of a fundamentally new photosensitive carrier-base prompted the transformation from one stage to the next. Usually one or two technically oriented persons who were either outside the industry or had only recently entered it concretized an abstract idea already in public circulation by making the key technological invention or discovery. The early key developments, wet collodion and dry gelatin, were employed by persons in the pursuit of noneconomic goals; the later key developments, roll film and cinematography, were created by persons committed to economic and entrepreneurial goals. The work in all cases initially received some public attention but largely in the context of the older dominating conceptions. Then the innovators, persons who took an invention or discovery and introduced it in a practical or commercial way, introduced the new developments in the framework of the older conception. In the case of the early developments, the innovators were distinct from the inventor-discoverers; in the case of the later developments, the innovators were inventor-discoverers. Once the innovations had been made, either the original innovators or their imitators began to recognize a different potential in the new technology; they committed themselves to it and devoted all their psychic and economic resources to its successful introduction and fuller exploitation.

In a very real sense, these key innovations led to the destruction of the older mind-set. In the newly emerging framework, the traditional fortifications and strategies of the dominating firms became less efficacious, and typically, the barriers to entry crumbled and fell. With the entry of new firms that began to challenge the established leadership, a new set of assumptions, conceptions, and strategies began to emerge in the industry. The destruction of the older business and technological mind-set, or at least a part of it, facilitated the reconceptualization of the technology and its relationship to business and the revision of strategies. Therefore, the demise of the old mind-set prompted the utilization of new ideas, strategies, and practices developed outside the industry. Often, however, the older leadership was less responsive to the new ideas than were the new and less security-conscious enterprises. The new entrants helped to re-create a market structure closer to the model of perfect competition than that which had prevailed toward the end of the previous stage. Then, with the aid of their new set of conceptions, assumptions, strategies, practices, attitudes, values, and ambitions, the leadership sought in the new setting to fortify and insulate their positions in the industry, and gradually the market structure moved once again toward oligopoly.

In interpreting the development of the American photographic industry in terms of this framework, we should note certain general, interrelated trends that spanned two or more of the stages. Among the most significant were: (1) the movement from decentralized to centralized factory production; (2) the separation and specialization of functions in producing photographs; (3) the emergence of a mass market for apparatus and materials; (4) the increase in scale and integration of enterprise; and (5) the increased awareness of technology and its sophisticated use as a business strategy. The attention to and employment of technology in the early days of the industry were casual and spasmodic. Gradually this attitude changed as entrepreneurs and businessmen began to conceive of (1) technological innovation as a means of entering the industry, and (2) technological innovation protected by patents as a means of gaining legal monopolistic control of certain products and sectors of the industry. Therefore, they began to pursue research and development in a systematized manner. In the later stages of this development, they saw technological change as having a dual potential. On the one hand, through the institutionalization of innovation by means of research and development laboratories, large corporations sought to control technological change as a means of protecting and fortifying their positions in the industry. Thus, research and development became part of the strategy to maintain the status quo in the industrial structure. On the other hand, fundamental technological change posed a threat to the status quo when technological innovation originated outside the leadership of the industry. Each of the first four trends noted above contributed to the creation of the organizational structure out of which the institutionalization of innovation grew, but the institutionalization merely climaxed and formalized the systematic utilization of technology and science as part of business strategy that had been adopted decades earlier.

The Frenchman here says the Plates cannot be made here & he calculates to make a Fortune by Importing them from France . . . we will try to disappoint him.

J. M. L. Scovill to W. H. Scovill
18 December 1839

I

Daguerreotype Period 1839 to 1855

1/The Daguerreotype Period

From the inception of the photographic industry in America, its structure, the ideas of its leaders, and the direction of technological innovation were strongly influenced by the character of the daguerreotype process. Two Frenchmen of limited technical training, Joseph Nicephore Niepce and Louis Jacques Mandé Daguerre, developed, through their cooperative endeavors, the first commercial photographic process: the daguerreotype. In August 1839, under the auspices of the leading men of science in Paris, Daguerre disclosed to the public the details of how to take a daguerreotype: (1) sensitize silver-coated copper plates with iodine vapors just prior to exposure in a camera (fig. 1.1); (2) expose the plates to a sharply focused optical image in the camera (fig. 1.2); (3) treat the plate with mercury vapors in order to develop the latent image created by exposure (fig. 1.3); and (4) "fix" the image (remove the remaining photo-sensitive salt) in a bath of "hypo" (sodium thiosulfate). The public in Paris responded to the daguerreotype with enthusiasm, and details of the process quickly spread to the United States.

The daguerreotype, with its sharply delineated image, was warmly received and quickly assimilated as part of American popular culture. In 1839 the United States, a relatively new nation still strongly influenced by European art and culture but separated from its ancestoral home by more than a two weeks' ocean journey, sought its own distinctive national culture by celebrating, with Romantic reverence for the past and for nature, its people, its frontiers, and its wildernesses. Hence, the daguerreotype, with its highly detailed and sharply focused image, helped to express America's distinctive culture. The process of production, which combined a "scientific" technology of optics and chemistry with the use of light, the "hand of nature," linked the quasi-objectivity associated with a positivistic conception of science and the Romantic mystical appeal to nature. Furthermore, the public willingly accepted the transition from hand techniques to optical-chemical methods of image production because artists had previously employed the camera obscura as a sketching aid (see fig. 1.4) and because the daguerreotype, like a painting, provided a unique image that could be replicated only with considerable difficulty. The great popularity of the daguerreotype in the United States reflected a democratization of many of the functions of the traditional painter through the broad acceptance of a new technology and a certain standardization of product in return for low cost and speed of production. The daguerreotype provided the population at large for the first time with unique personal images of the past at low cost. These widely adopted images embodied elements of the American conception of the origins of truth: the objectivity of science and the insight of the pencil of nature. The acceptance of these images represented an American popular response to the search for a distinctive culture in the context of European culture and the American socioeconomic setting.[1]

This new art form and technology and its accompanying demand for apparatus and materials arrived in America precisely at the be-

[1] A recent excellent study of the reception and influence of the daguerreotype in America is that of Richard Rudisill, *Mirror Image: The Influence of the Daguerreotype on American Culture* (Albuquerque: University of New Mexico Press, 1971).

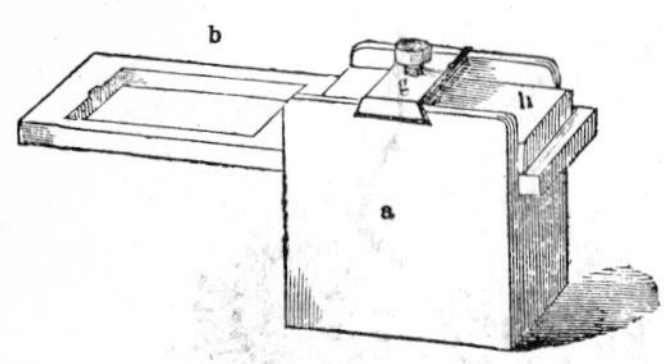

Fig. 1.1
Iodine box for sensitizing daguerreotype plates. From Henry Hunt Snelling, The History and Practice of the Art of Photography, *4th ed. (New York: G. P. Putnam, 1853).*

Fig. 1.2
Daguerreotype camera box and lens. Parts: a, *camera box;* b, *tube with lens;* c, *lens cap (early shutter);* e–f, *plate holder; and* g, *ground glass for focusing. From Snelling,* Art of Photography.

Fig. 1.3
Mercury developing box for daguerreotypes. From Tissandier Hist.

ginning of a sharp economic depression, which lasted from the fall of 1839 until early in 1843; however, from that point until late in the mid-1850s, the daguerreotype rode the wave of renewed economic growth and expansion in America.[2] The generation prior to the Civil War witnessed a marked shift, especially in the Northeast, from handicraft to factory modes of production. Although the organization of business still depended predominantly on the single proprietorship or partnership, the distribution system moved from the generalized merchant to the more specialized jobber or wholesaler.[3] These trends were further stimulated by the very rapid growth of the domestic markets for manufacturers. The U.S. population, which numbered about sixteen million in 1840, doubled during the next twenty years.[4] During this same period, major changes in communication and transportation stimulated business. The expansion of canal and railroad routes into the interior and the creation of specialized express companies cut sharply the time and cost of freight shipments.[5] Furthermore, communications vastly improved with the creation of major telegraphic networks in the late 1840s and early 1850s.[6] These changes helped to foster and mold the character of the early photographic industry in the United States, as it emerged from a variety of traditional industries that sought to meet the demand for cameras, chemicals, plates, and auxiliary equipment.

1839–45

Prior to Daguerre's disclosure of the details of his process, Americans like John W. Draper, a prominent chemist and physiologist, and Samuel F. B. Morse, a well-known artist and inventor, became interested in photography and anxiously awaited details of Daguerre's process. The first manuals from Paris describing this process arrived in New York on 20 September 1839. Soon Draper, Morse, and a few others equipped themselves and captured a few images of nature. Early in October details of the process were published in the American press.[7] The new knowledge swept like a wave from city to city, and soon pioneer daguerreotypists in Boston, Philadelphia, and elsewhere were proudly displaying their pictures. By late November the Parisian agent officially appointed by Daguerre to produce the authentic apparatus sent a representative, François Gouraud, to

[2] Douglass C. North, *The Economic Growth of the United States, 1790–1860* (New York: Prentice-Hall, 1961; reprint ed., New York: W. W. Norton & Co., 1966), pp. 194–215.

[3] Glenn Porter and Harold C. Livesay, *Merchants and Manufacturers: Studies in the Changing Structure of Nineteenth-Century Marketing* (Baltimore: Johns Hopkins Press, 1971), esp. p. 7.

[4] U.S., Department of Commerce, Bureau of the Census, *Historical Statistics of the United States, Colonial Times to 1957* (Washington, D.C., 1960), p. 8.

[5] Notably Adams Express, American Express, and Wells Fargo Express. See George R. Taylor, *The Transportation Revolution* (New York: Holt, Rinehart & Winston, 1951).

[6] Telegraph orders began to flow to New York from St. Louis in 1850.

[7] Beaumont Newhall, *The Daguerreotype in America* (New York: New York Graphic Society, 1968), p. 22.

Fig. 1.4
Artist using camera obscura to make sketch from nature. From Photog. N., *10 September 1858.*

New York, where early in December he presented an official display and lecture and shortly thereafter initiated a series of lecture-lessons. Late in the winter and early spring to 1840 Gouraud lectured in Boston and Providence, seeking to promote the apparatus of the French company.[8] Stimulated by such lecturers, professional daguerreotypists began to appear across America.

These technically oriented daguerreotypists performed the complex production function (see fig. 1.5). The photosensitive materials, the iodized silver plate, proved to be quite perishable and not only had to be exposed soon after sensitization but also required development and fixing shortly after exposure. Consequently production remained decentralized, with daguerreotypists working at the site of the exposure, which was often one of the numerous galleries established in cities across the nation but was sometimes the farm or village visited by an itinerant daguerreotypist.

One of the key supplies required by these new daguerreotypists was silvered plates, and almost at once, in the fall of 1839, the process of specialized photographic supply productions began in the United States with the entry of the J. M. L. and W. H. Scovill Company into their manufacture. The Scovill company of Waterbury, Connecticut, an old, established firm whose ancestry extended back to the beginning of the century, occupied a leading position in the production of brass and gilt buttons and brass and copper hardware items. This partnership of two brothers, James Mitchell Lamson and William Henry, also had considerable experience with the production of rolled plate metals, including silver plate. Although Samuel F. B. Morse had initially purchased for his experi-

[8] Ibid., pp. 27–32; François Gouraud, "Manner of Taking Portraits by Daguerreotype," *Boston Daily Advertiser and Patriot*, 26 March 1840.

Fig. 1.5
The daguerreotypist. From Godey's Lady's Book, *May 1849.*

Fig. 1.6
J. M. L. Scovill. From William G. Lathrop, The Brass Industry in the United States *(Mount Carmel, Conn.: Lathrop, 1926).*

Fig. 1.7
William H. Scovill. From Lathrop, The Brass Industry.

ments coiled silver-plated copper from New York hardware stores,[9] he soon placed an order with the Scovill agent in New York urging him to introduce specialized production of plates and predicting that a large quantity would soon be wanted.[10] Despite production difficulties attributable to the employment of impure silver and poor rolling techniques, the company was determined to compete with French manufacturers and did receive an order from a New York daguerreotypist for hundreds of plates. A New York customer who had seen plates rolled in Paris provided the valuable information that on the final roll of the silver on the copper plate the French rolled the plate double with the two silver surfaces facing each other (fig. 1.8). Also, in shipping the final product, the French covered the silver surface with tissue paper. Having adopted these suggestions and other improvements by early 1840, the Scovill company

[9] Samuel F. B. Morse to M. A. Root, 10 February 1855, in Marcus A. Root, *The Camera and the Pencil . . .* (Philadelphia: Root, Lippincott, Appleton, 1864), pp. 344–48. Scovill probably even produced this silver-plated copper.

[10] J. M. L. Scovill to W. H. Scovill, 15 October 1839, quoted in Theodore F. Marburg, "Management Problems and Procedures of a Manufacturing Enterprise, 1802–1852: A Case Study of the Origins of the Scovill Manufacturing Company" (Ph.D. diss., Department of Economics and Sociology, Clark University, 1945), p. 246.

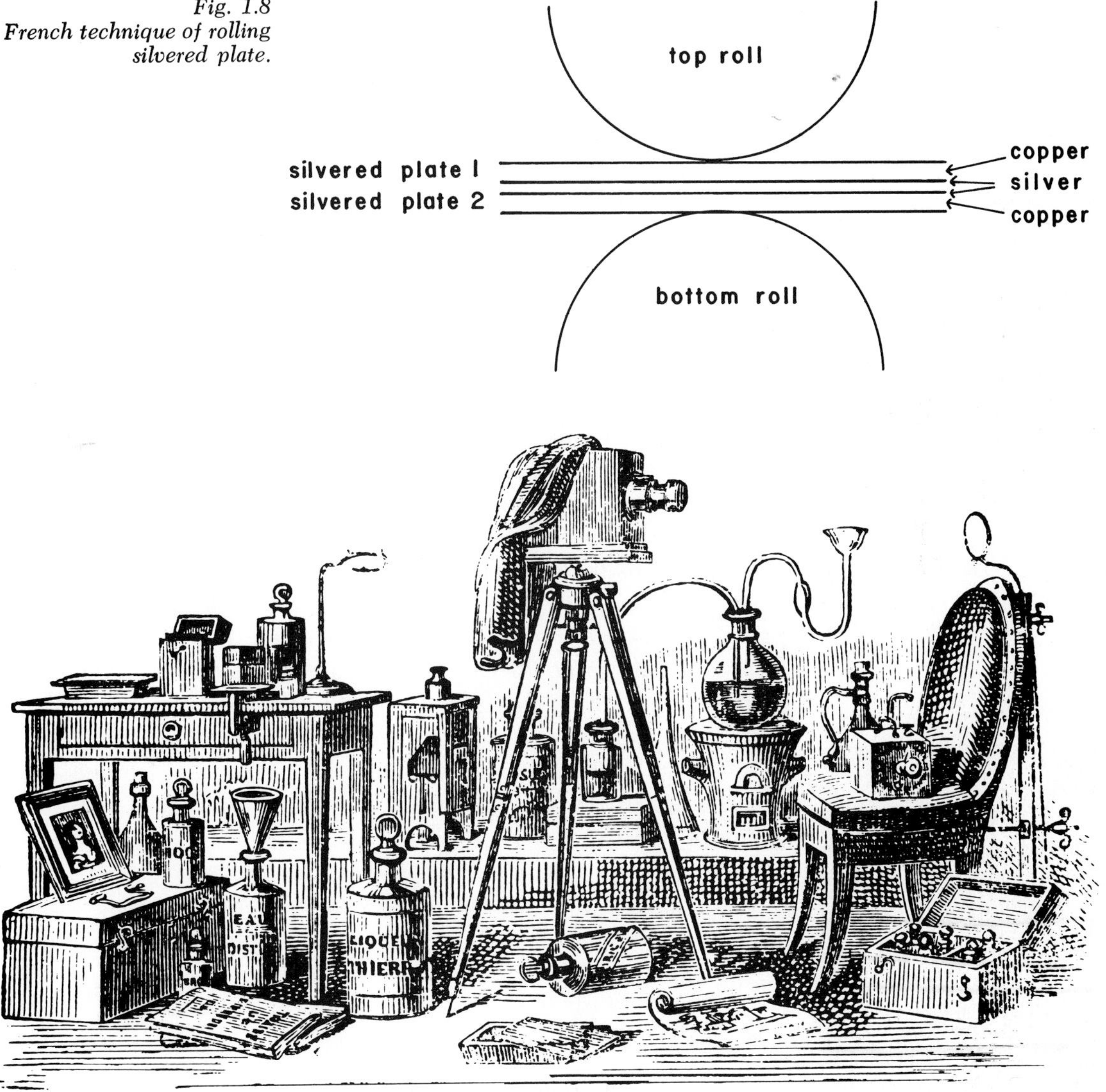

Fig. 1.8
French technique of rolling silvered plate.

Fig. 1.9
Daguerreotype outfit. From J. Thierry, Daguerréotypie, précédées d'une histoire générale abrégée de la photographie *(Paris: Lerebours et Secretan, [1847]).*

markedly improved its plates, one lot at least equalling in quality the plates shipped from Paris.[11]

In the next few years Scovill's production of plates expanded, and the firm established itself as the leading daguerreotype plate manufacturer in the United States. During 1840 the quality of the plates improved further as Scovill started importing copper from England and employing workmen from France. The expansion of production in the face of French competition and the receipt of a diploma for the Scovill plates and other daguerreotype apparatus at the Exhibition of Manufactures of the American Institute of New York in 1843 indicated the improvements made in product quality during the first years of production.[12] Sales were handled at first through the company's New York agent, but in the summer of 1841 the Waterbury

[11] Ibid., 18, 30, and 31 December 1839 and 24 January 1840, pp. 248–50.
[12] Ibid., 24 January 1840.

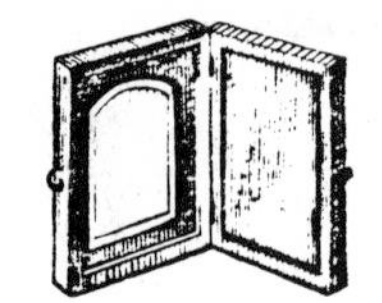

Fig. 1.10
Daguerreotype case. From J. J. Griffin, Catalogue of Photographic Apparatus, *May 1854.*

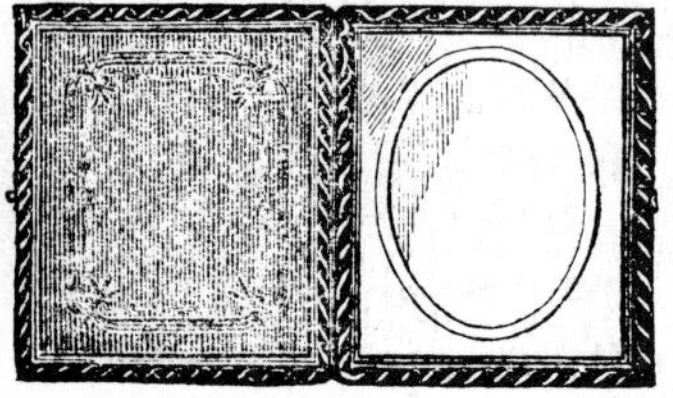

Fig. 1.11
Mats and case. From J. J. Payne, Descriptive Catalogue, *1857.*

office began opening accounts with a small number of individual daguerreotypists.[13]

By the mid-1840s the leading plates in the American market were those of the French, followed by Scovill's and then those of a few very small American producers.[14] Technical knowledge of metallurgical composition and rolling techniques served initially as a barrier to entry to competition in the manufacture of this important commodity. Nevertheless, as technical knowledge spread, the number of domestic producers increased, and plate prices fell sharply. In early 1840 French plates sold in the United States for $2.00 each. A year later American-made plates sold at two for $1.35 and in 1845 for $3.50 per dozen, reflecting a shift from imperfect competition toward perfect competition.[15]

Of equal importance as consumable items were the photographic chemicals, all of which would be classified as fine chemicals.[16] In the period prior to the mid-1840s, most daguerreotypists or their suppliers acquired the necessary chemicals directly from leading chemical firms or through pharmaceutical channels. The two principal sources for fine chemicals were Paris and Philadelphia, the principal chemical center of the United States. In the early 1840s the firm of Rosengarten & Denis of Philadelphia offered silver salts, halogens, and other photographic chemicals. In time, their production list expanded to cover all the requisite photographic chemicals.[17] The technical requirements of production served as a deterrent to many firms wanting to engage upon specialized photochemical production.

Soon after the introduction of the daguerreotype, special cases became important as means of transporting and protecting the fragile images. Initially, small jewelry or miniature cases were shaped to standard sizes of daguerreotypes. The shallow box and lid were hinged, with the lid carrying a piece of velvet or silk and the box carrying the daguerreotype plate (fig. 1.10). A glass cover was placed over the plate with a gilded "mat," or border, sandwiched between the plate and the glass to prevent the glass from scratching the surface of the plate (fig. 1.11). Then a "preserver," a metal foil border, enclosed the plate, mat, and glass cover (fig. 1.12). One of the first daguerreotype case manufacturers in the United States was Morse's former daguerreotype pupil Mathew Brady, who worked for a jewelry shop and then initiated his own production of cases in

[13] P. W. Bishop, "Scovill and Photography: The Tail That Almost Wagged the Dog" (unpublished manuscript), p. 428. Dr. Bishop of the Smithsonian Institution has generously shared the results of his researches in the Scovill Collection with me and given me permission to publish them here.

[14] Three New York producers were Joseph Corduan, L. B. Binsee & Co., and Edward White.

[15] Newhall, *Daguerreotype in America*, pp. 120, 165.

[16] Fine chemicals are chemicals of unusually high purity and are produced in small quantities. Customarily, their cost per unit weight or volume is high.

[17] N. F. H. Denis, a young French chemist, entered the firm in 1840. Williams Haynes, *American Chemical Industry*, 6 vols. (New York: D. Van Nostrand, 1954), 1: 214–15.

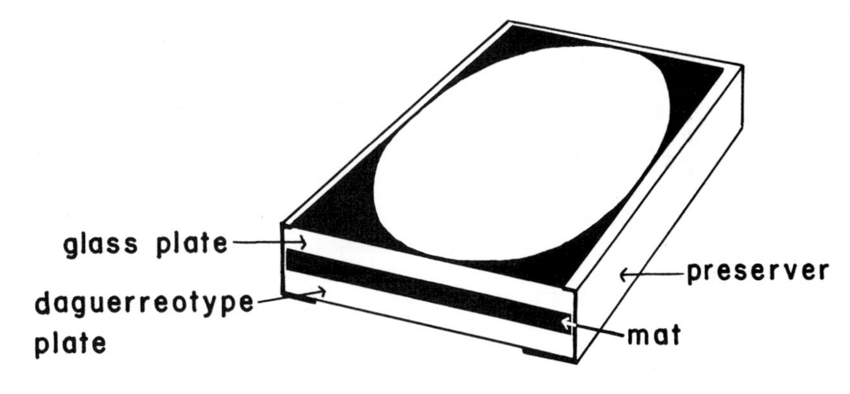

Fig. 1.12
Cross section of daguerreotype ready for insertion into case.

Fig. 1.13
Daguerreotype equipment, c. 1854. From Griffin, Catalogue.

New York in 1843, continuing production for two years.[18] Although in the early 1840s there were virtually no barriers to entry into case manufacturing, which was pursued on a handicraft basis, the inducements were still small, and consequently, there were only a few case makers.[19]

Opticians and instrument makers such as Henry Fitz and John Roach of New York made most of the requisite cameras and lenses. By the middle of the 1840s the high-quality apparatus from the Viennese optical firm Voigtländer & Son was readily available through agents in the United States. The French instruments of Gouraud, Chevalier, and Lerebours were also available. One New

[18] Newhall, *Daguerreotype in America,* p. 140; and James D. Horan, *Mathew Brady* (New York: Crown Publishers, 1955), p. 9.

[19] Sheldon & Co., *Business or Advertising Directory* (New York, 1845), p. 43; and Katherine M. McClinton, *A Handbook of Popular Antiques* (New York: Bonanza Books, 1946), p. 229. For a more detailed account of case makers and art see Floyd Rinhart and Marion Rinhart, *American Daguerreian Art* (New York: Clarkson N. Potter, 1967).

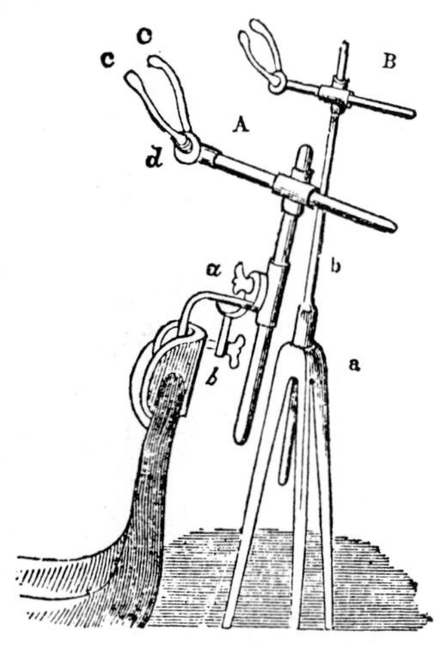

Fig. 1.14
Headrest. From Snelling,
Art of Photography.

York company, William and William H. Lewis, specialized at an early date in the production of sensitizing, developing, and camera boxes, camera stands, head rests, and plate holders and gradually developed a national reputation for quality and low cost apparatus (figs. 1.13 and 1.14).[20]

By the middle of the 1840s the production of photographic materials and apparatus had already begun to shift from companies and shops in other industries to specialized firms. Furthermore, a more competitive market structure had emerged with the appearance of several manufacturers in each sector. In the chemical and metal plate sectors the leadership of established firms in other industries decided to pursue specialized production for the photographic trade. In the case of the Scovill company, the photographic division soon became the most profitable sector of the business. Besides the skilled technical requirements of production, there were no substantial barriers to entry into any of the supply sectors; in the case sector, where the technical requirements were low, competition became quite keen during the next decade.

1845–55

The decade from 1845 to 1855 witnessed sustained prosperity and the rapid development of each of the three major divisions in the industry: daguerreotype producers, supply houses, and supply producers. With the return of prosperity in 1843 and the gradual training of persons in the techniques of daguerreotyping, the demand for daguerreotypes grew, and the ranks of daguerreotypists swelled. While decentralized production generally prevailed because of the character of the technological process, within those boundaries various operators sought to develop organizational and technological efficiencies. One notable entrepreneur, John Plumbe of Boston, created in the middle 1840s a chain of fourteen daguerreotype galleries located in the major cities of the nation.[21] The chain provided many potential advantages, including the sharing of technical and artistic improvements, collective purchasing of supplies, and joint national advertising. The Plumbe galleries also provided supplies to other daguerreotypists. The problems, however, outweighed the advantages. Apparently incompetent and perhaps even dishonest management at certain galleries brought about the demise of the ambitious enterprise in the late 1840s.[22]

As the number of operators and galleries grew and competition increased, particularly in the larger cities, gallery owners introduced various labor-saving techniques. One Broadway gallery, Reese &

[20] *Dag. J.* 1 (1851): 190; and *Photog. T.* 9 (1879): 212.

[21] New York; Boston; Washington; Philadelphia; Baltimore; Newport, R.I.; Saratoga Springs, N.Y.; Petersburgh, Va.; Alexandria, Va.; Harrodsburgh Springs, Ky.; Louisville, Ky.; Cincinnati; St. Louis; and DuBuque, Iowa. See advertisement in *Scientific American*, 11 September 1845.

[22] Robert Taft, *Photography and the American Scene: A Social History, 1839–1889* (New York: Macmillan Co., 1938), pp. 50–52; and Newhall, *Daguerreotype in America*, pp. 38–41.

Fig. 1.15
Interior of Meade brothers photographic gallery in New York City. From Photog. A. J., *February 1853.*

Company, introduced what Reese called the German system. It consisted of mass production of photographic portraits through the division of labor and movement of the product from specialized department to specialized department. Professor Reese, a recent immigrant from Germany, established departments for polishing, exposing, mercurializing, gilding, and coloring. The plates and customers moved from department to department. The company claimed to have taken between three hundred and one thousand portraits per day in the early 1850s.[23] In Boston in the late 1840s and early 1850s, the well-known gallery of John Whipple employed steam power and machinery for buffing and polishing the plates prior to iodizing, a procedure usually performed by hand. Whipple also installed a fan to cool his customers, who had to sit for long periods in the sun. Other Boston galleries, for example, Southworth & Hawes and Masury & Silsby, also employed machinery.[24]

As the popularity of the daguerreotype grew, the number of urban galleries increased rapidly and the competition became quite keen. In response to the competitive price cutting certain galleries—some of the large ones on Tremont and Washington Streets in Boston and those on lower Broadway in New York—began to cater to a more elite clientele. Daguerreotypists such as Mathew Brady, Charles and Henry Meade, Martin Lawrence, and Jeremiah Gurney turned their galleries into elaborate parlors with plush furniture and elegant trappings. They featured the qualitative and artistic element in their work, thereby trying to differentiate their work from that of the "factories." Of course, this quality justified a higher price for their products.[25]

A second response to the growing competition and consequent decline in prices and profits was the formation of associations of daguerreotypists and gallery owners created for the purpose of

[23] Newhall, *Daguerreotype in America*, pp. 63–65.

[24] Taft, *Photography and the American Scene*, p. 74.

[25] Newhall, *Daguerreotype in America*, pp. 55–66; and Horan, *Mathew Brady*, pp. 15–27.

maintaining prices. One such organization, the American Daguerre Association, met in New York in July 1851 and by mid-October had developed a constitution for its membership that included among its provisions the regulation that "no person shall be elligible [*sic*] to membership who publicly advertises low priced pictures by signs or other means."[26] In August 1851, the New York State Daguerreian Association held a meeting in Utica, at which its membership agreed to sell no daguerreotype for less than $1.50.[27] The emergence of product differentiation and associations of daguerreotypists in the early 1850s indicated the price competition that was emerging in the urban centers among the decentralized producers of photographs. Within the boundaries imposed by the technology, the operators sought methods and techniques of maintaining their profits.

Early in the history of the industry there emerged the supply house, a vitally important intermediary between the daguerreotypist and the manufacturer of the daguerreotype materials and apparatus. In the early 1840s a few large daguerreotype galleries in New York, Boston, and Philadelphia began ordering and stocking requisite supplies from various domestic and European firms and selling them to other daguerreotypists.[28] As the number of daguerreotypists grew in the late 1840s, specialized supply houses emerged. The three leading New York supply houses, Scovill, Anthony, and Chapman, through their influence in manufacturing and distribution commanded a dominant position in the American industry in the early 1850s.

The Scovill company in Waterbury, Connecticut, found in the middle 1840s that its accounts with daguerreotypists were expanding rapidly, and at the same time, it encountered difficulties with its New York agents. Consequently, in the late summer of 1846 the company established a store in New York that carried a complete stock of the company's products, including those for the daguerreotype trade.[29] The success of sales in that trade, however, required that the New York store carry a complete line of daguerreotype supplies, including competing lines of daguerreotype plates including those from France.[30] As a consequence, the managers of the store, George Mallory and Samuel Holmes, gradually adopted a mercantile as opposed to a manufacturing attitude.

Soon the Scovill New York store proved an immense success, showing substantial profits that soon exceeded those of any other division of the Scovill company (see table 1.1). However, despite its success, Mallory and Holmes became increasingly attentive to the markets in the interior of the country. After a competitor, Edward Anthony, sent a representative in 1850 on a circuit of major inland cities, Samuel Holmes and an assistant made an extensive trip dur-

[26] *Dag. J.* 2 (1851): 342–46.

[27] Ibid., pp. 248–49. In 1854 daguerreotypes sold far as little as 25¢ each.

[28] Edward White (N.Y.C.), William Lewis (N.Y.C.), Edward Anthony (N.Y.C.), A. S. Southworth (Boston), John Plumbe (N.Y.C., Boston, Philadelphia, etc.).

[29] Bishop, "Scovill and Photography," p. 431.

[30] Marburg, "Origins of the Scovill Manufacturing Company," pp. 464–67; and *Dag. J.* 1 (1851): 190 and 7 (1855): 6.

Table 1.1 Profits of Scovill New York Store, 1846–51

Year	*Net Profit*	*% of Net Profit of All Scovill Stores*
1846	$ 3,706	10
1847	8,283	14
1848	11,167	28
1849	25,840	52
1850[a]	30,872	
1851[a]	30,034	

SOURCE: The *Ledgers*, Profit and Loss Accounts, of Scovill, as reported in Theodore F. Marburg, "Management Problems and Procedures of a Manufacturing Enterprise, 1802–1852: A Case Study of the Origins of the Scovill Manufacturing Company" (Ph.D. diss., Department of Economics and Sociology, Clark University, 1945), pp. 519–21.

[a] In 1850, three different Scovill companies, including the New York Store, merged to form the Scovill Manufacturing Company, so the total net profits for 1850 and 1851 are not comparable with those prior to 1850.

Fig. 1.16
Edward Anthony. From Photog. A. J., *April 1853.*

ing the winter of 1851–52, stopping in most of the major cities east of the Mississippi.[31] The number and size of daguerreotype supply houses in the interior grew rapidly, and therefore they commanded special attention and prices from the New York supply houses.[32] Although the Scovill store, in contrast to its competitors, divided its loyalty between the hardware trade and the photographic supply business, it became the leading photographic supply depot in the country during the daguerreotype period, serving both as a retail supplier to daguerreotypists and as a wholesale jobber to interior supply houses.

In contrast to the Scovill company, which started as a manufacturer of daguerreotype materials and then went into marketing (*vertical integration forward*), the Anthony company reversed the process. The Anthony business began as a daguerreotype gallery, added the sale of supplies, then dropped the gallery work, operated a supply depot, and finally initiated the manufacture of certain supplies (*vertical integration backward*) (see chap. 8, n. 2, for a discussion of these terms). The founder of this important New York company was Edward Anthony, a civil engineer with formal scientific training at Columbia. At the time that the daguerreotype was introduced, he purchased lessons from Morse and soon engaged as a daguerreotypist on a survey of the northeast boundary of the United States. After his return to New York, he opened in 1842 a daguerreotype gallery.[33] From 1842 until 1847 he pursued daguerreotypy with a series of partners and established an excellent reputation in New York. Noting the increase in the sales of supplies at his gallery and at the Scovill store, he decided in 1847 to establish a depot devoted exclusively to photographic supplies (fig. 1.17). Anthony

[31] Marburg, "Origins of the Scovill Manufacturing Company," p. 474.

[32] Peter Smith of Cincinnati had sales of $50,000 in 1851. See Ibid., pp. 474–81.

[33] *Anthony Photog. Bul.* 19 (1888): 737–40; *Photog. N.* 32 (1888): 807; Henry Hall, ed., *America's Successful Men of Affairs,* 2 vols. (New York: New York Tribune, 1895), 1: 21–22; and Taft, *Photography and the American Scene,* pp. 52–55.

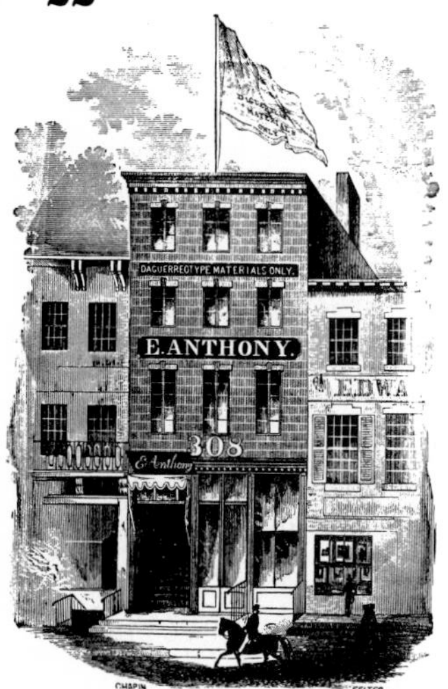

Fig. 1.17
Warerooms of E. Anthony, 308 Broadway, New York City. From Anthony Cat. *Courtesy of Tom and Elinor Burnside.*

Fig. 1.18
Henry T. Anthony. From Robert Taft, Photography and the American Scene: A Social History, 1839–1889 *(New York: Macmillan Co., 1938; reprint ed., New York: Dover Publications, 1964). Courtesy of Dover Publications, Inc.*

attained some measure of success in the late 1840s as sales grew rapidly; however, shortly thereafter, as a consequence of competitors' new production technology, the company was left with a supply of outmoded plates and became somewhat weakened financially.[34]

The introduction of new production techniques substantially changed the plate market in the United States. First, in the early 1840s Daniel Davis, a Boston maker of philosophical instruments, introduced a technique of electroplating a final silver surface on the rolled silver-copper plates, thereby insuring a fine surface. In the late 1840s daguerreotypists in the United States purchased small galvanic cells and widely adopted this method of improving the polish of the surface of the plate.[35] Warren Thompson of Philadelphia introduced the technique, which became known as the American process, into the factory production of plates for the French plate firm of Charles Christofle.[36] Second, in 1850 the Scovill company rebuilt its plate production facilities, introducing new techniques that the company expected would make the Scovill plates superior in finish to those of the French. At this time the firm introduced the planishing and factory buffing of plates, and it is likely that electroplating was also adopted. The surface quality of the Scovill plates showed a marked improvement after the installation of the new facilities.[37] The production changes in both Waterbury and Paris caught many of the smaller plate companies with stocks of old plates, a financial handicap that most could not overcome.[38]

This change in technology prompted Edward Anthony in the early 1850s to initiate some new strategies. First, his company started the production of daguerreotype cases, camera boxes, and photographic chemicals. Second, attentive to the growing domestic market in the West, Edward Anthony sent in the spring of 1850 a representative to visit suppliers and daguerreotypists in major interior cities as far west as St. Louis. The success of this effort was reflected by Scovill's imitation a year later. Third, as the business expanded, requiring additional working capital, Anthony acquired in 1852 two silent partners, one of whom was his brother.[39] Besides capital, Henry T. Anthony, with a background in civil engineering and banking,[40] brought to the business valuable technical and financial experience as he assumed special responsibility for the manufacturing phase of the company.[41] These strategies led to

[34] After entering the business, "for six years sales doubled the year before" (*Photog. N.* 32 [1888]: 807); the financial pressure of the plate problem was reported in "Edward Anthony: 30 Oct. 1851," D & B, N.Y.C. Series 8, p. 362.

[35] Newhall, *Daguerreotype in America*, p. 120.

[36] See the Christofle advertisement in *La Lumière*, 11 May 1851; see also Newhall, *Daguerreotype in America*, p. 120.

[37] Marburg, "Origins of the Scovill Manufacturing Company," pp. 251–53; and Bishop, "Scovill and Photography," p. 444.

[38] "Edward Anthony: 30 Oct. 1851," D & B, N.Y.C. Series 8, p. 362.

[39] "Edward Anthony: 13 July 1852," in ibid.

[40] He held an A.B. degree (1832) from Columbia University and had been a clerk for the Bank of the State of New York.

[41] *Brit. J. Photog.*, 26 September 1884, p. 613; and *Anthony Photog. Bul.* 15 (1884): 453–61.

Fig. 1.19
Anthony manufactory in New York City. From Anthony Cat. *Courtesy of Tom and Elinor Burnside.*

growth in sales, which reached a quarter of a million dollars in 1853. By 1855 the firm possessed a local business reputation as "one of the strongest in the trade."[42]

A third supply house of importance in the late 1840s and early 1850s was that of Levi Chapman of New York City. In the late forties Chapman manufactured razor strops and daguerrean materials, including cases, mats, preservers, and plates, but he did not enjoy any particular success until the early fifties. At that time, like Anthony, he began to develop and cultivate a jobbing trade with inland supply houses. Between 1852 and 1855 the Chapman company experienced considerable success and by 1855 devoted its major attention to manufacturing.[43]

Although there were other supply houses in New York, Philadelphia, and Boston and in the larger inland cities, Scovill, Anthony, and Chapman emerged as the principal national jobbers as well as suppliers to the daguerreotypists of the New York area because of their proximity to the supply producers who were concentrated along the coast from Boston to Philadelphia and because of their established relationship with foreign companies that exported supplies. The jobbing function was particularly critical to the photographic industry at this time because the market for supplies was large and broadcast and because a variety of supplies was needed from a large number of firms in several different industries. The output and sales of most of these supply manufacturers were so

[42] "Edward Anthony: June 13/54 and 18 June 55," D & B, N.Y.C. Series 8, p. 362.

[43] "Levi Chapman: Dec. 15/51 and Sept. 29/52 to Aug. 9/55," in ibid., p. 379; and S. Holmes to Mallory, 26 December 1851, in Marburg, "Origins of the Scovill Manufacturing Company," p. 479.

Fig. 1.20
Anthony mat and preserver factory. From Anthony Cat. *Courtesy of Tom and Elinor Burnside.*

Fig. 1.21
Anthony dipping room. From Anthony Cat. *Courtesy of Tom and Elinor Burnside.*

small that it was not practical for them to sell directly to the individual daguerreotypists. Furthermore, shortly after the birth of the industry, the newly created telegraphy and railroad networks and the major express companies began to link the markets of the interior closely with those of the East Coast. For light-weight, expensive commodities such as those in the photographic trade, a truly national and international market emerged at this time. However, the development of this national market also brought some complications. The considerable turnover in the ownership of galleries and the reputation of daguerreotypists as generally poor businessmen and financial risks meant that the supply merchant carried considerable risk in offering credit. Consequently, despite the development in the 1840s of national credit investigating companies and improved communications, the major New York supply houses were content to let local supply houses, which had more reliable information about the customer, develop the trade in the interior cities. Then, the New York houses anxiously solicited and cultivated the business of the interior supply houses from whom they reaped lucrative jobbing profits.

While the supply houses served the national market as jobbers, they also manufactured a variety of consumable photographic materials because they saw greater long-range opportunity for growth and profit in the producer consumable sector than in the producer durable sector of the supply market. Although Anthony moved from marketing into manufacture and Scovill and Chapman from manufacture into marketing, all three were at least partially integrated companies. When the two largest jobbing houses, Scovill and Anthony, discovered that they were pursuing similar strategies in jobbing and manufacturing, negotiations directed toward merging the ownership and operation of the two houses were begun; by 1853 the board of directors of Scovill Manufacturing specifically recommended a merger of the two firms. The final agreement, however, had to wait nearly half a century, when both companies found themselves in much different circumstances.[44]

With the growth of popularity of daguerreotypy in the late 1840s and early 1850s, many new, specialized companies entered the ranks of supply manufacturers; but, likewise, many old and new companies failed; by the early 1850s one or a few companies in each sector had moved into a dominant position. The daguerreotype plate sector illustrated this pattern well. While at least four American companies produced plates in the early 1840s, of these only Scovill remained in production by 1850.[45] In the meantime, at least five new companies in New York, Boston, and Philadelphia initiated production, but of those only two remained in business in 1855.[46] With the French, Scovill, and the new entrants in the market, prices fell in 1850 to $2 per dozen for French plates and less than $1.40 per dozen for American plates.[47]

In the early 1850s, with the raising of the technical and capital requirements for production, one American and two French firms dominated the American plate market: Scovill Manufacturing Company of Waterbury, Charles Christofle of Paris ("Scale" brand; fig.

[44] Bishop, "Scovill and Photography," p. 441.

[45] See n. 18 above.

[46] Anthony, Clark, and Co. (N.Y.C.), ca. 1846–51; Levi Chapman (N.Y.C.), ca. 1851–56; Louis L. Bishop (N.Y.C.), ca. 1851; Benjamin French (Boston), ca. 1848; and A. Beckers and V. Piard (Philadelphia or N.Y.C.), ca. 1851–55. Other companies of this period include: Holmes, Booth & Hayden (N.Y.C.), 1854–[?]; John W. Norton (N.Y.C.), ca. 1856; Jones and Co. (N.Y.C.); and Pemberton and Co. (Conn.).

[47] Newhall, *Daguerreotype in America*, p. 120.

Fig. 1.22 Trademark Charles Christofle et Ce "Scale" brand daguerreotype plates. From La Lumière, *11 May 1851.*

1.22), and Alexis Gaudin of Paris ("Star" brand).[48] In reflection of the new market structure, the prices on American plates were raised in January 1854.[49] In the related but less technically critical metallurgical products, mats, and preservers, the Scovill company maintained its dominant position. Although Chapman and other smaller producers also entered the market, few companies were better equipped technically and financially than the Scovill company for this production.[50]

In the area of photographic chemicals, the methods of production did not change notably during the 1840s and 1850s. Such requisite chemicals as gold chloride, the halogens, sodium thiosulfate ("hypo"), and silver nitrate were fine chemicals that required elaborate processing for relatively small yields. Iodine was obtained through the treatment of the mother liquors obtained from kelp. Similarly, bromine was obtained with treatment of salt brines.[51] One of the early methods of producing sodium thiosulfate was to treat sodium carbonate with oxides of sulfur; however, in the 1850s a much cheaper method based on the treatment of alkali residues from soda-making was introduced. The new method, which originated in the early 1850s in Great Britain, lowered the price of the sodium thiosulfate substantially and fostered the extension of its use to papermaking.[52]

In the early 1850s photography represented the largest market for most of the chemicals mentioned above. Although foreign supplies of such fine chemicals remained important in the American market, specialized producers of photographic chemicals, like Garrigues & Magee of Philadelphia and Andrews & Thompson of Baltimore,[53] came into being. David Alter and Edward and James Gillespie of Freeport, Pennsylvania, were among the first to produce bromine in the United States, and their production was devoted exclusively to the photographic market.[54] Edward Anthony of New York initiated the production of photographic chemicals; yet, many of the items on Anthony's extensive list were repackaged chemicals from American and European sources. Because of the limited demand and the complexity of production, the chemical sector of the industry retained an imperfect market structure. Yet, the few companies in this sector always faced the possibility of competition from other chemical companies that might seek to expand their product lines.

In the daguerreotype case sector of the industry, none of the early

[48] In 1850 Scovill sold 120,000 French plates; despite only forty weeks of production, nearly 185,000 Scovill plates were sold (Bishop, "Scovill and Photography," p. 444).

[49] Scovill Mfg. to J. M. Sunsquest, 8 March 1854, Letterbook, N.Y. Store, 1854, vol. 456, pp. 39–42, SC.

[50] *Dag. J.* 1 (1851): 189; and *Humphrey J.* 7 (1856): 9.

[51] F. Sherwood Taylor, *A History of Industrial Chemistry* (London: Heinemann, 1957), pp. 192–93.

[52] Thomas Richardson and Henry Watts, *Chemical Technology; or, Chemistry in Its Applications to the Arts and Manufactures*, 2d ed., 5 pts. (London: Baillière & Co., 1855–67), 4: 183–87.

[53] Edwin T. Freedley, *Philadelphia and Its Manufacturers . . .* (Philadelphia: Edward Young, 1858), pp. 209–11.

[54] *Dag. J.* 1 (1851): 345.

Fig. 1.23
Anthony case factory: covering and finishing room. From Anthony Cat. *Courtesy of Tom and Elinor Burnside.*

Fig. 1.24
Anthony case factory: gilding room. From Anthony Cat. *Courtesy of Tom and Elinor Burnside.*

producers survived beyond the middle 1840s, and during the subsequent few years the three major New York supply houses either directly undertook the production or gained control of producers of daguerreotype cases. In the early fifties Levi Chapman produced and marketed throughout the nation a large supply of low-priced cases.[55] At about the same time Edward Anthony entered upon production (figs. 1.23 and 1.24).[56] The Scovill company secured its supply of cases from Ogden Hall of New Haven beginning in mid-1847, but Hall encountered financial difficulties beginning in 1849,

[55] Marburg, "Origins of the Scovill Manufacturing Company," p. 479.
[56] Rinhart and Rinhart, *American Daguerreian Art*, p. 88.

and years later Scovill and Samuel Peck of New Haven became joint owners of the production facilities.[57]

In the early 1850s Samuel Peck made a significant innovation: the union case, an ordinary daguerreotype case made from a plastic composition of shellac and wood fibers pressed between hot rollers. In 1850 Halver Halverson, Peck's brother-in-law, had developed the plastic material, and Peck's case shop became one of the first plastic-molding plants in the United States.[58] Peck patented his union case but failed to prevent imitators. The advantages of the new plastic cases included their greater rigidity and strength and their greater attractiveness, since a variety of elaborate cover designs could be die cast in the plastic. The union cases soon attracted considerable popularity, and the Peck-Scovill company assumed a prominent position in the case sector, even exporting cases to England.[59] Many smaller companies entered the case business in the 1850s, but generally they were not commercially successful for any extended period of time.[60] In the early 1850s, Peck-Scovill, Chapman, and Anthony were the principal case producers.

In the camera and optics sector, like the photographic chemical sector of the industry, the number of American producers remained small, and imports from Europe were commonplace. The cameras and Petzval portrait lens (discussed below) produced by Voigtländer established the standard for the industry. However, these German optical products were quite expensive and therefore left a significant market for American products of lower cost and lesser quality. While individual instrument makers filled the demand for cameras and optics during the early 1840s, two large companies came to predominate. William and William H. Lewis, pioneer American producers of daguerreotype apparatus, specialized in camera boxes and camera stands, chemical boxes, baths, and headrests. As their business flourished, the company established a large factory outside New York. By 1851 the Lewises, employing from sixty to seventy men, boasted the largest daguerreotype apparatus factory in the world at Daguerreville, approximately a mile south of Newburgh, New York, on the Hudson; however, the expansion proved unsuccessful and within two years the Lewises sold the business and returned to a small shop in New York, where they continued for another quarter century to play a major innovative role in apparatus manufacturing. However, the position of the Lewises in the market at this time plummeted.[61]

Part of the problem the Lewis company encountered was the rapid rise in reputation of the optical and camera products of a new competitor, Charles C. Harrison of New York. Harrison possessed a

[57] Bishop, "Scovill and Photography," pp. 432–34.

[58] Williams Haynes, *Cellulose: The Chemical That Grows* (New York: Doubleday & Co., 1953), p. 226.

[59] W. B. Holmes to Samuel Peck & Co., 27 February 1854, Letterbook, N.Y. Store, 1854, vol. 456, p. 18, SC.

[60] Two that became prominent later were Holmes, Booth & Hayden (Waterbury; N.Y.C.) and Littlefield, Parson & Co. (N.Y.C.).

[61] *Dag. J.* 2 (1851): 212–14, 370; 3 (1851): 20; and 4 (1852): 11–12; and *Humphrey J.* 4 (1852): 28; 4 (1853): 287; and 5 (1854): 302.

Fig. 1.25 Anthony apparatus manufactory: cabinet shop. From Anthony Cat. *Courtesy of Tom and Elinor Burnside.*

Fig. 1.26 Anthony apparatus factory: machine shop. From Anthony Cat. *Courtesy of Tom and Elinor Burnside.*

strong background in optics, the more critical part of the "box" (the camera box) and "tube" (the lens). During the 1840s he operated his own daguerreotype gallery and studied optics and lens making with the optician and telescope maker Henry Fitz.[62] In the late 1840s Harrison started his own optical and camera shop in New York and soon won an enviable national reputation for his lenses,

[62] Newhall, *Daguerreotype in America,* pp. 144–45; and Julia Fitz Howell, "Henry Fitz," *Contributions from the Museum of History and Technology, Bulletin* #228 (Washington, D.C.: Smithsonian Institution, 1962), pp. 164–70.

Fig. 1.27
Anthony apparatus department: finishing room. From Anthony Cat. *Courtesy of Tom and Elinor Burnside.*

which were good quality and yet below the cost of the lenses imported from Germany.[63] During the middle 1850s, Scovill possessed the sole agency for Harrison cameras and lenses, the best-made in America at the time, thus assuring itself of a steady and reputable supply of optical and camera goods.[64]

Technological Conceptions and Strategies

During the first stage of growth of the photographic industry, the daguerreotype concept, i.e., the idea of a direct positive image produced on a silvered metal plate, and the related technology helped to mold the mind-set of the technical and business leaders of the industry, who in turn influenced the basic structure and character of the industry and the direction of technological innovations. Although many developments, such as the calotype, albumen, and wax paper processes, broke with the daguerreotype concept, these were not translated into popular innovations because of certain inherent technical or commercial limitations.[65]

After the introduction of the daguerreotype, improvements and modifications in optics and camera equipment came almost at once as an adaptive response to the special conditions of daguerreotypy. Since the daguerreotype provided a direct positive, the image pro-

[63] Marburg, "Origins of the Scovill Manufacturing Company," p. 466.

[64] Holmes to S. H. Walker, 27 February 1854, Letterbook, N.Y. Store, 1854, vol. 456, SC; and Bishop, "Scovill and Photography," p. 438.

[65] Calotype: image imperfections, soft focus, and licensing restrictions (Talbot patents); albumen and wax paper: insensitivity.

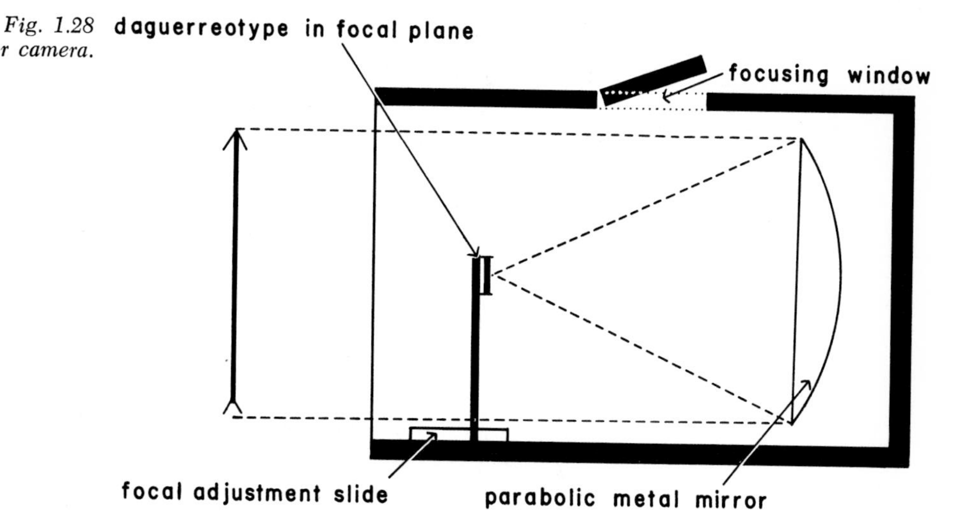

Fig. 1.28
Wolcott's mirror camera.

Fig. 1.29
Skylight gallery. From Photog. T., *19 (1889).*

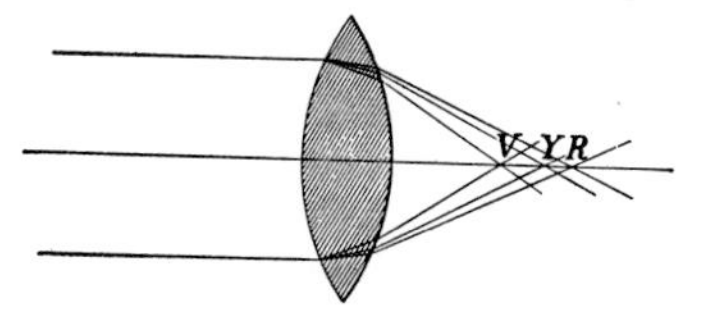

Fig. 1.30
Chromatic aberration in a simple lens. V *(violet), actinic or chemical focal point;* Y *(yellow), visual focal point; and* R *(red), red focal point.*

duced was the reverse of the original. Soon opticians introduced reversing prisms or mirrors to reinvert the daguerreotype image. For portraiture, reducing the exposure time was a primary goal. One method was to increase the light-gathering ability of the lens. An early effort in this direction was made by the American instrument maker, Alexander Wolcott, who designed a camera in which a large concave mirror replaced the lens, thereby increasing the light-gathering power and eliminating reversal of the image for direct positives (fig. 1.28).[66] Another and more significant early effort to improve light-gathering power was the specially designed and calculated daguerreotypic portrait lens of Joseph M. Petzval, an Austrian physicist. The Petzval lens, produced by Voigtländer and imitated by other optical producers, set the standard for the period. In a further effort to increase the light-gathering power of the portrait lens, German optical houses produced lenses of ever larger diameter. By the mid-1850s such lenses reached heroic size, with one twelve inches in diameter.[67] Reducing the period of exposure was also accomplished by increasing the sensitivity of the photosensitive materials. Successful efforts were devoted to the utilization of mixtures of halogen vapors in the daguerreotype sensitizing boxes. Opticians accustomed to producing camera obscuras for visual use had to recalculate the focal length of lenses for daguerreotypic use since the actinic focal length differed perceptibly from the visual focal length (fig. 1.30).

Other innovations that addressed problems of the daguerreotype included a variety of methods of improving the surface of the silvered plate, culminating in the adoption of electroplating and the tinting and coloring of daguerreotypes. The methods of display of a

[66] Helmut Gernsheim and Alison Gernsheim, *History of Photography: From the Earliest Use of the Camera Obscura in the Eleventh Century up to 1914* (London: Oxford University Press, 1955), p. 91 (a revised edition of this work, entitled *History of Photography, 1685–1914,* was published in 1969 by McGraw-Hill).

[67] Josef M. Eder, *History of Photography,* trans. Edward Epstean (New York: Columbia University Press, 1945), pp. 289–313.

daguerreotype also addressed its specific character: the assembly of mat, glass, and preserver addressed the fragile nature of the mercurized image and the necessity of protecting it from surface contact or dirt, and in turn, the daguerreotype case itself reflected the need for a protective case for this fragile "sandwich."

While the daguerreotype concept strongly influenced much of the innovation in the industry from 1839 to 1845, new ideas and developments during the first decade of daguerreotypy reflected the scientific origins of the process. Just as leaders of French science had acted as mentors in the introduction of photography, these and other leaders from science, intrigued by the theoretical implications of the photochemical reactions, continued to shower attention on the new process. The scientific journals carried a steady flow of articles on experiments and new developments and thereby served as the chief organs of dissemination of new information in the field. Because new ideas and developments were generally shared freely, attempts to restrict the free use of processes or new developments were deeply resented.[68] Furthermore, the small scale and decentralized character of the production of daguerreotypes reinforced this attitude, inasmuch as there was relatively little commercial advantage in patenting some new improvement.

At mid-century the attitudes and relationships began to change. After a decade the photochemical process no longer represented a frontier problem for science.[69] At the same time that photography was becoming more commercial, the scientific journals published fewer papers, and those they published were not usually of direct interest to the ordinary daguerreotypist. Soon photographic societies began to emerge and with them specialized photographic journals, which, while retaining a link with prominent scientists, served as trade journals to the daguerreotypists and gallery owners. With this change in relationships also came changes in attitudes. With the emergence of supply houses and specialized supply manufacturers, patent and trade secrets assumed a place in business strategies; yet, the attitude toward new ideas remained ambivalent. In general, the daguerreotypist accepted patents and trade secrets on producers' supplies, but he deeply resented any attempt to patent processes or improvements in processes that he might employ in his production. In any case, patenting of processes for use by daguerreotypists proved highly ineffective because of the difficulty of enforcement.[70]

The New York supply houses that also served as jobbers secured an increasingly important position in the industry and developed some significant attitudes and strategies. The leadership of the supply houses became committed to the daguerreotype concept: Chapman and Scovill because of their manufacturing interest in daguerreotype plates and related metallurgical supplies, Anthony

[68] Note the reactions to the English patents of Daguerre and Talbot. See Gernsheim and Gernsheim, *History of Photography*, pp. 159–60.

[69] For example, the fields of thermodynamics, kinetic theory, electromagnetic theory, and organic chemistry were much more attractive to the scientific community.

[70] None of those who sought to exploit their patents through the sale of licenses profited substantially: Talbot, Daguerre, Langenheim, Whipple, and Cutting.

because of his linkage to chemical production; and all three because of their involvement in case production. But during the 1850s the management also became increasingly committed to the jobbing and marketing function, treating production of supplies as simply a part of the strategy of the supply function. They acquired the assumptions, attitudes, and strategies of the merchant. These houses handled products that competed with those of their own production, encouraged the sales of those products for which there were high profits, and sought to acquire sole agencies from producers of popular items.

In the area of innovation, these supply houses basically pursued a strategy of externalizing the risk of innovation. They conceived of the process of creation of new products as unpredictable, arising from genius fortuitously inspired. They believed that though concentrated effort on a problem might produce moderate improvement, systematic pursuit within the business was unlikely to produce genuinely new ideas or solutions to major problems. Therefore, they pursued a policy of allowing new ideas to be generated outside the firm, keeping alert to the ideas of genius wherever they occurred, and seeking to gain control of new ideas when they appeared. If a new product sold well, then the supply house sought a sole agency or possibly an interest in the new company, but even if these efforts failed, they would be assured of the jobbers' profits. They assumed, of course, that the innovating firm would always be smaller than the supply houses and dependent on supply houses for its marketing. In the case of a really novel and apparently revolutionary development —for example, Levi L. Hill's claim to have developed a color photographic process—the leaders of the supply houses (in Hill's case, Mallory of the Scovill company) sought to invest in the innovation, but even then they were seeking an interest in an idea that had arisen outside the business.[71] Instead of stimulating new ideas within their firms, they scoured Europe for novelties and kept alert to the American daguerreotypic scene.

In the early 1850s, when the daguerreotype was at the height of its popularity, a new photographic process was introduced. A major change in the photosensitive carrier and base technology soon tested the strategies of each of the three dominant firms. The influence of this testing and the response to it helped create a new mind-set, which then dominated the industry for the next quarter century.

[71] Bishop, "Scovill and Photography," p. 446.

Collodion Period 1855 to 1880

The boy who gets a "nibble" when fishing is carefully watched by his companions, and if he catches a fish, the waters in which he angles, is soon a perfect network of lines, and the fish which he hoped to carry home on his own string, are divided up among so many fishermen, that no one has a meal, and the fishing is spoiled for all. Men in their habits toward one another, in the business world, are only overgrown boys fishing; with the same impulses, the same selfishness, and the same disregard of right, only covered by greater knowledge and increased power to deceive.

Victor Moreau Griswold, c. 1871

2/Glass Plate and Paper Photography

Introduction

While the American cultural scene did not alter radically during the middle of the nineteenth century, the economic setting underwent substantial change. The Civil War disrupted business relations between the North and the South, created a high demand for manufactured products in the North, and left in its wake economic lethargy in the North and devastation in the South. A major depression occurred in the middle and late 1870s that accelerated a general trend of falling prices. The completion of transcontinental railway lines opened the region west of the Mississippi River to settlement. This settlement of the western land and the growth of the national population between 1850 and 1880 from less than 25 million to more than 50 million created a large and broadcast domestic market.[1] At the same time, however, American cities were experiencing unprecedented growth, with New York reaching more than a million and Philadelphia a half million by 1860.[2] This urban growth partially reflected the increasing tempo of industrial development of the nation, particularly in the traditional sectors of textiles, leather, and food processing and the newer sectors of steel and metal fabrication.

Organizationally, after the middle of the century manufacturing concerns began to shift from single proprietorships and partnerships

[1] U.S., Department of Commerce, Bureau of the Census, *Historical Statistics of the United States, Colonial Times to 1957* (Washington, D.C., 1960), p. 8.
[2] U.S., Census Office, *8th Census, 1860 Population of the United States* (Washington, D.C., 1864), pp. 337, 432.

to the new, limited-liability corporate form of enterprise, reflecting the desire for organizational stability and longevity and the increased level of capital requirements in mechanized manufacturing. While participating in this economic environment, the photographic industry underwent an important technological change that helped reshape its structure and character. The collodion photographic processes, the use of which characterized the second stage of the history of the industry, influenced the ideas and concepts of business and technical leaders and helped to mold the structure of the industry from the middle 1850s to the late 1870s.

1855–65: Toward a More Competitive Market Structure

The development of the new collodion carrier for photosensitive silver salts originated outside the commercial activities of the photographic industry. In 1846 a German-Swiss chemist, Christian F. Schönbein, discovered that raw cotton treated with an appropriate mixture of nitric and sulfuric acids possessed explosive properties. Shortly thereafter, other chemists observed that certain types of this nitrated cotton, called guncotton, were soluble in sulfuric ether. The dissolved guncotton, which was transparent and sticky, was called collodion, meaning "adhere" in Greek.[3] In the fall of 1848 Frederick

[3] James R. Partington, *History of Chemistry*, 4 vols. (London: Macmillan & Co., 1961–70), 4: 190–96.

Fig. 2.1
Dark room in the time of the wet collodion process. From Tissandier Hist.

Fig. 2.2
Printing frame. From J. J. Payne, Descriptive Catalogue, *1857.*

Scott Archer, an English sculptor interested in photographing his models, initiated experiments with collodion and by the following summer was employing glass plates with a collodion surface in his camera. His published account of this wet-collodion-on-glass photographic process in the March 1851 issue of the *Chemist* initiated the movement to adopt collodion as the carrier for the photosensitive material.[4]

The technical complexity of this wet collodion process, like that of the daguerreotype, limited its use to professional photographers and the most ardent of amateurs. Yet, the collodion process possessed some decisive advantages. Most notably, it enabled photographers to produce multiple images because it was a negative-positive rather than a direct-positive process, as was the daguerreotype. The transparent collodion served as a carrier for the halogen salts while adhering to the glass plates. Just prior to exposure in the camera, the photographer treated the collodion plate with silver nitrate, thereby creating on the plate a photosensitive silver halide surface. Upon exposure in the camera, the light reduced a small number of silver halide grains to metallic silver, creating a latent negative image. Subsequently, the photographer employed certain reducing agents, such as gallic acid, or protosalts of iron, to develop the latent image through the accumulation of additional silver about the light-reduced silver halide grains. He then washed the developed plate with the fixing agent, sodium thiosulfate ("hypo"), which removed the remaining unexposed silver halide grains. What remained was a visible negative image composed of dark areas of opaque silver where light had struck and transparent areas where light had not struck the plate; from this negative image multiple positive prints could later be drawn (fig. 2.3).

The photographer prepared photographic paper for positive printing by coating paper with a silver halide salt, often silver chloride. When he wanted a positive print he placed the negative glass plate immediately above the photosensitive paper and exposed it for long periods to sunlight. The sunlight would strike the transparent portions of the negative and be transmitted to the paper where, through prolonged action, it reduced the silver salt to metallic silver. Beneath the opaque portions of the negative, the photosensitive salt remained unaffected. Upon the appearance of the sunlight-developed image on the paper, the photographer removed the paper from the sunlight and treated it with the fixing agent, sodium thiosulfate. He thus removed the unexposed silver halide grains and obtained a positive image consisting of whitened areas where light had originally affected the negative and darkened silver areas where light had not originally struck the negative. In order to execute this new collodion process, the photographer needed, in addition to his traditional camera and lens, glass plates, collodion, halide salts, silver nitrate, such developers as protosalts of iron or gallic or pyrogallic

[4] Helmut Gernsheim and Alison Gernsheim, *The History of Photography: From the Earliest Use of the Camera Obscura in the Eleventh Century up to 1914* (London: Oxford University Press, 1955), pp. 152–55.

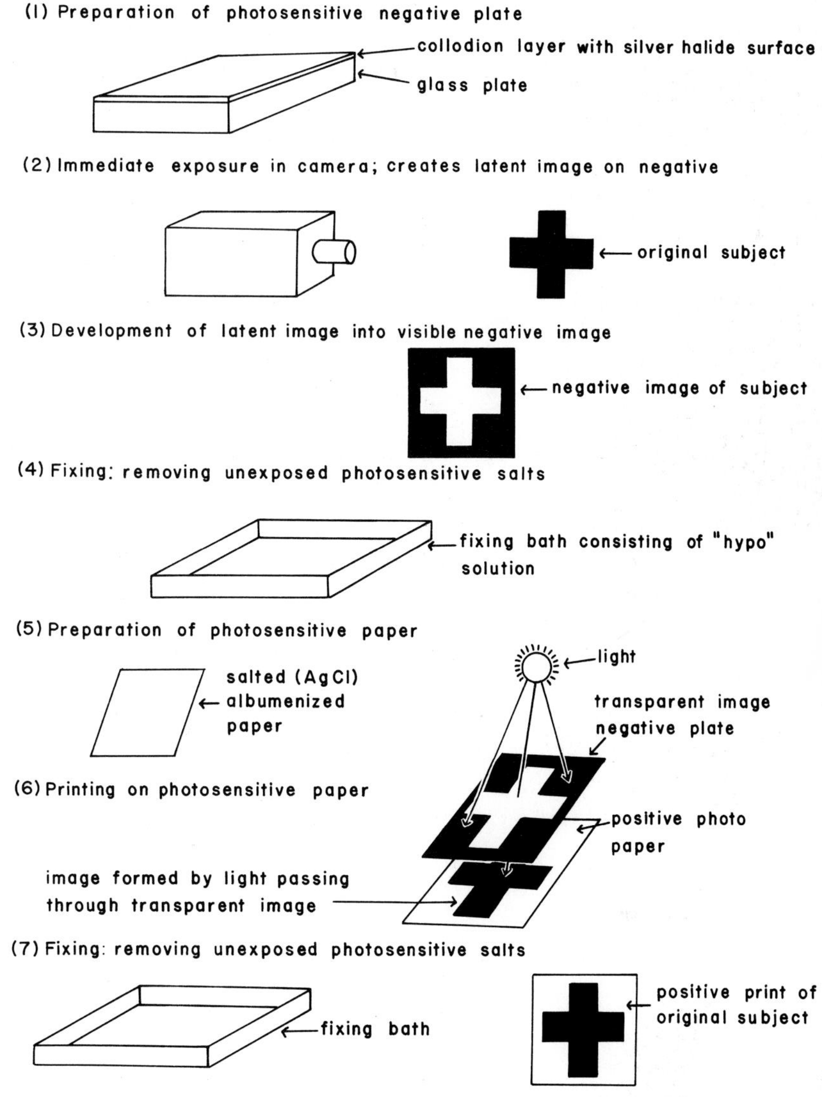

Fig. 2.3
The Collodion Process.

acids, hypo, special photographic papers, varnishes (to preserve the surface of the photograph), bath pans, and ancillary chemical apparatus. Moreover, the field photographer needed to be both technically oriented and also sufficiently muscular to bear the burden of his apparatus and small laboratory (figs. 2.4 and 2.5).

The collodion process possessed disadvantages and advantages compared with the daguerreotype. The new positive-negative process required more steps of production, did not attain the fineness of detail, and produced a heavier, breakable negative. However, it came to replace the daguerreotype because of its potential for multiple prints, greater sensitivity, and lower cost. While all the advantages of the collodion process were not initially realized, its immediate and potential advantages led to its general acceptance.

In the years between 1853 and 1855, a few of the most prominent daguerreotype artists in America adopted the paper print because of its desirable artistic qualities, and they served as the agents of

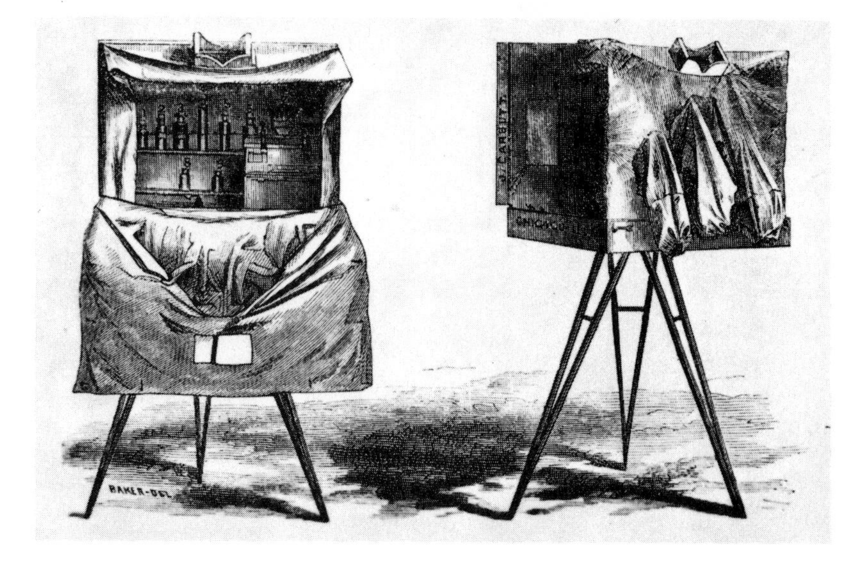

Fig. 2.4
Dark tents. From Philadelphia Photographer, *1866; and Hermann W. Vogel,* La photographie et la chimie de la lumière *(Paris: Baillière, 1876).*

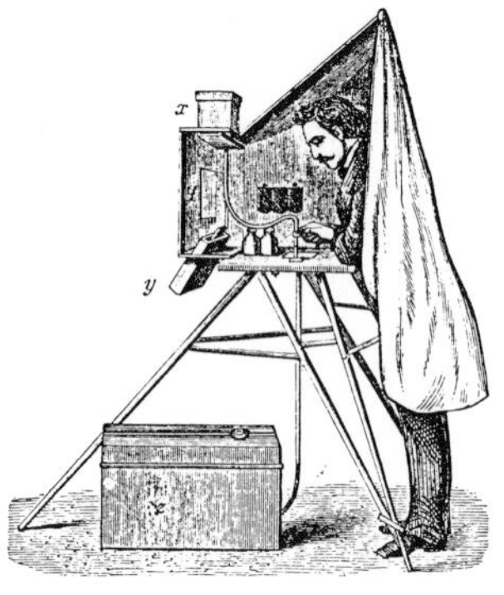

Fig. 2.5
Field photographer and his equipment, wet collodion period. From Tissandier Hist.

change.[5] In general, however, most daguerreotypists and most patrons of the less elite galleries did not at once accept the new form of photography.[6] The shift from direct positive to negative-positive photography, the change to a new carrier for the photosensitive material, and the adoption of a new base for the carrier represented three major conceptual and technical modifications, the combination of which was too radical for immediate acceptance. Consequently, during the 1850s most photographers, who were naturally wedded to the older techniques, first adopted two intermediate direct-positive collodion processes, the ambryotype and the tintype, before gradually adopting the more radical wet collodion negative-positive photography.

The *ambryotype*, which became popular in the middle and late 1850s, consisted of an ordinary wet collodion glass negative with either a black background behind the glass or black varnish on the back of the negative. The negative image ordinarily seen with transmitted light became with the black background a positive image seen by reflected light (fig. 2.6). Hence, ambryotypes were

[5] Boston: Whipple; New York: Gurney, Lawrence, and Brady; Philadelphia: McClees and Germon; Chicago: Hesler; and St. Louis: Fitzgibbon.

[6] Robert Taft, *Photography and the American Scene: A Social History, 1839–1889* (New York: Macmillan Co., 1938), pp. 119–22.

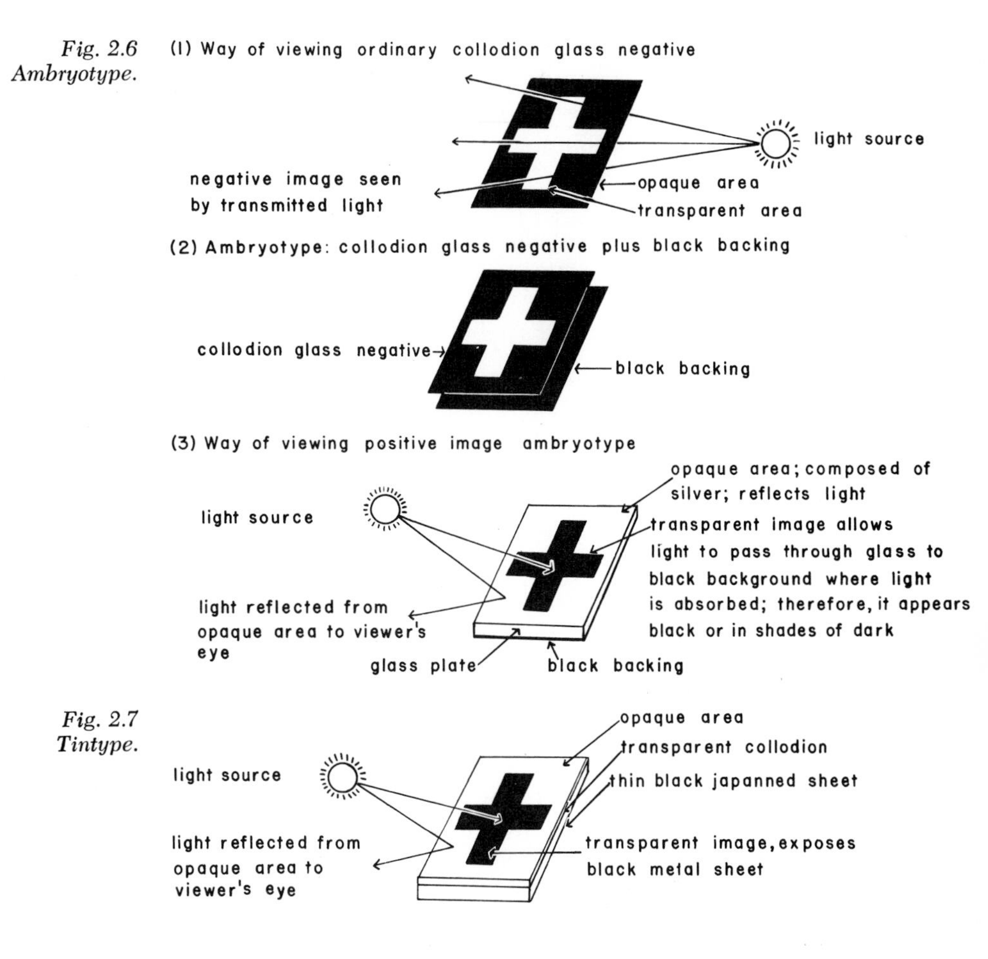

Fig. 2.6 Ambryotype.

Fig. 2.7 Tintype.

direct positives like the daguerreotvpe, only on collodionized glass plates rather than on silver metal. A second direct-positive form of collodion photography was the *tintype*, which consisted of wet collodion on thin, black lacquered or japanned iron sheets (fig. 2.7).[7] This style of photography, which was introduced in 1857, endured for three quarters of a century because tintypes were much cheaper than any other form of photography. By the late 1850s the positive print on paper had been widely adopted, and during the sixties and seventies nearly all photographs were made on albumenized paper or, for the cheaper trade, on the tintype plate.

The first decade of the collodion period witnessed numerous innovations. Many daguerreotypists ignored the new techniques, arguing that the new process was uncertain but the old was well understood and could be reliably employed. Those who refused to change eventually left the business, but the majority joined the growing ranks of photographers and tintypists. Moreover, the Civil War provided opportunities for itinerant photographers. Also, the development of a new and popular entertainment, the stereoscope, provided some photographers with the opportunity to devote at least part of their time to taking field photographs for the production of stereoscopic view cards. In general, the change to the collodion tech-

[7] Also known as the ferrotype, melainotype, melanotype, etc.

Fig. 2.8
Peter Neff. From the
Bulletin of the Geological Society of America.

nology did not strongly influence the structure of the photograph production sector of the industry (the photographers); it simply presented new opportunities. With the leaders of the manufacturing and jobbing sectors, the change had a more profund influence.

A New Manufacturing Sector: Tintype Plate

One of the most important new manufacturing opportunities was the production of tintype plates. The popular acceptance and mass production of the cheap and almost indestructable tintype began in the mid-1850s and lasted until the close of the century. Entry into and dominance in this field was independent of previous large-scale photographic materials and manufacture, although in most cases manufacturers came either from the ranks of photographers or from the patent-leather business, where processes were similar to those employed in tintype plate production. The basic problems confronting the tintype manufacturers were price competition and oversupply; ultimate success in solving the problems derived from coordinating strategies in technology, marketing, and supply of raw materials.

The first discussion of a tintype-style process occurred in 1853 when a French natural philosopher, Adolphe Alexandre Martin, presented a memoir to the Academie des Sciences in Paris. The first published details appeared shortly thereafter in the Academie's *Compte rendus.* As in France, the scientific community introduced the process in the United States. Apparently without knowledge of Martin's work, Hamilton L. Smith,[8] professor of natural science at Kenyon College, Gambier, Ohio, experimented with collodion processes and direct positive images in 1853–54, with the assistance of a young seminary student, the ambitious Peter Neff, Jr. (fig. 2.8).[9] Both men continued their experiments independently in 1855 and soon succeeded in producing a successful tintype. At the urging of Neff, who prepared the patent specification and paid the fees, Smith applied for a patent on the process of making tintype photographs. Upon the issue of the patent in February 1856, Smith assigned it to Neff and his father, a wealthy Cincinnati businessman.[10]

Reflecting a new attitude toward innovation and patents, Neff

[8] Smith graduated from Yale with both B.A. (1839) and M.A. (1842) degrees, lectured in the early 1840s on chemistry at the Western Reserve Homeopathic Medical College in Cleveland, and later became a world-renowned microscopist and geochemist at Hobart College (*Cleveland News,* 1 June 1896, Western Reserve Historical Society Biographical Files, Cleveland, Ohio).

[9] Neff graduated Phi Beta Kappa with B.A. (1849) and M.A. degrees from Kenyon and later from Bexley Hall, an Episcopal seminary in Gambier, Ohio. See obituary notice in the *Cleveland Plain Dealer,* 12 May 1903; and "Peter Neff, Jr., Questionnaire, 31 January 1902," Archives, Kenyon College, Gambier, Ohio. Mr. Thomas B. Greenslade, college archivist at Kenyon, kindly supplied me with this valuable questionnaire. Also see his son's article: Thomas B. Greenslade, Jr., "The Invention of the Tintype," *Kenyon Alumni Bulletin,* 29 (July–August 1971): 16–23.

[10] U.S. Patent #14,300 (19 February 1856); and "Peter Neff, Jr., Questionnaire."

Fig. 2.9
Victor M. Griswold. From Robert Taft, Photography and the American Scene, *A Social History, 1839–1889* (New York: Macmillan Co., 1938; reprint ed., New York: Dover Publications, 1964). *Courtesy of Dover Publications, Inc.*

initially conceived of the Smith patent as the key to successful commercial exploitation of the new photographic process. In the spring of 1856 he opened a tintype gallery in Cincinnati and promoted the new process through the publication and gratis distribution of his pamphlet, "The Melainotype Process Complete." In the summer of 1856, after obtaining thin charcoal iron plates from England, Neff established a plant in Cincinnati for the japanning of the thin plates.[11] The following summer fire destroyed the factory, and he relocated it at Middleton, Connecticut, nearer the Eastern markets. When production difficulties led to his failure to supply the leading New York supply houses and when the patent on the tintype process proved unenforceable, Neff sold the Connecticut factory to its local manager and spent the remainder of his life engaged in a series of unsuccessful business ventures.[12]

Another Ohioan, Victor Moreau Griswold, an enterprising, relatively young painter and photographic portrait artist with a definite experimental inclination, explored the new collodion processes, introduced some new ideas, and in July and October of 1856 obtained his first patents (fig. 2.9). One patent covered an improvement in collodion and the other the introduction of a bromine-iodized bitumen carrier for photographs, similar to the japanning material on tintypes.[13] In late 1856 or early 1857, a few months after Neff's introduction of the manufacture of tintype plates, Griswold borrowed a small sum of money from his father and began producing tintype plates in Lancaster, Ohio. During the late fifties and early sixties facilities were expanded several times, and he opened a new factory of even greater capacity in Peekskill, New York. By 1862 the company had two factories with a dozen ovens and between forty and fifty employees.[14]

Such rapid expansion reflected an initially successful marketing strategy. Handling his own marketing, Griswold sent samples of plates to the leading stock dealers of the country with the request that the plates be given to prospective customers among the professional photographers. The demand increased soon thereafter. In the late 1850s he also published a small trade paper called *Camera*.[15] With the opening of an Eastern plant in 1861, the firm also opened a salesroom and office in New York City.[16]

Combined with successful marketing techniques, Griswold carefully pursued product innovations. He enjoyed experimenting and was able thereby to improve the endurance of the japanned surface; by 1857 he had successfully produced japanned surfaces tinted in blue, green, red, chocolate, and white. Despite his efforts to introduce tinted plates to the national market, the conservatism of pro-

[11] In his patent Smith described how to japann the surface: #14,300.

[12] The factory was sold to James O. Smith. "Peter Neff, Jr.," D & B, Ohio, vol. 79, p. 249; "Peter Neff, Jr., Questionnaire"; and Edward M. Estabrooke, *The Ferrotype and How to Make It* (Cincinnati: Gatchel & Hyatt, 1872), pp. 99–100.

[13] U.S. Patent #15,336 (15 July 1856) and #15,924 (21 October 1856).

[14] Victor M. Griswold's own account is quoted in Estabrooke, *Ferrotype*, pp. 80–82, 86–87.

[15] A decade later Scovill and Anthony imitated this business tactic.

[16] Griswold's own account in Estabrooke, *Ferrotype*, pp. 80–87.

fessional photographers doomed this effort to failure. Somewhat later he also introduced tintype plates that were coated on both sides, and he introduced the two styles of surface coating, glossy and eggshell, that became standard in the industry for the remainder of the century.[17] In the late 1850s Victor Griswold reduced production costs by cutting the plates after coating and baking rather than beforehand, by introducing a process requiring a much shorter baking time, and by employing heavy cardboard boxes instead of the costlier and heavier wooden crates previously used for packing and shipping.

The two firms producing tintypes in the early 1860s showed a marked contrast in the internal administration of their business, perhaps reflecting differences in the personalities of the businessmen who managed them. The Griswold firm, although lacking capital, had the advantage of Griswold's fifteen years' successful experience in the photographic industry, whereas the Neff operation, with substantially more capital than Griswold's, was led by a man who possessed less business creativity and stamina and who had considerably less photographic experience. Griswold was persistent and creative in both marketing and technology; Neff was not. Griswold closely integrated the marketing, production, and administrative functions; Neff delegated such functions in a more decentralized style of operation.

During the first five years of the companies' existence, the market structure was duopolistic (that is, there were only two sellers of the commodity in question), but the behavior of the two firms only partially reflected the predicted theoretical behavior. The introduction of the tintype was slow compared to that of the daguerreotype and the ambryotype because of patent restrictions on the practice of the process and because, initially, the two plate manufacturers operated their own marketing departments independent of the established New York jobbers. Griswold's initial advantage of several months head start was maintained for a couple of years. Neff responded to Griswold's entry into the business by threatening lawsuits against both him and individual supply houses for infringing upon the Smith patent. Griswold, recognizing the weakness of the Smith patent—it did not cover the production of japanned plates—ignored the threats, confident that Neff would not make the weakness of his position public by bringing the matter to court.[18] Griswold cut his prices as he improved production methods. Neff, who had sought to sell licenses to practice the process, responded in 1858 by making the licenses gratis, hoping that removal of practice restrictions would increase his plate sales. Recognizing Griswold's threat, Neff also sought to purchase the competing firm in order to regain his initial monopoly. Griswold, unwilling to forego the prospect of future profits, counteroffered that they merge, but on terms that were indignantly rejected by Neff.[19] Thus, the market structure for the first five years of tintype manufacture remained duopolistic.

[17] Ibid., pp. 81–83.
[18] Ibid., p. 83.
[19] Ibid., pp. 79–84.

The Civil War in 1861 brought a great demand for the popular tintype as the soldiers sent inexpensive pictures to loved ones at home. Because the two established firms failed to meet the growing demand for plates and because they initially conducted much of their marketing independent of the largest jobbing houses, the jobbing and supply houses, seeking the jobber's profits but not anxious to assume the risks of engaging in manufacture, actively encouraged new manufacturers to enter the market.[20] The experience of the first five years in tintype manufacture had indicated that weak patents were worthless, especially when they were so regarded by a rival. During the Civil War, about half a dozen small companies began production of plates. Most of these companies, because of their previous association with patent-leather manufacture, which employed processes similar to japanning, were able to overcome the technical barriers to entry into tintype production.[21] Therefore, the center of the tintype industry shifted eastward from Ohio to the center of the patent-leather industry, Newark, New Jersey, close to the New York City supply jobbers, who were interested in freeing themselves from reliance on Griswold and Neff.[22] Accordingly, the tintype-plate-manufacturing sector moved from a duopolistic structure to one more closely approaching perfect competition.

Alteration in Industry Leadership: The New York Jobbers and Allied Manufacturers

While the technological change from daguerreotype to collodion processes helped to create a new manufacturing sector—the producers of tintypes—it also profoundly influenced the three leading supply houses, which between 1855 and 1865 underwent internal reorganizations and shifts in their relative positions of leadership. The weakest of the three New York jobbing houses prior to the advent of the collodion processes was that of Levi Chapman. Chapman's strong commitment to manufacturing of daguerreotype plates, cases, mats, and preservers, combined with a conflagration of his factory in 1856, placed the firm in a vulnerable position at the time of major technological change.[23] Chapman did not respond to the new processes and filed bankruptcy early in 1857. Reorganized under the name of Levi's teen-aged son, the firm continued as a local supply depot.[24] Later Chapman directed attention to production of cameras and albumenized paper, but the company was never

[20] Ibid., pp. 90–91.

[21] Charles H. McDermott, ed., *A History of the Shoe and Leather Industries* (Boston: Demehy, 1918), pp. 206–7.

[22] Ibid., p. 206; *National Cyclopedia of American Biography*, s.v. "Seth Boyden"; Charles T. Davis, *The Manufactures of Leather* (Philadelphia: H. Carey Baird, 1885), p. 70; J. Leander Bishop, *A History of American Manufacturers from 1608 to 1860*, 3rd ed., 3 vols. (Philadelphia: E. Young, 1868), 1: 443; Victor S. Clark, *History of Manufactures in the United States*, 3 vols. (1929; reprinted, New York: Peter Smith, 1949), 2: 466–67; and Estabrooke, *Ferrotype*, p. 94.

[23] "Levi Chapman: March 15/55 and March 31/56," D & B, N.Y.C. Series 8, p. 379.

[24] "Levi Chapman: Feb. 19/57, Feb. 24/57, and Dec. 18/57," in ibid.

able to regain either the manufacturing or jobbing status it had held in the early 1850s.

Like Chapman, Scovill Manufacturing Company had a deep commitment to the production of metallurgical products related to daguerreotypy, especially the silvered plate, which the advent of collodion soon made obsolete. Moreover, at the time of the technological change, the two Scovill brothers died and Mallory, manager of the New York store, left. Consequently, new leadership entered simultaneously with the new technology. Samuel Holmes, Mallory's assistant, became manager of the New York store[25] and consolidated the company's strengths, redirected its strategy, and succeeded in maintaining profits in the late 1850s.[26]

As plate orders fell precipitously in the middle 1850s, nothing could be done to regain the immense profit loss represented by this key sector of the Scovill business;[27] however, the case business, conducted by Samuel Peck & Company, continued strong because of the demand of both ambryotypists and tintypists. Yet, as case producers became numerous once again in the mid 1850s, competitive pressures increased.[28] Difficulties in obtaining a reliable flow of cases from New Haven prompted Scovill to acquire in 1857 all the stock in Samuel Peck & Company and to relieve Peck of his managerial duties.[29] Other companies—such as Littlefield, Parson & Co. (Mass. and N.Y.) and A. P. Critchlow & Co. (Northampton, Mass.) —sought to employ patents, as Peck had done, to restrict competition, but no litigation ensued.[30] Therefore, patents played no important role in this sector. After 1860 the potential for profits in case production decreased because only tintypes, which were directed to the low-priced trade, continued to be mounted in cases.

During the period of major technological change in the middle 1850s, the ownership and business organization of the optical-camera sector of the industry increased in complexity. At the same time this sector grew in domestic and international reputation. In 1855 Edward Anthony company combined its camera factory with that of Charles C. Harrison's optical shop.[31] Scovill's forfeiture of its sole agency with Harrison represented an initial step toward that union then under consideration by the Anthony and Scovill companies. During the late 1850s Scovill exercised more control over the Harrison operation than did Anthony, but early in 1862 the two firms inexplicably transferred ownership of the factory to a Mr. Nelson Wright of New York. Scovill's withdrawal may have been a

[25] *Humphrey J.* 7 (1856): 8 (advertisement).

[26] "Scovill Mfg. Co., New York City: Jan. 23/53, Mar. 14/56, April 23/58, and Nov. 29/59," D & B, N.Y.C. Series 7, p. 133.

[27] Inventory Book B (1858–73), SC.

[28] New entrants included Edward G. Taylor (N.Y.C.) in 1855, John Barnett (N.Y.C.) in 1855, and DeForest Bros. (New Haven County) in 1855–57.

[29] Edward Atwater, *History of the City of New Haven, Connecticut* (New York: Munsell, 1887), p. 625; and Bishop, *History of American Manufacturers,* 1: 441.

[30] Beaumont Newhall, *The Daguerreotype in America* (New York: New York Graphic Society, 1968), pp. 131–32.

[31] P. W. Bishop, "Scovill and Photography: The Tail That Almost Wagged the Dog" (unpublished manuscript), p. 438.

Fig. 2.10
Genealogy of Harrison Optical Company.

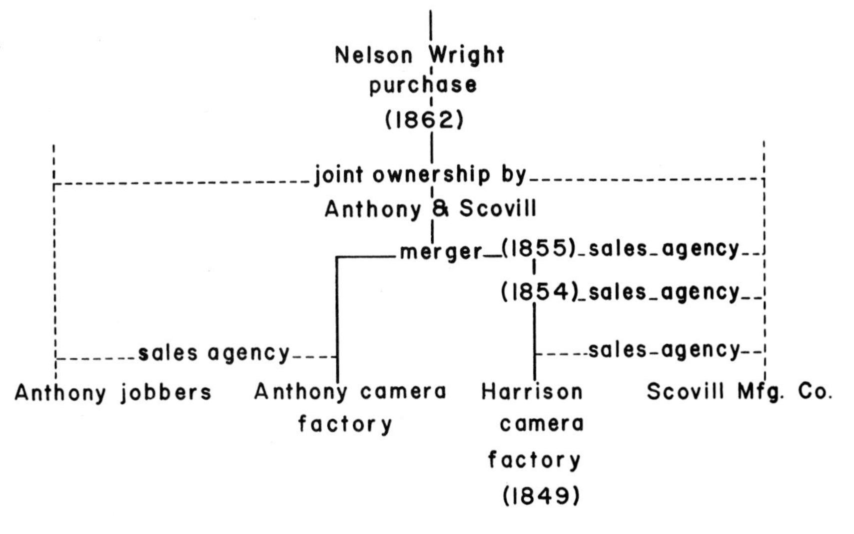

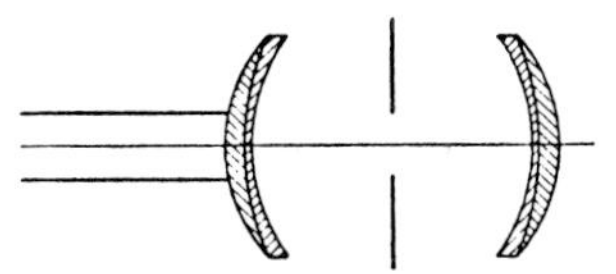

Fig. 2.11
Globe lens. From Moritz von Rohr, Theorie und Geschichte des photographischen Objektivs *(Berlin: Julius Springer, 1899).*

part of the more general change in Scovill business strategy, which was prompted by a response to the major technological change of the mid-fifties: a gradual withdrawal from the manufacturing sectors of the business and an increased emphasis on the jobbing of apparatus and materials produced by smaller, outside firms that carried the risk in the event of economic depression or further major technological change. The introduction of a new and very successful lens may also have contributed to the change in ownership. In 1860 the talented Harrison and an associate, I. Schnitzer, designed a new landscape lens called the Globe lens (fig. 2.11). This new symmetrical doublet soon acquired an international reputation as a fine quality and relatively inexpensive lens.[32] Under Wright's managership the optical and camera firm retained Harrison as supervisor of production and lens designer.[33] Harrison's world-renowned ability as an optical designer and his willingness to innovate helped to maintain the company's position of leadership among American optical and camera manufacturers.[34]

Meanwhile, a number of new companies had entered the market,[35] but they faced difficulties in competing with established firms in sectors little affected by the major change in technology. Moreover, they also had difficulty making the transition in sectors directly affected by the technological changes. The entry in 1853 of the firm of Holmes, Booth & Haydens well illustrates such difficulties. A group of men previously associated with either the Scovill company or with brass manufacturing in Waterbury established this well-capitalized company with manufacturing facilities in Waterbury and a supply depot near the Scovill company in New York. In direct imitation of Scovill, the new firm manufactured products in both the

[32] Moritz von Rohr, *Theorie und Geschichte des Photographischen Objektivs* (Berlin: Julius Springer, 1899), pp. 174–77.

[33] Bishop, "Scovill and Photography," p. 439; and advertisement in John Towler, *The Silver Sunbeam* (New York: Ladd, 1864).

[34] Rohr, *Photographischen Objektivs*, pp. 174–77.

[35] In New York: Henry J. Lewis, John Stock & Co., George Chapman, E. M. Corbett, and E. Gordon; in Philadelphia: Woodward.

hardware and the daguerreotype trades.[36] For the photographic trades it produced mats, preservers, cameras, and plates. Entering the plate business at a late stage, the company secured the services of August Brassart, a French plate maker and former associate of Daguerre.[37] In following the same basic product line as Scovill, the fledgling firm soon felt the impact of the new collodion technology, and its entry into metallurgical phases of photographic materials was not a sustained success. It shifted its emphasis to camera production in the late 1850s and later to tintype plate production. In camera production the company's initial reputation for excellence allowed it to compete with Harrison; however, by the early to mid sixties, the superiority of Harrison's new optics weakened the Holmes, Booth & Haydens position in the camera market, too. In addition several smaller camera companies also entered upon production and had some small initial success.[38]

The destructive effects of the technological change were minimal for the firm of Edward Anthony. Anthony, as one of the largest jobbing houses in the country, was less committed to manufacturing than were Chapman or Scovill, and the areas of manufacture it adopted—daguerreotype cases and chemicals—were not immediately affected by the change. Instead, the chemical sector experienced a rapid increase in demand because of the many chemical needs of the new process. Soon the company produced guncotton, collodion, iodides, bromides, silver nitrate, gold chloride, various varnishes, and many other photographic chemicals.[39] However, the Anthony company had competition in the production of new chemicals: the fine chemicals companies of the Philadelphia area expanded their production of photographic chemicals; French chemicals continued to be imported; and a large number of very small, specialized chemical companies arose in the New York area.[40] Many of the supply houses across the country also began to produce their own salted collodion and other chemical specialties, featuring them under their house names. Nonetheless, the Anthony company was the largest and most vigorous specialized photographic chemical producer in the United States. The change to collodion expanded the company's opportunities for and commitment to chemical production.

Soon the Anthony company began to produce albumen paper. Paper negatives and positive paper prints had been employed since the early 1840s when Talbot introduced his calotype process, but the material used for positive prints was simply well-sized paper that had been salted with the appropriate silver salt. The resulting

[36] The new firm was organized on 2 February 1853 with capital of $110,000, which was soon raised to $200,000, the nominal capitalization of the Scovill Manufacturing Co. (Joseph Anderson, ed., *The Town and City of Waterbury, Conn.,* . . . 3 vols. [New Haven: Price & Lee, 1896], 2: 352).

[37] Ibid., 3: 1041–42.

[38] *Humphrey J.* 6 (1854): 137 and 249, and 7 (1855): 2.

[39] Advertisement in Towler, *Silver Sunbeam.*

[40] Daguerre Manufacturing Co. (N.Y.C.), Seely & Boltwood (N.Y.C.), R. A. Lewis (N.Y.C.), and S. D. Humphrey (N.Y.C.). Powers & Weightman of Philadelphia began production of photographic chemicals at this time. See advertisements in ibid.

images were usually deep in the surface of the paper and diffused; however, in 1850 Louis-Désiré Blanquart-Evrard introduced a paper with a thin coating of salted (but unsensitized) albumen on its surface.[41] This surface helped to retain the sharpness of detail from the negative by keeping the image on the surface. With the advent of the collodion negative-positive process, the use of albumenized paper grew rapidly, and with it the consumption of egg whites. The presence in the raw paper of impurities that interfered with the photosensitive salts posed one of the principal problems in the use of paper in photography. Production of quality photographic raw stock required not only high-quality paper with a high rag content but also considerable care in avoiding contamination by impurities, for example, metal buttons or clasps in the rags or reactive minerals in the water. Because mineral-free water was required, during the nineteenth century only mills in two locations in the Western world succeeded in producing satisfactory photographic raw stock: Steinbach of Malmedy, near the Belgian-Prussian border, and Blanchet Freres & Kleber of Rives, in northeastern France.

Because of their proximity to these sources of relatively pure paper stock, specialized German photographic paper producers arose in the late 1850s and early 1860s and established a worldwide reputation for outstanding quality and reliability. In the late fifties the papers of these companies began to reach the American photographic market.[42]

The demand for albumen paper increased substantially during the Civil War period because of the great poularity of the *cartes de visite* photographs, which were printed on albumen paper. At first Anthony imported raw paper from Germany and France and sold it along with collodion and other chemicals to photographers, who did their own albumenizing. Then, in the late 1850s and early 1860s, the company imported from Germany paper that had already been albumenized. Having tested the market and found a demand for the product, Anthony then decided to move into production. By 1863 the New York company had several factory rooms devoted to albumenizing raw paper imported from Europe.[43] Anthony assumed the leadership of albumen paper production in the United States but soon had many smaller imitators in New York and Philadelphia.[44]

The demand for albumenized paper prints originated not only from the gallery patron's desire for multiple prints of his image but also from the sale of multiple prints of famous personages and from a new and most important form of mass entertainment, the stereoscopic card. Attempts to create the illusion of three-dimensional views grew out of the new subjectivist approach to optics in the first

[41] Gernsheim and Gernsheim, *History of Photography*, p. 151.

[42] Reese V. Jenkins, "Some Interrelations of Science, Technology, and the Photographic Industry in the Nineteenth Century" (Ph.D. diss., Department of the History of Science, University of Wisconsin, 1966), pp. 229–31.

[43] Oliver W. Holmes, "Doings of the Sunbeam," *Atlantic Monthly*, December 1863, pp. 1–15.

[44] James Wilcox (N.Y.C.), George Chapman (N.Y.C.), John H. Simmons (Philadelphia), and George Dabbs & Co. (Philadelphia).

Fig. 2.12
Stereoscopes. From Payne, Descriptive Catalogue.

half of the nineteenth century. Charles Wheatstone, an English natural philosopher, developed a stereoscope in the late 1830s, which, during the next decade, represented a novel and curious device mostly of interest to men of science. The Langenheim brothers of Philadelphia, German emigrant photographers who were deeply interested in transparencies and projection, introduced in the late 1840s stereoscopic transparencies and prints. These attracted the attention of the members of the Franklin Institute in Philadelphia, and with the financial support of some of its members, the Langenheims issued in 1854 a series of stereoscopic landscape views.[45] At about the same time the London Stereoscopic Company and Ferrier of Paris introduced stereoscopic views to the public.[46] In the late 1850s, with some foreign view cards being imported and with the Langenheims continuing to produce view cards, one of the early major mass consumer items was born. The stereoscope viewer and box of view cards were as common a feature of the post-Civil War American home as is the television set, today (fig. 2.12).

In 1859 Edward Anthony launched the manufacture and sale of stereoscopic view cards with a series of 175 views, and soon the company became the leading producer of stereo cards in the country. However, a very large number of photographers also adopted the business as a profitable sideline to gallery operation. Consequently, nearly every city and large town had at least one stereoscopic slide photographer and producer. But Anthony had the advantage of possessing the best distribution channel in the country.[47]

In addition to having entered into production of new chemicals, albumen paper, and stereoscopic view cards, Anthony also inaugurated production of photographic albums and such plush goods as gallery furniture, backdrops, and photographers' curtains. This major shift to production while maintaining the commitment to jobbing, combined with the stimulus of the Civil War, caused Anthony's sales to climb from $100,000 in 1859 to $600,000 in 1864.[48] Expansion of the company came in the early 1860s, when sales and profits were very high and when a major technological change in the industry was providing new opportunities. Aiding in Anthony's expansion was new capital brought to the company by H. Badeau, an employee of the firm for several years, who joined the company as a limited liability partner. Soon thereafter he left for Paris, where he served as the European agent for the American company, seeking new markets for Anthony products and searching for European novelties for the American market.[49] During that period, too, many other firms entered into manufacturing in each of the new sectors of supply production, thereby creating a market structure approaching

[45] Josef M. Eder, *History of Photography*, trans. Edward Epstean (New York: Columbia University Press, 1945), p. 340; and William Culp Darrah, *Stereo Views: A History of Stereographs in America and Their Collection* (Gettysburg, Pa.: Times and News, 1964), pp. 27–32.

[46] Taft, *Photography and the American Scene*, p. 184.

[47] Darrah, *Stereo Views*, pp. 35–41.

[48] "Edward Anthony: July 29/59 and Nov. 2/64," D & B, N.Y.C. Series 8, p. 400B.

[49] "Edward Anthony: March 9/57; June 30/58; May 20/61; July 8/62; Nov. 28/63; Jan. 1/64; and Jan. 15/64," in ibid.

Fig. 2.13
Stereoscope viewer and cards. From Tissandier Hist.

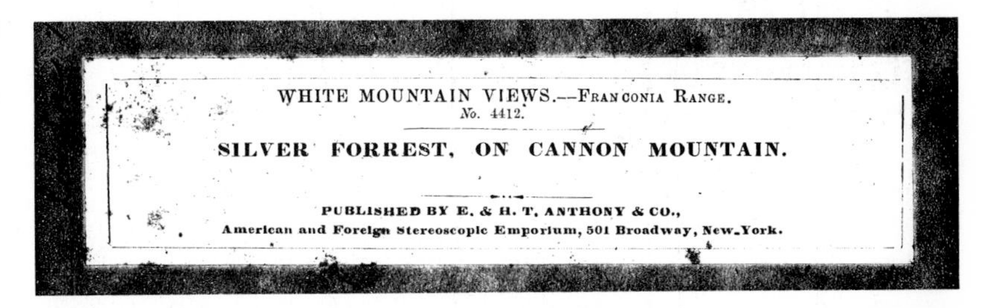

WHITE MOUNTAIN VIEWS.—Franconia Range.
No. 4412.

SILVER FORREST, ON CANNON MOUNTAIN.

PUBLISHED BY E. & H. T. ANTHONY & CO.,
American and Foreign Stereoscopic Emporium, 501 Broadway, New-York.

Fig. 2.14
Anthony stereoscope card: front and back views.

perfect competition in those sectors. Yet, the combination of leadership in jobbing and in most of the sectors of supply production gave the Anthony company a uniquely powerful position in the industry.

During the period 1855 to 1865 the technological mind-set was less rigid than it had been earlier. Many new processes were tried—albumen, wet collodion negative, ambryotype, and tintype—but businessmen and technical men became attached to the collodion negative-positive process and devoted themselves to improving it; in the early sixties Henry T. Anthony introduced ammonia sensitizing.[50] With increased interest and recognition of new commercial potential in field and landscape photography, the sixties witnessed

[50] Gernsheim and Gernsheim, *History of Photography*, p. 259.

technical developments such as Harrison's Globe lens and also numerous efforts to produce a dry collodion plate for field work.[51]

1865–80: Return to Oligopolistic Market Structure

From 1865 to 1880 the wet collodion negative process continued to dominate the photographic industry, but in contrast to the first decade of the collodion period, which witnessed the opening of many new opportunities for both photographers and supply manufacturers, the later period coincided with a time of depressed prices and sales not only in the photographic industry but in the economy generally. A contraction and consolidation of supply manufacturers accompanied the business decline in the industry.

Tintype Plate Sector

One of the most important supply manufacturing sectors, tintype plate, moved during the 1860s from a duopolistic structure to a more competitive structure and then back to a duopolistic structure. Out of the large group of small companies that derived from the patent-leather industry in Newark, the most important in terms of long-range influence was that of Horace Hedden and his son, Horace M. The son, wanting to expand the business, made arrangements in 1863 to join in partnership with John Dean, a native Englishman, and Samuel P. Emerson, producers of ambryotype mats and preservers in Worcester, Massachusetts.[52] The new firm, Dean, Emerson & Company, continued to produce mats and preservers but also engaged in the production of tintype plates. Hedden brought to the company key technical information and knowledge, while Dean and Emerson contributed the capital and experience in the photographic business. They marketed their new plates under the name "Adamantean" and granted E. & H. T. Anthony & Company the sole agency for its distribution. The Hedden transfer to Massachusetts set the stage for the eventual shift of the center of tintype manufacture to Worcester.[53]

Changes in partners in the company prompted several significant reorganizations between 1865 and 1870 (fig. 2.15). Significantly, in 1866, when Victor M. Griswold closed his business, John Dean & Company purchased his formulas and production information.[54] Shortly thereafter the partners separated, and two distinct Worcester companies emerged: Phenix Plate Company (Emerson & Hedden) and John Dean & Company (Dean & Morgan). These two Worcester companies inherited both the product and the production

[51] Ibid., pp. 256–61.

[52] Estabrooke, *Ferrotype*, p. 96.

[53] Ibid.; "John Dean," D & B, Mass., vol. 95, p. 300E; *Anthony Photog. Bul.* 13 (1882): 55; and *Worcester Evening Gazette*, 8 February 1882, p. 2.

[54] See E. & H. T. Anthony & Co. advertisement in Estabrooke, *Ferrotype*.

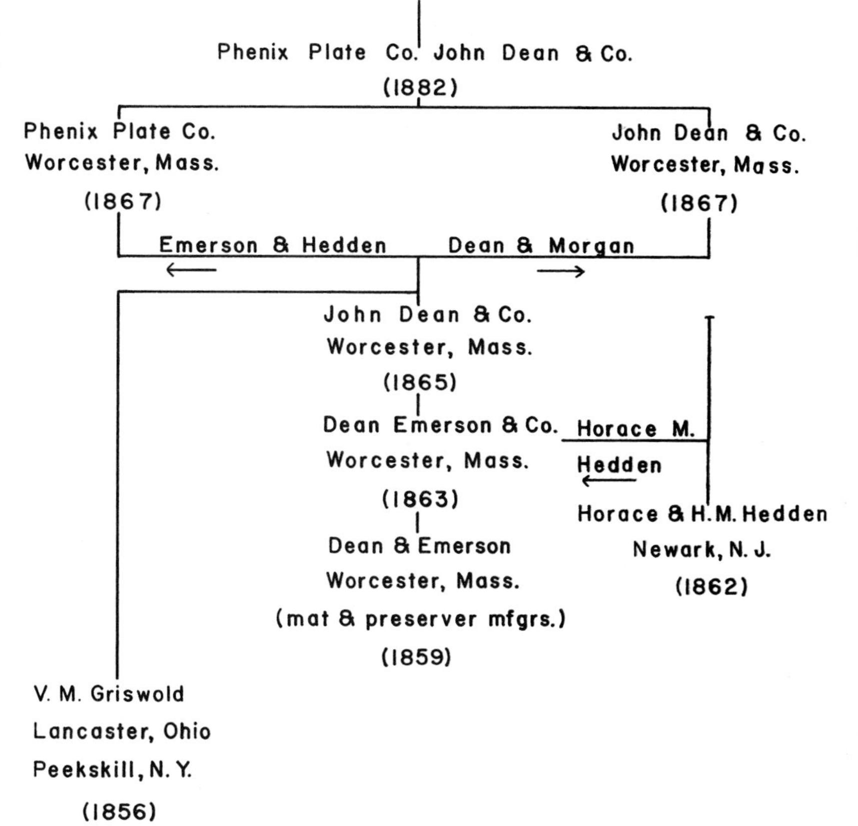

Fig. 2.15
Worcester tintype business genealogy.

technology from the two major branches of the industry: the Ohio–New York branch through Griswold, the patent-leather group of Newark through Hedden. After the changes of the late sixties and early seventies, a long period of stable ownership reflected the general stability within the industry.[55]

A major new entrant appeared in the industry in 1870. The Scovill Manufacturing Company erected an extensive plant in Waterbury, Connecticut, and commenced production of the Sun plate in March 1870. Undoubtedly, Scovill had not ventured into this area of metal processing earlier because it had little experience in either fabricating iron or japanning metals. Its policy had been to externalize the risk by allowing others to do relatively small-scale manufacture and to profit from the jobber's margin. The difficulty Scovill had in marketing its plates, despite offering glossy, egg-shell, and tinted plates and uncut sheets of tintype metal, confirmed the wisdom of that policy. During 1872 Scovill discontinued production and negotiated an agreement with the Phenix Plate Company to act as the general sales agency for Phenix plates.[56]

[55] "John Dean," D & B, Mass., vol. 95, p. 300E, and "Phenix Plate," in ibid., vol. 104, pp. 710, 787J7, and 883; Charles Nutt, *History of Worcester and Its People*, 5 vols. (New York: Lewis Historical Publishing Co., 1919), 4: 507–8. Mr. Charles A. Hill of Worcester also kindly shared with me some of his knowledge of his grandfather and the Phenix Plate Company.

[56] Bishop, "Scovill and Photography," p. 450; Estabrooke, *Ferrotype*, pp. 99–101; and *Photog. T.* 1 (1871): 37.

By 1872 business and technological superiority in the tintype industry was centered in the two companies in Worcester, Massachusetts. The best quality plates were made by these firms, and a popular new fad was introduced by Horace M. Hedden: tinted tintypes, particularly ones of a chocolate tint.[57] Hedden patented this development in 1870, but the patent was not enforced. A contemporary observer pointed out the advance in productivity from the Neff-Griswold period to the Dean-Phenix period, indicating that "by the improved process of manufacture, one man [could] now make as many plates in a week as the two then existing factories did in a month."[58] During the decade of the 1860s, when the demand for tintypes had grown and the number of plate suppliers had increased, a number of new improvements in the product and the methods of production had been introduced; however, during the seventies and the early eighties, when the demand and market structure stabilized, the number of changes in technology diminished drastically.

With the emergence of two dominant firms in the industry, duopolistic behavior characterized the interfirm relations. Price competition had been severe during the 1860s, and the late years of the decade had not been very profitable for either of the two Worcester companies; but the early years of the 1870s saw the companies move toward close cooperation on a number of fronts.[59] Each became identified with one of the two national photographic jobbers—Dean with Anthony and Phenix with Scovill—thus providing price and demand stability.

Control of the iron supply became another major method of maintaining stability. The thin, polished-surface iron plates required for tintype plates were imported from Britain. The tinplate industry of South Wales, through its possession of technological secrets, pursuit of technological innovation, and proximity to raw materials, had come to dominate the international tinplate market in the middle of the century.[60] During the 1860s efforts had been made—with some limited success—by combinations of large manufacturers to purchase the entire importation of this metal and thereby deny this vital raw material to smaller producers. By the early 1870s a New York firm had control of the importation of this iron and agreed to give the Worcester firms an exclusive contract for the tintype trade.[61] Then, Dean and Phenix moved to a mutual price stabilization understanding, which resulted in several marked increases in retail plate prices in 1872 and 1873. These cooperative moves resulted in

[57] U.S. Patent #100,291 (1 March 1870), issued to H. M. Hedden.

[58] Estabrooke, *Ferrotype*, p. 95.

[59] "John Dean," p. 952, and "Phenix Plate," in ibid., pp. 710 and 787J7.

[60] The imported sheet metal was Taggers iron, Pontimeister number 38. See W. H. Dennis, *A Hundred Years of Metallurgy* (Chicago: Aldine Publishing Co., 1964), pp. 219–21; Leslie Aitchison, *A History of Metals* (New York: Interscience Publishers, 1960), pp. 526–27; Charles L. Mantell, *Tin: Mining, Production, Technology, and Applications*, 2d ed. (New York: Reinhold, 1949), pp. 384–86; and John W. Oliver, *History of American Technology* (New York: Ronald Press, 1956), pp. 325–26.

[61] Estabrooke, *Ferrotype*, pp. 102–3.

substantial gain in profits for both companies in the early and middle 1870s.[62]

The Phenix Company looked beyond the national market as that market began to stabilize. Hedden obtained a British patent on his chocolate-tinted tintype in August 1871 and soon thereafter sent agents to England to introduce the plate there. The next spring the company sent a representative to London to establish the first English gallery devoted exclusively to the tintype; however, such marketing endeavors were not particularly successful at that time.[63]

The continued cooperation of Hedden-Hill and Dean-Morgan during the depths of the depression in the middle and late 1870s, when demand and profits fell, was insured by a joint ownership of both properties. By 1877 the financial operations of the two firms had merged; the firms were run from the same office but under their own company names and established brand names. The merger was formalized when, upon the death of Dean in 1882, the management of both production facilities was turned over to Charles A. Hill and Horace M. Hedden, with Morgan continuing as a silent partner. Their products continued to be marketed under the established names and brands.[64]

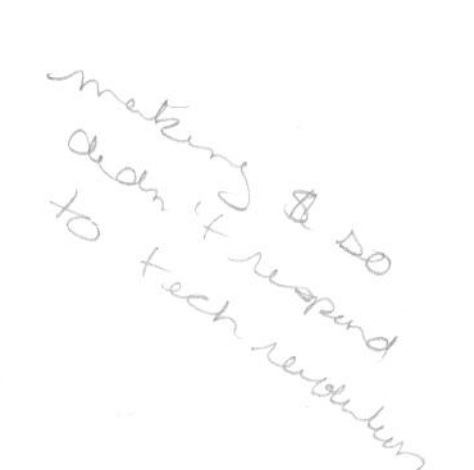

The introduction in the early 1880s of gelatin dry plates revolutionized the entire photographic industry. However, because the Worcester concern continued to make a good profit, it made no serious effort to respond creatively to the new opportunities of the technological revolution. Hill and Hedden, both in their fifties and not seeking their fortunes in a new line of business, continued into the mid-1880s to pursue a single product line: japanned materials.

The Style of Scovill and Anthony Leadership

Just as the two principal jobbing houses played a role in shaping the market structure of the tintype sector, Scovill and Anthony, despite policy and personnel changes, exercised considerable influence over the entire industry. The Scovill Manufacturing Company, which had slipped to second place in the jobbing sector of the industry, continued during this period to have stable management and administration. Responsibility for the photographic division rested principally with Samuel Holmes, who continued as manager of the New York store until 1872, and with his successor, Washington Irving Adams.[65] Both men developed close personal relations with photographers across the nation. Although Scovill's business did not grow as rapidly during the collodion period as did Anthony's, the photographic division during the quarter century from 1850 to 1874 made earnings in excess of one million dollars, a figure that nearly equaled

[62] Ibid.; and "John Dean," and "Phenix Plate."

[63] Estabrooke, *Ferrotype*, pp. 98–99; and *Photog. T.* 2 (1872): 139, 147, and 178.

[64] "John Dean," pp. 952, 992, and 1078.

[65] "Samuel Holmes to the Photographic Trade and Fraternity," *Photog. T.* 2 (1872): 78 and 28 (1896): 65–66.

that of all the other divisions of the Scovill company in the same period and that indicates the general success of the policies and strategies pursued by Holmes and Adams.[66]

From 1865 to 1880 Holmes and Adams maintained the earlier policy of concentrating on jobbing and marketing. Even though the company did engage in some manufacturing outside its traditional areas during the last years of Holmes's leadership, the problems it experienced in the quality of its product and the organization of production and the risks of technological change prompted the maintenance of the traditional policy. In marketing, Holmes and Adams dispatched full-time traveling agents to represent the company across the nation. Imitating the innovation of the Anthony company, Holmes introduced a house trade journal, *Photographic Times*, to provide advice to professional photographers and to promote orders for supplies. In addition, sales offices in Boston, Chicago, and Liverpool, England, were coordinated with the New York headquarters of the photographic division. Some of these marketing strategies led the company into direct conflict with its principal rival, Anthony; however, in the 1870s relations between the two firms were more cooperative than they had been during the 1860s. They jointly ordered foreign supplies, such as photographic paper, and had informal understandings on price levels and discounts, reflecting behavior in accord with the oligopolistic market structure that had emerged in both jobbing and manufacturing.[67]

In the apparatus sector of production, the Scovill company encountered increasing difficulty in obtaining a reliable supply of camera box and optical apparatus, especially after the loss of the sole agency relationship with the Harrison company in the early 1860s (see figs. 2.10 and 2.16). However, in 1867 the Scovill company acquired the American Optical Company, which had acquired the former Harrison plant and the rights to the important Harrison optical designs.[68] The optical and camera assembly operations remained in New York, but many of the wooden case parts were prepared for assembly at the New Haven case plant. During the next decade the American Optical Company apparatus, which included stereoscope viewers in addition to camera boxes and accessories, developed an international reputation and market and ranked first in America. This position derived fundamentally from the generally good quality and low cost of American wood craftsmanship. In the early 1870s American Optical Company received the highest medal for photographic apparatus at the Universal Exposition at Vienna and thereafter established small but growing markets for apparatus in Western Europe, Asia, and Africa.[69]

The sequence of changes and mergers associated with the Harrison factory and American Optical were indicative of a pattern among camera and apparatus manufacturing firms in the United

[66] Bishop, "Scovill and Photography," p. 453.
[67] Ibid., p. 441; and *Photog. T.* 6 (1876): 135.
[68] Rohr, *Photographischen Objektivs*, pp. 176–77.
[69] *Photog. T.* 3 (1873): 169 and 7 (1877): 158 and 195.

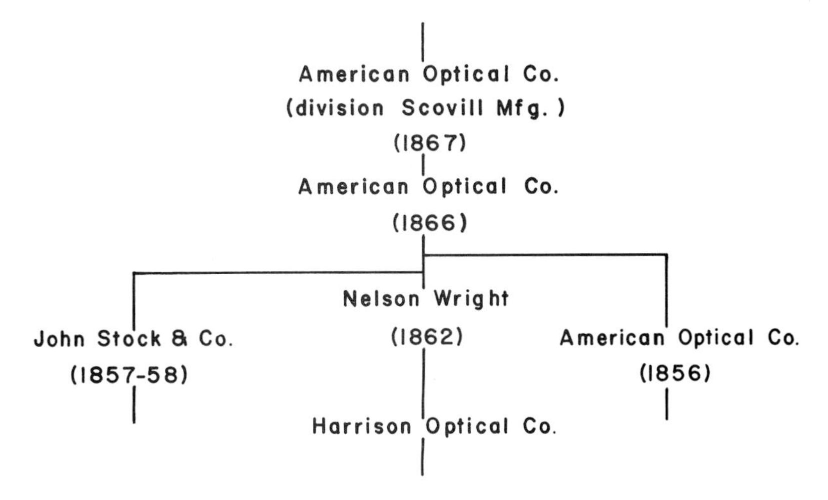

Fig. 2.16
American Optical Company genealogy.

Fig. 2.17
Photographic bellows camera. From Tissandier Hist.

States in the immediate post–Civil War years.[70] The contraction in the number of manufacturers reflected the general business decline but also, significantly, the growing importance of the quality of the optics associated with the camera. Several new European optical firms, such as Ross, Dallmeyer, and Grubb in England and Steinheil in Germany, introduced important new photographic lenses. The most significant development was the collaboration between Ludwig P. Seidel, a German professor of mathematics, and H. Adolph Steinheil, the photographic lens designer in the German firm of Steinheil & Son. Seidel's studies of optical theory and of deviations of rays from the ideal paths they should follow through a lens led to his enunciation of formulae for correction of five important kinds of aberrations, or lens imperfections.[71] Adolph Steinheil employed results from this work and adopted mathematical ray-tracing techniques in the design of a new landscape lens called the aplanat (fig. 2.19).[72] This lens was to landscape photography for the next thirty years what the Petzval lens had been to portrait photography: the

[70] *Humphrey J.* 17 (1865): 222 and 18 (1866) (advertisement).

[71] Ludwig Seidel, "Zur Dioptrik . . . ," *Astronomische Nachrichten* 43 (1856): 289–332.

[72] The lens had a relatively low *f* number and was corrected for spherical and chromatic aberrations as well as for coma.

Fig. 2.18
Twin-lens, or stereo, camera.
From Tissandier Hist.

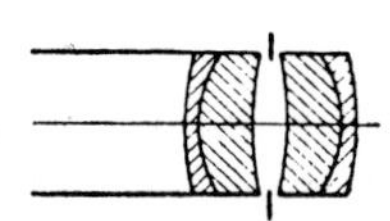

Fig. 2.19
Steinheil's aplanat landscape lens. From Rohr, Theorie und Geschichte.

standard of excellence. As a result of this lens system and others designed from similar principles, Steinheil captured the leadership in photographic optics from Voigtländer. Unable to patent the new lens in England, however, Steinheil lost the English market to John H. Dallmeyer, who introduced in the middle sixties the rectilinear lens, which was similar to that of Steinheil. With the change in Europe in the techniques of lens design from trial-and-error methods to mathematical ray-tracing techniques, the quality of European optics improved markedly, hurting most of the American optical and camera makers, who then competed principally in the moderate and low-cost segment of the optical market.[73]

One new piece of apparatus developed during the collodion period, the stereoscopic viewer, proved to have a large sale. In the 1850s individual instrument makers copied designs from Europe, including table viewers, but in the early 1860s Oliver Wendell Holmes designed a hand viewer consisting of a card holder and two viewing lenses on an adjustable slide mounted on a rigid axle with a small handle underneath (fig. 2.20). This simple device, which Joseph L. Bates of Boston initially produced, became immensely popular and, because Holmes did not patent it, was imitated by American Optical, Anthony, and several smaller producers.[74]

E. & H. T. Anthony & Co. consolidated its leadership position in jobbing between 1865 and 1880 while maintaining its activities in those manufacturing sectors it had entered earlier. As an organization, Anthony experienced only two significant changes in ownership and management. In 1870 Vincent M. Wilcox, a clerk in the Anthony store for several years, entered the partnership and during the next two decades came to assume increasing responsibility for the conduct of the business (fig. 2.21). Badeau, the partner residing in Paris, retired from the business in 1875 and gradually withdrew his

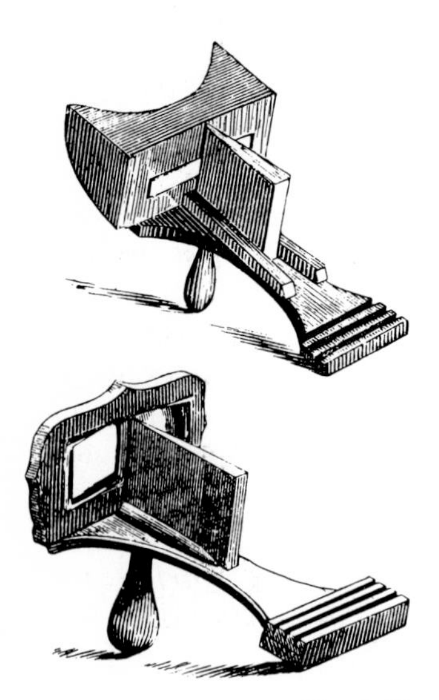

Fig. 2.20
Hand stereoscopes: Holmes and Bates. From Philadelphia Photographer, *January 1869.* Top, *stereoscope designed by Oliver Wendell Holmes in 1859;* bottom, *stereoscope based on Holmes design and produced by Bates of Boston.*

[73] Jenkins, "Some Interrelations of Science, Technology, and the Photographic Industry," pp. 105–15.

[74] Darrah, *Stereo Views,* pp. 10–11.

Fig. 2.21
Colonel Vincent M. Wilcox. From Taft, Photography and the American Scene. *Courtesy of Dover Publications, Inc.*

capital; because of difficulties in the depression years, the company was able to acquire his share only by mortgaging the firm to European interests. The remaining three partners decided to incorporate in 1877 to insure that the firm would not be dissolved upon the death of one of them.[75]

The Anthony company, which during the 1860s had developed a much stronger commitment to manufacturing than had Scovill, maintained its manufacturing facilities and even expanded them; but in several important manufacturing sectors its leadership declined. In the late 1860s, when numerous supply and manufacturing businesses either merged or dissolved, Anthony acquired the supply house and albumen-paper manufacturing capacity of Chapman & Wilcox, a firm descended from Levi Chapman, and acquired the camera business of William Lewis.[76] Anthony's position in camera production did not, however, seriously rival that of Scovill's American Optical Company.

Anthony maintained a strong position in chemical manufacturing but in this sector shared the market with a small number of domestic fine-chemicals manufacturers. Three Philadelphia companies, Rosengarten & Sons, Powers & Weightman, and J. F. Magee & Company, represented not only key firms in pharmaceutical and industrial fine-chemicals production but also important suppliers of nearly the full range of photographic chemicals.[77] One of the most important chemical developments of this period was the discovery of new domestic sources of bromine. Discovery of brine deposits in the Ohio and Kanawha River valleys stimulated the appearance shortly after the Civil War of a number of small bromine plants.[78] One firm that exploited these new sources, the Mallinckrodt Chemical Works of St. Louis, soon became a major producer of bromides for photography.[79] Although in the late 1860s the Germans initiated production of bromides as a by-product of their exploitation of the vast potash deposits at Stassfurt, the American firms continually lowered their prices and, ultimately, challenged the Germans, even abroad. The domestic price of a pound of potassium bromide fell from more than six dollars in the 1840s to thirty cents in 1875.[80] Also at this time, German chemical companies, most notably Schering of Berlin, entered the American photochemicals market with a broad range of quality products.[81] As earlier in the collodion period, photographic supply houses across the country continued to market their specialty items, such as salted collodion, developers, and varnishes. Consistent

[75] "E. & H. T. Anthony & Co.: May 14/70, Sept. 30/75, Jan. 18/77, and April 12/77," D & B, N.Y.C. Series 8, pp. 301DD, 400°, and 400EE.

[76] "James Wilcox: Feb. 15/66 and March 14/66," in ibid., p. 400D; and Estabrooke, *Ferrotype*, p. 101.

[77] Estabrooke, *Ferrotype*, advertisement; *Photog. T.* 1 (1871): 31; and Edwin T. Freedley, *Philadelphia and Its Manufactures . . .* (Philadelphia: Edward Young, 1858), p. 209.

[78] Williams Haynes, *American Chemical Industry*, 6 vols. (New York: D. Van Nostrand, 1954), 1: 324.

[79] Ibid., pp. 324–25; and *Dictionary of American Biography*, s.v. "Edward Mallinckrodt."

[80] Haynes, *American Chemical Industry*, 1: 324.

[81] *Photog. T.* 5 (1875): 171.

with its general policy, Anthony marketed these specialties along with its own chemical products.

In the photographic paper sector, Anthony was one of few American firms to survive the 1860s, but other companies came to replace the early ones and ultimately to challenge Anthony. Three Philadelphia companies developed considerable markets for their paper,[82] one of which, Willis & Clements (associated with the English Willis firm), introduced in the late 1870s platinum paper, which eventually became very popular among the leading art photgraphers. In addition, the firm of Douglas Hovey, founded in Rochester immediately after the Civil War by a Philadelphia photographer, assumed the name American Albumenizing Company and became a small but leading producer in America.[83] The control of raw paper from the two European paper mills facilitated the cartelization of the German photographic paper industry. After the merger of most of the Dresden albumen paper producers in 1875, a single firm imported Dresden albumen paper into the United States. Hence, during most of the sixties and seventies four domestic companies and one foreign company supplied the American albumen paper market.[84]

The stereoscopic view sector of the industry became less attractive to the Anthony company in the 1870s, perhaps as a result of the company's inability to dominate the sector owing to the competitive market structure. After 1873 the company clearly lost interest in the field and in 1880 issued its last series of cards; however, a new company appeared at about the same time, the Kilburn Brothers of Littleton, New Hampshire; it became the most prominent company in stereoscopic view production and marketing.[85] The popularity of stereoscopic views declined during the late 1870s and early 1880s but revived in the late 1880s and early 1890s with the introduction of mass production methods. The revived popularity continued until just before World War I,[86] when movies, the automobile, and later, radio provided alternative forms of entertainment. The truly oligopolistic phase of stereoscopic view production occurred only with the advent of mass production methods after the introduction of gelatin emulsions.

Despite its withdrawal from the stereoscopic view sector, Anthony continued to dominate the photographic jobbing trade in the United States during the late 1870s.[87] Scovill held the secondary position

[82] D. U. Morgan (succeeded by John Howarth), John R. Clemons, and Willis & Clements; see *Photog. T.* 1 (1871): 67 and 130; 2 (1872): 8; 3 (1873): 23–24; 5 (1875): 132 and 218; 9 (1879): 237 (advertisement); 19 (1889): 84; and 27 (1895): 216–20.

[83] Ibid. 1 (1871): 24; 6 (1876): 129; and 11 (1881): 181; *Rochester Morning Herald*, 10 February 1886, p. 7; and "Douglas Hovey," D & B, N.Y., vol. 162, p. 72, and vol. 166, p. 326.

[84] Paper from Trapp & Munch (Friedberg, Hessen) and Marion & Co. (England) also continued to be imported in small quantities (*Photog. T.* 5 [1875]: 134).

[85] Darrah, *Stereo Views*, pp. 35–45, 50–53. Note that Anthony issued more than 10,000 different titles between 1859 and 1880.

[86] Ibid., pp. 109–14.

[87] "E. & H. T. Anthony & Co.: Nov. 18/71, April 17/72, June 14/74, April 12/80," D & B, N.Y.C. Series 8, pp. 301EE and 400°.

with strengths in apparatus production. The firm of Gustav Gennert, which was founded in New York in the middle 1850s and developed a small jobbing business and some limited manufacture in photographic chemicals, ranked a far distant third. From the late 1860s the two jobbing and manufacturing firms of Anthony and Scovill maintained substantial influence over the photographic business. Through their national marketing function, they performed a critical and highly profitable role as intermediaries between the photographers who were broadcast across the nation and the few, small-scale manufacturers who operated in duopolistic or oligopolistic markets.[88]

The attitudes toward innovation held by Holmes and Adams at Scovill and by the Anthony brothers and Wilcox at Anthony remained similar to those held during the latter part of the daguerreotype period. These men saw in product novelty an intrinsic sales stimulant and, when patented, a deterent to gross imitation.[89] Hence, the interest in product invention and innovation increased during the collodion period. Yet, the leadership of the Scovill and Anthony companies continued to depend primarily upon external sources of invention. When an improvement was created, the managers of Scovill and Anthony sought to acquire it outright through purchase or to control it through a sole agency for its sale. Moreover, following the European pattern in which societies offered prizes for resolution of specific technological problems,[90] Holmes sought in the early 1870s to encourage invention by awarding prizes and medals. He clearly stated that his intent in offering such awards was "to induce experiment and useful discovery."[91] In addition, the American leaders sought new products in Europe. The Anthony brothers, of course, assigned their new partner, Badeau, to Europe with the specific task of seeking novelties for the American trade. It is not clear to what extent the management of the two leading American firms encouraged invention by their own staffs; however, an examination of photographic patents issued from 1841 to 1873 indicates that less than 4 percent of the more than four hundred patents originated with the employees and owners of Scovill and Anthony. Their staffs generated only seven and eight patents respectively, Henry T. Anthony being the most prolific patentee, with five. The patents of both companies tended to be on widely scattered features, from chemical trays and plate holders to cameras and photograph albums. None of them proved to be of major financial consequence.

Patents produced by staffs of companies in the photographic business tended to come from proprietors of small manufacturing companies rather than from Anthony or Scovill. Among these companies, which accounted for about 16 percent of all photographic

88 Richard A. Anthony testimony, *U.S.* v. *EKCo.*, pp. 556–58; Gustave Gennert [the son] testimony and Alexander C. Lamoutte testimony, in ibid., pp. 175 and 540, respectively.

89 *Photog. T.* 11 (1881): 221.

90 Ibid. 7 (1877): 17, 134, 135, 158, and 159.

91 Ibid. 2 (1872): 82.

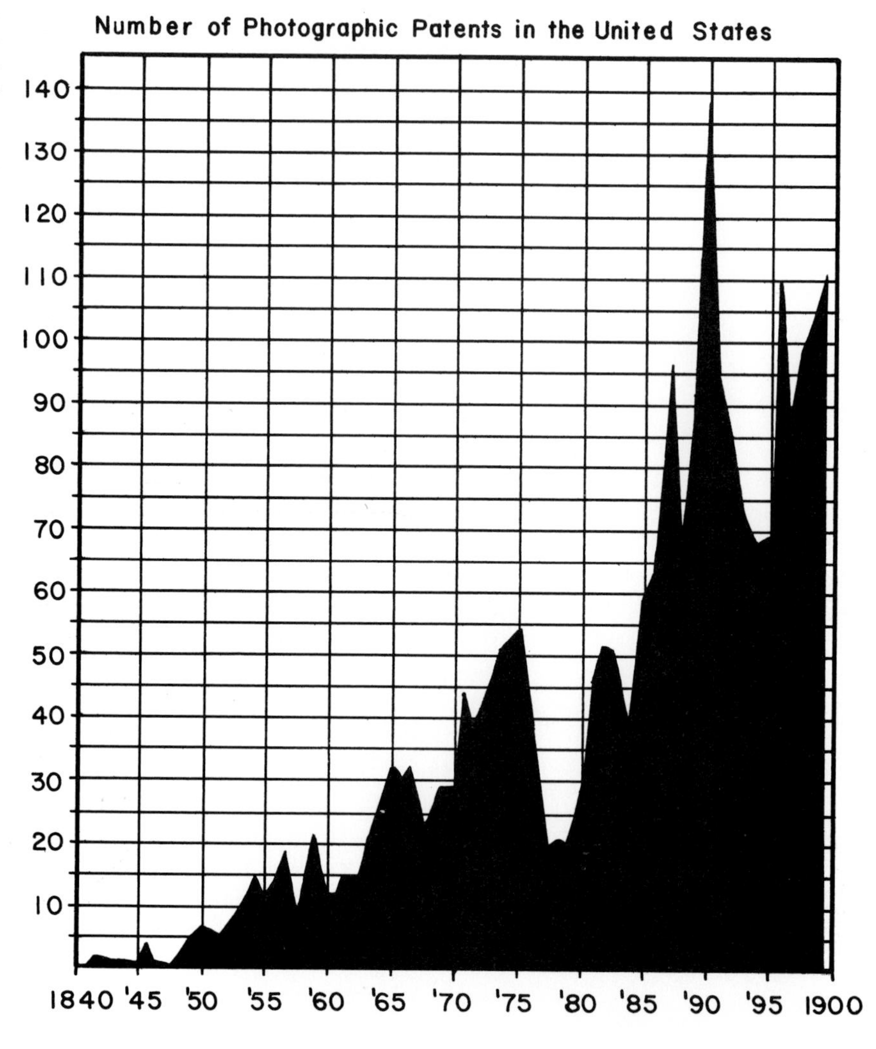

Fig. 2.22
Number of photographic patents in the United States (items indexed under camera, daguerreotype, photographic, portable camera, solar camera, and stereoscopic camera in General Index of Patents, U.S. Patent Office, 1790–1873, *and* Patent Gazette, *1874–99*).

patents, American Optical or its predecessor companies (with sixteen patents) and the Lewis family company (with thirteen patents) were the most prolific patentees and showed the greatest tendency to sharply concentrate their inventive and patent endeavor. Staff members of both companies, which were at times associated with the Scovill and Anthony companies, patented principally cameras, camera features, and lenses. Three circumstances may help account for the greater interest in patents among the small-scale producers. First, because of the small size of the firms, the owners were themselves directly involved in both the technical details of production and the demands of the market place. Therefore, they were exposed to a wide range of opportunities and incentives for technical innovation. Second, despite the substantial production commitment of Anthony and Scovill, their leaders had come to conceive of the central activity of their enterprises as the jobbing of photographic materials and apparatus. Hence, they developed marketing strategies in which internally developed product and production innovation played a subordinate role. While they sought to exploit novelty as a marketing tactic, they sought to control it through sole agencies and contract terms with the originator of the novelty rather than by taking the inventive initiative themselves.[92] Third, the negative attitudes toward patents prevalent among photographers at the time may also account for the apparent antipathy to internally derived invention, patenting, and patent enforcement. The intensity of these feelings toward patents, which the process patents of Daguerre and Talbot in England and of Cutting in the United States engendered, no doubt influenced the strategies of the leaders of the major supply houses in the United States.[93]

Hence, interest in invention and patents increased during the collodion period, but because of the difficulty of enforcement and the reluctance of the large firms to engage in litigation, patents did not play a commanding role in business strategy. In the technical and business contexts, technical secrets, control of critical supplies, and exclusive marketing agreements proved more efficacious in maintaining the oligopolistic structure of the industry during the last half of the collodion period. Yet, in the 1870s technical developments were under way which in the next decade were destined to alter the technological pattern of the collodion period and to disrupt once again the established structure of the industry.

[92] For example: Lewis glass plate holders, ibid. 1 (1871): 6; Fitzgibbon's patent adhesive picture mounts, ibid. 3 (1873): 154; Vaughn's patent photograte, or shutter, ibid. 4 (1874): 5; and Gordon's patent reversible double corner, ibid. 2 (1872): 178 and 9 (1879): 141.

[93] In 1839 Daguerre and the French government ostensibly donated Daguerre's process "to the world," but Daguerre nonetheless patented his process in France's greatest rival, Great Britain. Then, in the early 1840s, Talbot patented his paper photography processes in Britain and, much to the consternation of British photographers, sought to extend these patents to cover the collodion processes in the early 1850s. A couple of years later in the United States, the Boston photographer, James A. Cutting, patented the use of the highly sensitive mixture of silver bromide and silver iodide in collodion photography. The Cutting patent, which was unevenly but persistently enforced from 1854 to 1868, incensed American photographers generally (Taft, *Photography and the American Scene*, p. 129) and prompted statements in the Scovill trade journal such as "we all know how grasping and greedy patentees sometimes are . . ." (*Photog. T.* 8 [1878]: 16).

The photographic business from 1880 down to the time I ceased to be connected with it [c. 1900], was more or less dependent upon patents. These patents were owned by the different concerns who were engaged in the trade. They, at times, in the opinion of lawyers, seemed to conflict very much, and brought about a state of affairs where one company was suing or threatening to sue the other from time to time.

Leo Baekeland

Gelatin Plate Period 1880 to 1895

3 / The Gelatin Revolution, 1880-95

Introduction

The American economic and business climate between 1880 and 1895, an important period of transformation that fell between the depressions of the 1870s and the 1890s, was characterized by expanding markets and falling prices and culminated in the beginnings of the merger movement whose attendant internal and external changes have been termed the Corporate Revolution. A number of forces converged during the period preceding the Corporate Revolution that enabled businessmen to explore new strategies for dealing with the rapidly changing technical and business environment. The railway network had basically reached its geographical limits by 1880; a growing national and urban market reflected new tastes and improved incomes; and high tariffs and the traditional ocean barrier protected the domestic market.[1] At the same time, improved production technology promoted economies of scale, which resulted in falling prices and increasing competitive pressure among manufacturers. Borrowing heavily from the new developments in science and technology from Europe, the American manufacturer experimented with new products and production techniques in his protected domestic business laboratory.

In the growth of the American photographic industry, the interaction between the forces of the market and the forces of technology were critical. The mode of production, the product line, and the

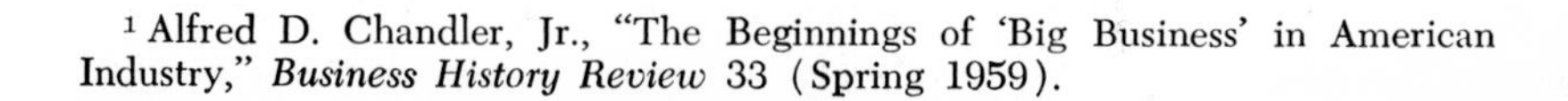

[1] Alfred D. Chandler, Jr., "The Beginnings of 'Big Business' in American Industry," *Business History Review* 33 (Spring 1959).

market were all intimately interrelated. The broad technological configuration—both product and production—defined the boundaries within which the business sector could operate. The market potential likewise defined the technological potentials that could be exploited. In the case of photographic apparatus and materials, a national—and even an international—market had existed from the middle of the nineteenth century. The barrier to expansion of photographic supply production was not the lack of a market but two internal technological restraints: (1) the rapid perishability of photosensitive materials; and (2) the complexity of photography for the average person. Thus, initially, the broad boundaries within which occurred the development of photography and the photographic industry were technological. Once these restraints had been removed, the forces of the national market could come into play.

The introduction in the late 1870s of factory-sensitized, gelatin-coated glass plates prompted a fundamental change in the business-technological mind-set in the United States. The change from the wet collodion to the dry gelatin process meant that photosensitive materials were no longer highly perishable. It set in motion fundamental product, production, marketing, and organizational changes that affected the entire industry. These changes promoted the introduction of new business ideas, ideas that ultimately revolutionized the photographic industry. The creation and modification of the roll film system of photography followed from this gelatin revolution. The introduction of the roll film system provided the bases for a further simplification of photography and, therefore, prompted the creation and successful exploitation of the economically explosive mass amateur market. Technical and economic prerequisites to the development of the roll film system were two key, interrelated changes: the introduction in the late 1870s of gelatin dry plates for negatives and the introduction half a decade later of gelatinized paper for positive prints. In the gelatin plate and photographic paper sectors, the changing market structures and technology provided the historical setting for the eventual introduction of the roll film system.

Gelatin Dry Plates

The origin and subsequent development of American dry plate manufacturing depended heavily upon European-derived photographic technology. Moreover, persons outside the established photographic industry played key roles in the introduction and improvement of dry gelatin emulsion plates. Since the introduction of the wet collodion process, photographers had employed numerous substances in an effort to preserve the photosensitivity of collodion plates. Although some of these efforts met with limited success, none had produced a plate with the sensitivity of ordinary wet collodion plates. In 1871 Dr. Richard L. Maddox, an English physician interested in photomicrography, grew dissatisfied with the high cost of wet collodion plates and the unhealthful fumes attending their use. In his search for a substitute he eventually dispersed silver-bromide salts through gelatin, which he then used to coat glass plates and paper. Maddox reported upon this first, moderately successful gelatin-silver-bromide emulsion in the *British Journal of Photography* in 1871 and received commendation from the journal's sympathetic editor.[2] Although this article apparently attracted little immediate attention, two years later other Englishmen improved the process and initiated commercial production of gelatin emulsions.[3]

During the 1870s a few British pioneer gelatin-emulsion makers committed themselves to the process and then gradually helped to unfold the potential of gelatin emulsions. They recognized at once the advantages of the dry carrier and the less perishable photosensitivity of the silver-halogen salts. Soon they noted the increased contrast of the gelatin negative as compared with the collodion negative, a characteristic they attributed to the lesser porosity of the gelatin, which enabled it to better protect the photosensitive salts from the developers. Yet, Maddox's original emulsion was less sensitive than the typical collodion plate. Through a series of empirically derived modifications of the process, the photosensitivity of gelatin emulsions was so greatly increased that it far exceeded that of the collodion plate. Later, this remarkable increase was attributed to the

[2] R. L. Maddox to W. Jerome Harrison, 19 August 1887, in W. Jerome Harrison, *A History of Photography . . .* (London: Trubner, 1888), pp. 129–32.

[3] Helmut Gernsheim and Alison Gernsheim, *The History of Photography: From the Earliest Use of the Camera Obscura in the Eleventh Century up to 1914* (London: Oxford University Press, 1955), pp. 262–65.

catalytic effects of trace quantities of sulfur compounds in some gelatins and to the control of the size of the salt grains. By the late 1870s the advantages of increased contrast and sensitivity as well as those of dryness and less perishability were evident, and the details of the improvements in the process were being freely published in the British photographic journals.[4]

Production of a highly sensitive emulsion was a complex operation, which lent itself to specialized manufacture. Commercial interest in the manufacture of gelatin plates increased markedly during the late 1870s, and because of the initial technical interest in the process in Britain, innovators there made Britain the principal supplier to the world. While centralized production of other photosensitive materials began at this time, handicraft methods dominated all stages of production in Britain.

Typical of the response to the introduction of new technologies, the professional photographer, whose livelihood and reputation depended upon the uniform and reliable production of photographs, was reluctant to adopt the new process. The use of gelatin plates required the re-education of most photographers because the plates were considerably more sensitive and, therefore, required new handling procedures and special care to insure that no stray light prematurely exposed them. Photographers also had to learn new procedures for the timing of exposures. Furthermore, gelatin dry plates were initially more expensive than collodion plates. Hence, the dry gelatin plate, despite its immediate advantages of factory preparation and increased sensitivity, only gradually won the loyalty of professional photographers in England and the rest of the world.

Pioneer Producers and the Trend toward Perfect Competition, 1879–85

During the late 1870s American photographic journals paid increasing attention to the English reports on gelatin emulsions and plates. The three established American jobbing houses responded to this new technology in their traditional manner. Initially, Anthony imported plates from England, and Scovill imported them from Holland.[5] During 1878 and 1879, Scovill acted as agent for plates produced locally and then in the summer of 1879 initiated its own production. Anthony also inaugurated production of its own Defiance plate in May of 1880.[6] However, both Anthony and Scovill encountered difficulty in meeting the standards and prices of imported and domestic plates, so they reverted to their traditional strategy of acting as jobbing agents for other more competitive manufacturers.

Three of the earliest and most important American pioneers in gelatin dry plate manufacturing were Carbutt of Philadelphia, Cramer and Norden of St. Louis, and Eastman and Strong of

[4] Ibid., pp. 264–65.

[5] Concerning the Anthony company, see the testimony of George Murphy, *U.S.* v. *EKCo.*, p. 81; Scovill imported from Wegner & Mottu, Amsterdam, Holland (*Photog. T.* 9 [1879]: 208).

[6] *Anthony Photog. Bul.* 11 (1880): 150.

Fig. 3.1
John Carbutt. From Fritz Wentzel, Memoirs of a Photochemist *(Philadelphia: American Museum of Photography, 1960). Courtesy of Minnesota Mining and Manufacturing Company.*

Rochester. Two of the three companies were inaugurated by immigrant professional photographers who had strong interests and backgrounds in gelatin technology or chemistry. As a youth John Carbutt (1832–1905) had emigrated from Sheffield, England, and he practiced in the 1850s as a professional photographer in the Chicago area. Within fifteen years he had attained the position of superintendent at the American Photo-Relief Printing Company in Philadelphia, where over many years he acquired practical experience working with gelatin in the manufacture of photo-relief plates.[7] An innovative, middle-aged man who was characteristically alert to new technologies and products, Carbutt founded the Keystone Dry Plate Works in 1879 and employed his experience with gelatin in the production of gelatin dry plates.[8] In a somewhat parallel pattern, Gustav Cramer (1838–1914) and Hermann Norden had immigrated from Germany, where they had gained some education in elementary chemistry. During the 1860s and 1870s they had served as successful professional photographers in St. Louis. In the late 1870s both men sought to produce gelatin plates for their galleries, and in 1879 they collaborated in commercial plate production

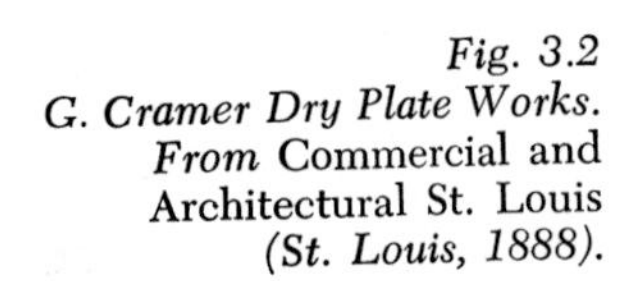

Fig. 3.2
G. Cramer Dry Plate Works. From Commercial and Architectural St. Louis *(St. Louis, 1888).*

under the name of Cramer & Norden. Norden soon withdrew from the business, and Cramer employed his own chemical talent to build a large, specialized enterprise in St. Louis (fig. 3.2).[9] Cramer's

[7] *Dictionary of American Biography,* s.v. "John Carbutt"; and *Photog. T.* 5 (1875): 5.

[8] *Photog. T.* 11 (1881): 206 and 235; D & B, Pa., vol. 142, pp. 435 and 461.

[9] Rudolph Schiller, *Early Photography in St. Louis* (St. Louis: W. Schiller & Co., [n.d.]), [pp. 8–10]; William Hyde and Howard L. Conard, *Encyclopedia of the History of St. Louis,* 3 vols. (St. Louis: Southern History Co., 1899), 1: 511–12; Charles van Ravenswaay, "The Pioneer Photographers of St. Louis," *Bulletin of the Missouri Historical Society* 10 (October 1953): 55; Ernst D. Kargau, *Mercantile, Industrial, and Professional Saint Louis* (St. Louis: Nixon-Jones, [1902]), pp. 409–13; and *Commercial and Architectural St. Louis* (St. Louis: Jones & Orear, 1888), pp. 293–95. I am grateful to Mr. Schiller for his generosity in sharing with me his extensive researches on the photographic industry in St. Louis.

Fig. 3.3
Gustav Cramer. Courtesy of Rudolph Schiller, St. Louis.

Fig. 3.4
George Eastman, c. 1884. Courtesy of Eastman Kodak Company.

Fig. 3.5
Henry A. Strong. Courtesy of Eastman Kodak Company.

warm, genial personality and the active role he played in national photographic conventions soon helped him to create both personal and product loyalty among many professional photographers and supply-house operators across the nation. Eventually his efforts in behalf of the "fraternity" won for him the endearment of the profession, which was expressed in the nickname Papa Cramer.

In contrast, Eastman and Strong were native Americans without previous professional photographic experience or scientific training and with little or no personal rapport with the "fraternity." Their eventual success depended more on the exploitation of the impersonal elements of product quality and market strategy. George Eastman (1854–1932) was a young assistant bookkeeper in a Rochester bank whose father, who had died when George was quite young, had owned and administered a successful business college in Rochester. With seven or eight years of public and private schooling and some experience as a clerk in an insurance office, the diminutive George Eastman became involved in 1878 as a serious amateur photographer.[10] Too much of a novice to be committed to the old wet collodion technology, the ambitious bachelor avidly pursued the new developments in gelatin emulsions as reported in the English journals and soon initiated limited production of gelatin plates.[11] In 1881 the socially shy entrepreneur attracted the interest, warm personal affection, and financial participation of Henry A. Strong (1838–1919), a local, highly successful buggy whip manufacturer. The fatherly and amiable Strong respected the austere, calculating, but scrupulously honest Eastman, and he gradually bestowed upon him the autonomy and unswerving loyalty that allowed Eastman to stamp his own goals of continual sales and profit growth on the firm on which he staked his high ambitions.[12]

The three major jobbing houses, through their dominance of the national photographic supply market, influenced the structure of the dry plate manufacturing sector of the industry. Each of the houses established a sole agency with one of the early producers of gelatin emulsions: Scovill with John Carbutt of Philadelphia (Keystone Dry Plate), Gennert with Gustav Cramer and Norden of St. Louis (Cramer & Norden), and Anthony with George Eastman and Henry Strong of Rochester (Eastman Dry Plate). Keystone and Cramer & Norden had received national recognition of their plates at the convention of the Photographer's Association of America in 1880. The reputation of Eastman's plates in the Rochester area by the summer of 1880 recommended them to Anthony. The effect of the sole agencies was to concentrate the major portion of the nation's dry plate production in the hands of three companies, leaving only local markets for the many urban producers across the country who had initially found the barriers to entry quite low.

[10] Carl W. Ackerman, *George Eastman* (London: Constable, 1930), pp. 25–28; and George Eastman to George Monroe, 10 May 1901, (GEC) (hereafter George Eastman will often be referred to as GE).

[11] GE to George H. Johnson, 18 January 1880, and GE to Fry, 1 April 1880 (GEC).

[12] GE to William H. Walker, 9 October 1890 (GEC); and Ackerman, *George Eastman,* pp. 30–32.

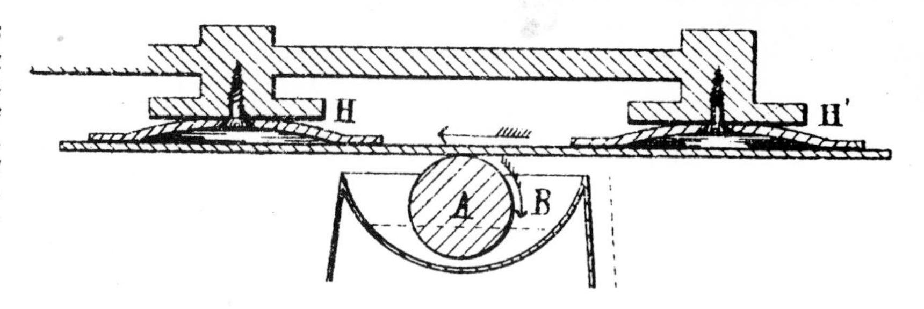

Fig. 3.6
Eastman plate coating machine. Roller A *in trough* B, *containing the photosensitive emulsion, coats the plate held by plate-holders* H. *From U.S. Patent # 226,503.*

The introduction of gelatin plates under the aegis of the dominating jobbing houses quickly shifted the mode of production of photosensitive materials from the decentralized hands of photographers to centralized manufacturers, who produced for a national market consisting largely of local professional photographers. The immediate mechanization of manufacture accompanied this centralization of production. The earliest and most notable example was that of George Eastman. He developed machinery for coating of plates, thereby gaining the economies of large-scale production. He patented his machinery in the United States and Europe,[13] hoping to gain an advantage over other American plate manufacturers and to sell either licensing rights or the patents themselves in Europe in order to provide capital for his initial production venture in the United States.[14] The Eastman plant featured an extensive range of machinery, including stamping machines, a large ventilating system, and machinery for cleaning the glass and for coating the plates with emulsion.[15] Stimulated by the high demand created by the national jobbing houses, both Cramer and Carbutt followed Eastman's example.

Although patents were acquired on some of the early production machinery, they did not play an important role in the competition among the leading firms because they could so easily be circumvented. However, the processes for the production of gelatin emulsion did play a key role in that competition and in the erection of technical barriers to entry in the industry; these processes, therefore, became carefully guarded trade secrets. Because of the technical character of emulsion making, most of the early plate companies were founded by experienced photographers, such as Cramer and Norden, or by those previously familiar with gelatin, such as Carbutt. The early American gelatin plate pioneers relied upon the information derived from published British and German sources. Most notable was the influence of Josef M. Eder, a leading German photochemist, on Cramer and Norden,[16] and the influence of the foreign photographic journals on George Eastman. While the relatively free flow of technical information in Europe and the United

[13] GE to E. & H. T. Anthony & Co., 18 August 1880, and GE to Mawson & Swan, 13 October 1879 (GEC).

[14] U.S. Patent #226,503 (applied for 9 September 1879 and issued 13 April 1880); Ackerman, *George Eastman*, p. 28; and GE to R. Talbot, 14 July 1880 (GEC).

[15] "Eastman Dry Plate Company," *Anthony Photog. Bul.* 12 (1881): 342–43.

[16] Schiller, *Early Photography in St. Louis,* [pp. 8–10].

Fig. 3.7
Eastman Dry Plate Company, Martin Building, State Street, Rochester, New York. Courtesy of Eastman Kodak Company.

Fig. 3.8
Early dry plate emulsion kitchen in Europe. From Wentzel, Memoirs. *Courtesy of Minnesota Mining and Manufacturing Company.*

Fig. 3.9
Early dry plate coating in Europe. From Wentzel, Memoirs. *Courtesy of Minnesota Mining and Manufacturing Company.*

States in the late 1870s served the pioneer American gelatin plate producers, with the emergence of centralized, mass production of plates in the United States, the patterns of transmission and diffusion of technological information changed. As the economic value of production information increased, technical communication about emulsion making and plate manufacture disappeared from public sources. Although technical journals, societies, and conventions continued to play a role in the circle of professional photographers, patents on new products, complex methods of production, expensive production machinery, and trade secrets all erected barriers against the diffusion of manufacturing information, leaving mobility of employees and acquisition of companies as the principal channels of diffusion (note the increase in patents issued in photography, fig. 2.22).

Characteristically, large national producers and small local manufacturers alike committed themselves almost exclusively to the production function and pursued a policy of manufacturing a single product. The only exceptions to this single-product orientation were Eastman and Carbutt and only Eastman diversified extensively, manufacturing bromide paper, photographic enlargers, and sundries during the first four years of the 1880s. Significantly, the most innovative companies in terms of improved and new products during the decade were also Eastman and Carbutt.

Although most metropolitan areas of the country witnessed the entry of many small dry plate producers in the early 1880s, St. Louis and Rochester emerged during the next two decades as the principal plate production centers of the Western Hemisphere. The large profit margins in plate production attracted many manufacturers in the early 1880s; however, most of them stayed in business only until their invested capital had been consumed.

After 1881 new producers faced three principal problems: (1) how to acquire sufficient knowledge to produce a good quality emulsion that would meet standards of quality control; (2) how to develop machinery, an organization, and a scale of production that would deliver a profit despite falling prices and increased competition; and (3) how to find a reliable market for the product. Producers such as Cramer, Carbutt, and Eastman, who had entered upon production very early, had an initial advantage because of their sole-distributor relationships with the three leading jobbers. Their access to the national market allowed them to focus attention upon improving their products and production methods. The other entrants had not only to overcome this marketing edge and the "established" reputation of the early companies but also to solve the quality and production cost problems with a smaller sales base. Most of the new firms confronted these problems with limited capital resources and limited technical knowledge, manufacturing experience, and national marketing experience, and they did not, therefore, meet with success.

In addition to the problems of the multitude of plate makers, the short economic recession in 1884–85 highlighted conditions of overproduction and overcapacity in the industry.[17] A number of readjustments indicating changes of strategy occurred in the industry. The marketing relationships with the leading jobbers began to change. A number of the leading dry plate makers established the Dry Plate Manufacturers Association. Eastman, Cramer, Norden, and Walker-Reid-Inglis were among those representatives attending the first association meeting in Cleveland, on 15 May 1883. Carbutt, although not in attendance, participated in the formation of the group. The purpose of the meeting was to establish a scale of prices for plates—to avoid price cutting as an increasing number of competitors entered the field.[18] The group met in Cleveland again on 9 and 10 January 1884, and a permanent association was formed with Carbutt as president, Cramer as vice-president, and Eastman as secretary-treasurer.[19] Further meetings were held in 1884, and the association continued for a couple of years. The combination of ease of entry into the industry and economic recession prompted this attempt to control prices. In the late eighties, as the leaders of the industry came to accept and advocate administered prices on plates

[17] From the time of the introduction of gelatin emulsions, the price trend was downward. By mid-1880 quarter plates were selling at 80¢ per dozen. By late 1881 Eastman had lowered the price to 65¢ per dozen, and in the year 1883 price cutting became a serious problem. Indicative of the extent of cutting is the price of the Beebe plate at 33¢ per dozen. See *Anthony Photog. Bul.* for the years 1880 to 1886.

[18] Ibid. 14 (1883): 160. Recommended quarter plate prices of 60¢ per dozen.

[19] Members included Cramer, Carbutt, Eastman, Inglis and Reid (Rochester), Monroe (Rochester), Hub (Providence), and Seed (ibid. 15 [1884]: 30–31); and *Rochester Union and Advertiser,* 15 January 1884, p. 2.

Fig. 3.10
Stanley Dry Plate Company trademark.

and as economic prosperity returned, prices stablized and the formal association disappeared.[20]

New Entrants, New Technology, and Emerging Oligopolistic Structure, 1885–95

Many of the smaller companies did not have the resources to weather the economic recession of 1884–85 and were forced, therefore, to quit the industry. The companies that remained increased their sophistication and began to put a greater emphasis on advertising and trademarks. Several companies also stressed improvements and innovations. Moreover, a number of new and important firms entered the field, making major inroads into the market between 1885 and 1895. The entry of new firms led to a restructuring of the industry.

Fig. 3.11
Miles A. Seed. Courtesy of Rudolph Schiller, St. Louis.

The new firms included three significant challengers: Stanley Dry Plate of Lewiston, Maine; M. A. Seed Dry Plate of St. Louis; and Hammer Dry Plate of St. Louis. The twin brothers Francis Edgar Stanley (1849–1918) and Freelan Oscar Stanley (1849–1940) founded the Stanley firm in Lewiston, Maine, about 1884. The twins (well-known now for their later introduction of the Stanley Steamer automobile) were born in Maine and received normal-school training. Francis Stanley, the more aggressive of the two, practiced in the 1870s as a crayon portrait artist and then as a photographer. In the early 1880s he prepared gelatin plates for his own gallery work. When a Boston supply house encouraged him to undertake commercial production of plates, Freelan joined him in the enterprise. Although their firm met with considerable success, the Stanley brothers were not deeply committed to long-term goals of growth, technical innovation, and imaginative business practice in the photographic industry.[21]

Fig. 3.12
Ludwig F. Hammer. Courtesy of Rudolph Schiller, St. Louis.

Two immigrant photographers assumed the leadership in inaugurating two important St. Louis dry plate companies. Miles Anscow Seed (1843–1913), a native Englishman, joined in 1883 with two other St. Louis men to found the Seed Dry Plate Company. While Seed's initial efforts in producing gelatin plates and albumen paper met with little success, the firm eventually became the largest and one of the most technically innovative plate makers in the country.[22] Ludwig F. Hammer (1834–1921), a native German, had long experience as a photographer and gallery owner in St. Louis. In the middle 1880s he was part of an unsuccessful dry plate company, but in 1891 he joined with a trained chemist and

[20] GE to G. Cramer, 30 November 1886 (GEC).

[21] Francis Edgar Stanley, *Theories Worth Having and Other Papers* (Boston, 1919), pp. xi–xiv; R. W. Stanley, *The Twins* (Boston: Privately printed, 1932); *National Cyclopedia of American Biography*, s.v. "Francis Edgar Stanley"; and *Newton Graphic* (Mass.), 2 August 1918, pp. 1 and 8; and F. E. Stanley testimony, *U.S.* v. *EKCo.*, p. 78.

[22] Kargau, *Mercantile, Industrial, and Professional Saint Louis*, pp. 414–15; John William Leonard, *The Industries of St. Louis* (St. Louis: Elstner, 1887), p. 142; Schiller, *Early Photography in St. Louis*, [pp. 10–11]; and "M. A. Seed Dry Plate Co." (GEN).

Fig. 3.13
Hammer Dry Plate Company trademark.

Fig. 3.14
Harvard Dry Plate Company trademark.

Fig. 3.15
Wuestner's Eagle Dry Plate trademark.

two former employees of Cramer Dry Plate Company to establish a highly successful plate company.[23]

Smaller but important new companies included Standard Dry Plate of Lewiston, Maine; Harvard Dry Plate of Cambridge, Massachusetts; Record Dry Plate of East Milton, Massachusetts; Wuestner Dry Plate of St. Louis and, later, Jersey City; and a group of small and interrelated companies in Rockford, Illinois. The emergence of these new companies after the period of overcapacity in the middle 1880s reflected (1) the success of the Dry Plate Association in maintaining prices and profits; (2) the return of prosperity to the American economy during the late 1880s and early 1890s; and (3) the growth of the relatively small yet significant serious-amateur market in photography. The factory production of photosensitive negative material removed the most critical technical operation in photography from the hands of photographers and, thereby, made photography attractive to an ever larger group of amateurs. Of course, these amateurs still had to develop and print their pictures and often had to produce their own sensitized printing materials. Both Anthony and Scovill gradually sought to exploit this new market through the introduction of light cameras that utilized dry plates. The market thus created promoted the entry of new firms and the deterioration of the market position of the leading dry plate manufacturers.

New markets, new companies, and the new strategies of old companies led to a new alignment and structure in the industry. Seed entered the industry in 1883 and after two difficult years succeeded in creating a national distribution network independent of the three major jobbers. In 1885 Eastman severed relations with Anthony, as the Rochester firm had begun to shift its focus from dry plates to roll film photography. Anthony then established a sole agency with Stanley, and in response, the Maine company's sales rose rapidly; yet, in the late 1880s strained relations led to a severance of the relationship, and Stanley, like Seed and Eastman, developed its own marketing department, which worked directly with supply houses and professional photographers. When Hammer entered the market in the early 1890s, it also mounted its own marketing department. Consequently, with the development of the new serious-amateur market, the market for dry plates reached the critical size that made it profitable for the larger companies—old or new—to perform independently the marketing function. Accordingly, the once powerful position of the national jobbing houses began to decline.

Although certain new companies could handle the marketing barriers because of the expanded market, the technical barriers remained. A few companies, such as Stanley, Seed, and Hammer,

[23] Schiller, *Early Photography in St. Louis,* [pp. 11–12]; Kargau, *Mercantile, Industrial, and Professional Saint Louis,* pp. 413–14; Hyde and Conard, *History of St. Louis,* 2: 977–78; *Historical and Descriptive Review of St. Louis: Her Enterprising Business Houses and Progressive Men* (St. Louis: John Lethem, 1894), p. 163; Fritz Wentzel, *Memoirs of a Photochemist* (Philadelphia: American Museum of Photography, 1960), pp. 19–20, 124; *Photog. T.* 34 (1902): 34 and 42; Salzgeber testimony, *U.S.* v. *EKCo.,* pp. 83–84; and *Philippi's Rockford City Directory, Including a Directory of Winnebago County,* 1891–92.

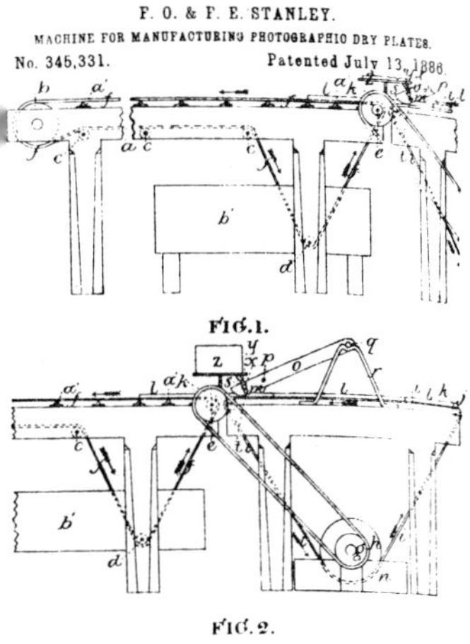

F. O. & F. E. STANLEY.
MACHINE FOR MANUFACTURING PHOTOGRAPHIC DRY PLATES.
No. 345,331. Patented July 13, 1886.

FIG. 1.

FIG. 2.

Fig. 3.16
Stanley coating machine. From U.S. Patent # 345,331.

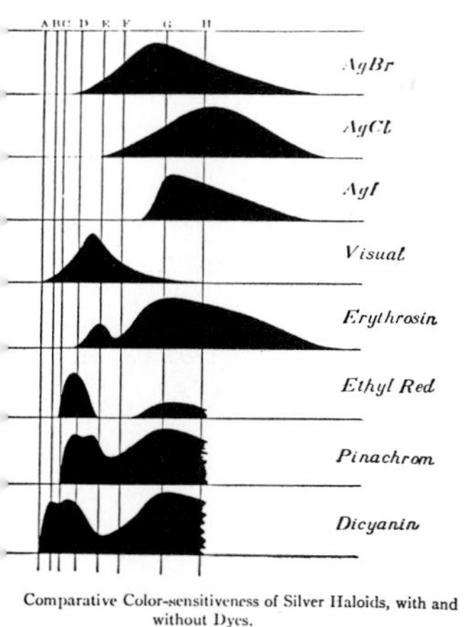

Fig. 3.17
Comparative color sensitiveness. From Louis Derr, Photography for Students *(New York: Macmillan Co., 1919).*

were able to circumvent these barriers successfully. The Stanley brothers experimented over a period of several years with their emulsion and by 1885 were able to produce a competitive one. Equally important, they developed, patented, and introduced an efficient plate-coating machine (fig. 3.16).[24] Miles A. Seed succeeded after two years of experimentation in producing a high quality emulsion that by the early 1890s had captured a significant share of the market. Hammer, starting operations in 1891, entered the industry relatively late but brought to the new company his own experience with the defunct St. Louis Dry Plate Company, the experience of two of his partners as office manager and salesman for Cramer, and the technical expertise of another partner who had been educated as a chemist at the Stuttgart Polytechnique Institute. Moreover, the Hammer company soon acquired the emulsion-making services of Rochester's George Monroe, the man who had originally taught George Eastman wet plate photography and who had operated his own dry plate company during the middle 1880s. Monroe soon left Hammer to become emulsion maker for Eastman and was replaced by Frank Pratt, an emulsion maker with previous experience at three dry plate companies in Iowa City and with a company in Rockford, Illinois (see appendix). Thus, the Hammer company soon captured a significant share of the dry plate market by exploiting one of the remaining methods of diffusion of technical information: employee mobility.[25]

As the new gelatin emulsion technology became widely adopted and commercially institutionalized, technological innovation tended to focus upon problems posed within the new technological framework. During the decade beginning in 1885, innovations from Europe continued to influence the American dry plate industry. One of the most significant problems addressed during this decade was that of the color sensitivity of gelatin emulsions. Silver salts are highly sensitive to violet light and relatively insensitive to the remainder of the visible spectrum. Dr. Hermann Wilhelm Vogel, world-renowned photochemist at the Charlottenburg Technische Hochschule, demonstrated in 1873 that silver salts could be sensitized to nonviolet portions of the spectrum by treating the photosensitive material with certain aniline dyes. With the introduction of gelatin emulsions, new dyes were required. Josef M. Eder discovered in 1884 that erythrosin added to gelatin emulsions would extend color sensitivity to the yellow and green portions of the spectrum. This information was published in the same year in the *Photographische Korrespondenz*, and plates so sensitized, called orthochromatic plates, were introduced in Vienna shortly thereafter.[26]

[24] F. E. Stanley testimony, *U.S.* v. *EKCo.*, p. 78; and D & B, Maine, vol. 9, p. 260, and vol. 10, pp. 120–21.

[25] Schiller, *Early Photography of St. Louis*, [pp. 11–12]; Kargau, *Mercantile, Industrial, and Professional Saint Louis*, pp. 413–14; Hyde and Conard, *History of St. Louis*, 2: 977–78; *Historical and Descriptive Review of St. Louis*, p. 163; Wentzel, *Memoirs of a Photochemist*, pp. 19–20 and 124, nn. 70 and 73; *Photog. T.* 34 (1902): 34 and 42; Salzgeber testimony, *U.S.* v. *EKCo.*, pp. 83–84; and *Rockford City Directory*, 1891–92.

[26] Josef M. Eder, *History of Photography*, trans. Edward Epstean (New York: Columbia University Press, 1945), pp. 457–78.

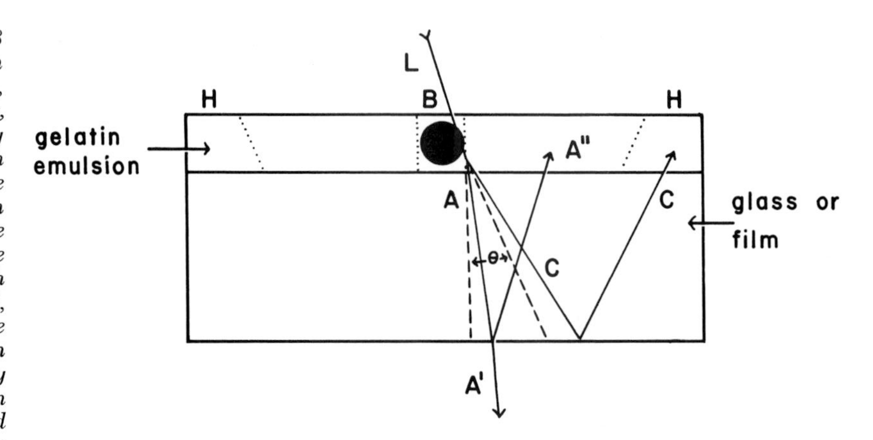

Fig. 3.18
Cause of halation. When an intense beam of rays, L, *strikes a photographic plate, the final photograph may show a bright spot,* B, *with a surrounding halo,* H. *The halo results from irradiation of rays from particles in the emulsion (the size of the particle here is very much exaggerated). Ray* C, *irradiated at an angle greater than a certain critical angle,* Θ, *totally reflects from the bottom boundary of the plate and exposes the emulsion from the back; ray* A, *irradiated at an angle less than* Θ, *partially passes through the boundary* (A′) *and partially reflects* (A″). *The intensity of the partially reflected ray is much less than that of the totally reflected ray and therefore produces in the final photograph a relatively dark area between the bright spot and the bright ring of the halo.*

Orthochromatic plates were first introduced in the United States in 1886 by John Carbutt of Philadelphia.[27] Cramer, in St. Louis, after a year of experimenting (1888), went to Germany to learn more about sensitizing dyes. While in Europe, he placed under contract for ten years the Swiss photochemist Edward V. Boissonnas and brought him back to St. Louis. Experiments continued, and in 1889 Cramer built a new plant especially for the production of orthochromatic plates. But when Boissonnas had been in St. Louis only eight months he contracted typhoid fever; he died in January 1890. Cramer next engaged B. J. Edwards of London for $10,000 to come with his son to St. Louis to assist in producing the appropriate emulsion. Edwards had been successful in producing orthochromatic plates in England but was unable to meet with success in St. Louis.[28] When Edwards returned to England, Cramer employed his former partner, Norden, and the two men worked together and succeeded in producing a satisfactory plate by June 1891. The plate was marketed under the name "Isochromatic," the term used by B. J. Edwards & Company in England.[29]

While orthochromatic plates partially solved one problem, the range and equality of sensitivity, another serious problem remained: *halation*, a blurring effect caused when very bright lights or brightly lighted objects were photographed against a dark background (fig. 3.18). It is now understood that halation is caused by light striking the thin emulsion and some of the light penetrating to the glass base where it is reflected back into the emulsion, either from the front or back surface of the plate. The degree of halation is a direct function of the sensitivity of the plate; hence, the introduction of gelatin plates, with their substantially increased sensitivity, aggravated the problem. German photochemists and physicists, for ex-

[27] Schiller, *Early Photography of St. Louis*, [p. 10]; and Joseph Jackson, *Encyclopedia of Philadelphia* (Harrisburg: National Historical Association, 1931), 2: 367–68.

[28] Schiller, *Early Photography of St. Louis*, [p. 10]; and Wenzel, *Memoirs of a Photochemist*, p. 19.

[29] Schiller, *Early Photography of St. Louis*, [p. 10]; *Anthony Photog. Bul.* 22 (1891): xxxv; and *American Annual of Photography Almanac* 5 (1891): 135 (advertisement).

ample, Dr. Adolf Miethe and Dr. Franz Stolze, first addressed the problem of halation with gelatin plates. In the late 1880s Thomas Sandell introduced a nonhalation plate that was marketed in England by R. W. Thomas & Company of London. The Sandell nonhalation plates were of two varieties: one was coated with two layers of emulsion of different sensitivity, the other with three layers. The top layer was the most sensitive and the lower layer, or layers, was less sensitive, acting as an absorber of admitted light.[30]

One of the new firms entering the dry plate market in the late 1880s, that of Edward Wuestner, pursued an innovative course by introducing orthochromatic plates in 1890 and nonhalation plates in December 1892. The Imperial Non-Halation Plate employed orthochromatic emulsion in the substratum as a further insurance against halation, following the specifications of Sandell's plate.[31] Nonhalation plates were also introduced in the United States by Seed of St. Louis at the same time that Wuestner entered the market. The larger plate companies, such as Cramer and Eastman, soon followed the pioneers in the production of orthochromatic and nonhalation plates.[32]

From the middle 1880s to the early 1890s John Carbutt introduced a number of product innovations besides the orthochromatic and nonhalation plates. Keystone Dry Plate produced a special plate for lantern slides, various engraving plates, developing tablets, and a variety of products of European origin. Carbutt kept well-informed of European developments by reading foreign journals, and he was quite responsive to innovations.

Reflecting the national trend in business strategy, the growing scale of operations, and increased capital needs, the styles of business organization among dry plate producers changed significantly between the early eighties and the early nineties. Where the pioneers, Carbutt and Eastman, had begun as single proprietorships, and Cramer & Norden as a simple partnership, by the mid 1880s all the larger companies, with the exception of Carbutt, had moved to partnerships with two or more members. By the early 1890s Seed (1883), Eastman (1884), Stanley (1887), and Hammer (1891) had organized as corporations.

The newer companies—Stanley, Standard, Hammer, and Wuestner—and the older companies—Cramer, Seed, Eastman, and Carbutt—pursued quite different strategies during the period 1885 to 1895. Cramer and the newer companies continued to focus on a single-product strategy, conceiving of themselves exclusively as dry plate manufacturers. While Wuestner and Cramer made some product innovations, they were confined to dry plates. The remaining older companies, perhaps challenged by the new entrants, en-

[30] Charles Scolik, "The 'Thomas' Sandell Plate," *Photog. T.* 22 (1892): 586–88.

[31] *American Annual of Photography Almanac* 4 (1890): 27 (advertisement), 5 (1891): 135 (advertisement); *Anthony Photog. Bul.* 22 (1891): xxxv; and *Photog. T.* 22 (1892): 586–88, 602, and 656; 23 (1893): 376; and 24 (1894): 146, 225–26, and 291.

[32] Wentzel, *Memoirs of a Photochemist*, p. 19; and *Photog. T.* 22 (1892): 602 and 24 (1894): 146, 225–26, and 291.

Fig. 3.19
Changing leadership in production of gelatin dry plates.

Early to middle 1880s	*Early to middle 1890s*
1. Cramer (St. Louis, Mo.)	1. Seed (St. Louis, Mo.)
2. Carbutt (Philadelphia, Pa.)	2. Cramer (St. Louis, Mo.)
3. Eastman (Rochester, N.Y.)	3. Hammer (St. Louis, Mo.)
	4. Stanley (Boston, Mass.)

larged their conceptions of their enterprise. Carbutt not only produced a very large variety of plates; he also introduced cut-celluloid films, filters, and sundry photographic supplies. Seed produced some photographic paper, cut-films, and much later, some cinematographic film. Eastman had minimal commitment to dry plates and switched not only to film photography but eventually to production of nearly a full line of both amateur and professional photographic materials and apparatus.

By the mid 1890s the market structure of the dry plate sector of the photographic industry had become oligopolistic, and its leadership had changed since the mid 1880s, when Carbutt, Cramer, and Eastman had dominated the market. In St. Louis, Cramer's position was eroded by Seed, which in the middle 1890s held approximately one-third of the national market, and by Hammer. In the East, the leadership of Carbutt and Eastman was gradually eroded by Stanley and Wuestner. In terms of marketing, the hold of the jobbers on the market had been substantially reduced during the late 1880s and early 1890s. Stanley, Hammer, and Eastman built their own marketing departments and thereby became independent of the traditional industrial leaders. Anthony and Scovill both continued to focus on the jobbing function and ignored the telltale implications of the growing independence of some manufacturers of photographic materials. The economic depression of the middle nineties stimulated the revival of the association approach of the middle eighties. The ultimate failure of this approach soon signaled the decline of influence of the three old-line houses.

During this period the technically oriented individual dominated the firms. All the major companies, with the exception of Eastman, were founded by professional photographers. Emulsion making remained an empirical art, and it was mechanically or chemically oriented persons with a background in photography and practical photographic chemistry who were attracted to positions of technical responsibility. The employment in the industry of the college-educated chemist or engineer was still the exception, with only Punnett (Case) at Seed, Sellner (Stuttgart) at Hammer, and Reichenbach (Rochester) and de Lancey (M.I.T.) at Eastman.

Photographic Papers, 1880–95

Although factory sensitized gelatin dry plates were widely adopted in the early 1880s, factory sensitized photographic papers were not introduced commercially in the United States until the middle 1880s and did not become widely adopted until the early 1890s. In general, the companies that led in dry plate manufacture did not become the leaders in photographic paper manufacture because, although sensitized photographic papers employed silver salts, the emulsions possessed entirely different characteristics and the methods of coating were also quite different. As with the dry plate manufacturers, only a few firms initially produced sensitized photographic paper; however, in contrast, the acceptance of sensitized paper by professional photographers took longer, and wide adoption did not come until the early 1890s, at which time a fairly large number of producers appeared.

Before the early 1880s, photographic papers were not generally sensitized until immediately prior to printing. Consequently, as with the negative photosensitive materials, each photographer made his own photosensitive print material; however, unlike the negative materials, unsensitized photographic papers were prepared, prior to 1880, in centralized factories. As indicated above, the most common photographic paper of the late collodion period consisted of high quality raw paper stock made with mineral-free water and coated with albumen that contained halogen salts. Special albumenizing companies located in Dresden, New York, Philadelphia, and Rochester coated raw paper stock from two special paper mills in Europe. The American photographer usually purchased albumen paper and then floated it on a silver nitrate bath for a short time prior to printing the paper. Then the collodion glass negative was placed above the sensitized paper and exposed to sunlight for a long period of time. The print was then produced directly, without the aid of a latent image. The photographer watched the direct production of the print, and when the print appeared to be finished, he removed it from the sunlight and fixed it. This type of paper became known as *printing-out paper* (POP).

The earliest and most significant innovation in factory-produced photosensitive papers was the gelatin silver bromide paper. As with the negative material, gelatin served to preserve the photosensitivity of photographic papers, thereby making possible factory production of sensitized papers. The gelatin silver bromide paper, unlike the albumen papers sensitized by photographers, employed the latent image in the production of prints. The glass negative was placed above the gelatin bromide paper and exposed to light for only a very short time; then the paper was removed in a dark room, *developed*, and fixed. This bromide paper, which employed a latent image that required development, was called *developing-out paper* (DOP).

The gelatin bromide papers (DOP) had advantages and disadvantages. The advantages included the possibility of enlargement—the production of large prints from small negatives—and the acceleration and mechanization of the printing process. With printing-out paper, the printing was often a time-consuming process, the length of which depended upon the intensity of sunlight on a particular day. With developing-out paper, artificial light of a specific duration was used, allowing for some standardization. Moreover, wheels or drums carrying large numbers of negatives and prints were used to process the prints mechanically. The major disadvantage, as with the gelatin negative materials, was the professional photographer's reluctance to adopt a new process where the sensitivity factor was so different. Careless exposure to light was relatively unimportant with printing-out papers, but considerable care and experience were required in the use of the more sensitive developing-out paper. Furthermore, with contact printing (POP), the photographer could judge precisely when he wished to stop printing because he was able to watch the slow process; but with developing-out paper, such "eye control" was impossible. If the photographer did not time the exposure accurately, he did not discover his error until the print had been developed. Consequently, professional photographers were more reluctant to adopt gelatin bromide papers than gelatin bromide negative plates. Gelatin bromide papers were not produced in any quantity in the United States until a few years after the introduction of gelatin dry plates, and the shift from the use of albumen paper to factory sensitized DOP was slow and not universal.

Pioneering Production of Sensitized Papers, 1880–89

The commercial introduction of factory sensitized gelatin photographic paper in the United States came from outside the traditional photographic paper sector. While Anthony, Haworth, Clemons, and Hovey had considerable experience of coating raw stock with unsensitized albumen, the technological change from unsensitized to sensitized paper was too great to be met by the established firms. However, one of the new dry plate companies, the Eastman Dry Plate & Film Company, in cooperation with the Anthony company, introduced gelatin bromide sensitized print papers in 1881 and 1882. Eastman inaugurated production of the bromide paper utilizing the old hand production methods employed in albumenizing paper. In 1882 and 1883 the Anthony company marketed and promoted the Eastman paper under the Anthony brand name; however, the paper was not a success because the methods of production were slow and costly, and the emulsion coatings lacked the necessary uniformity. Soon the production and marketing of the paper ceased.[33]

During this period George Eastman sought, as he had in dry plate manufacture, to develop a system of mechanized production. During the late 1870s several persons interested in coating papers developed and introduced mechanized methods of continuous coating. In Britain, Sarony and Johnson developed a machine for coating carbon and pigment paper; in France, Delaunay constructed a machine for coating emery paper with glue; in Germany, Colas developed a machine for coating blueprint paper, and Bertsch built a machine for the production of solar print paper. In the early 1880s Allen &

[33] *U.S.* v. *EKCo.*, pp. 3, 558, and 563; GE testimony, in ibid., pp. 267–68; and *Anthony Photog. Bul.* 14 (1883): 59.

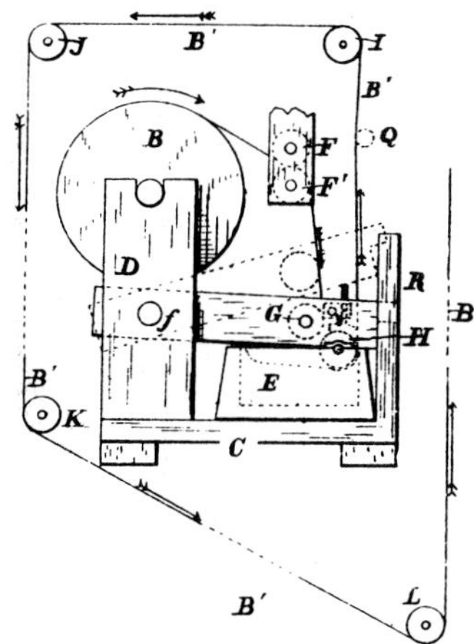

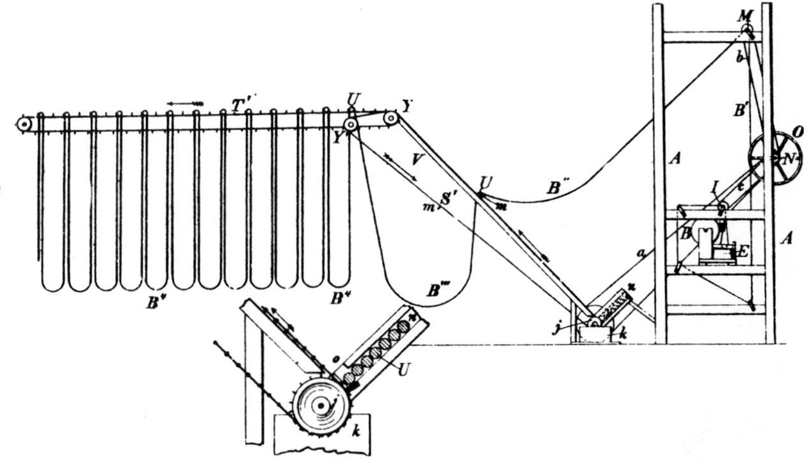

Fig. 3.20
Eastman-Walker paper coating machine. From U.S. Patent # 358,848.

Rowell, a new, small Boston photographic paper producer, and Anthony employed machines in their coating of photographic papers.[34] In the early 1880s George Eastman and William H. Walker, Eastman's co-experimenter, investigated the production methods and machines in use in certain American and European photographic paper companies. Eastman visited Anthony's paper factory in New York, where he observed that company's albumen coating machine. He returned to Rochester and made a replica of the machine but could not make it operate satisfactorily. In 1882 he spent two weeks at the factory of Mawson and Swan in Scotland, where he probably observed the Sarony and Johnson machine in operation.[35] Eastman's model of the Sarony and Johnson machine also proved inoperative. William Walker in the early 1880s visited the production facilities of Allen & Rowell and carefully observed their coating machine.[36]

In 1883 and 1884, Eastman and Walker, working together, devised a coating machine that successfully produced paper. They designed the machine initially for production of continuous roll paper film for film photography, but as Eastman observed, "it worked so well that we immediately decided to put out bromide paper, the manufacture of which we had abandoned."[37] The machine, for which applications for patents were filed in October 1884,[38] was similar in part to the machines of Sarony and Johnson, Allen & Rowell, and the Anthonys. It contained the basic elements of rollers, coating trough, and appropriate heating devices; however, the Eastman-Walker machine differed from the Sarony and Johnson coater by eliminating the belt that carried the paper (fig. 3.20). It

[34] *Eastman* v. *Getz*, 77 *Fed. Rep.* 412–20 (1896).

[35] Personal correspondence of Beaumont Newhall: D. Burton Payne, Director, Mawson & Swan, Ltd., to Beaumont Newhall, Director, George Eastman House, Rochester, N.Y., 23 December 1957.

[36] *Eastman* v. *Getz*, 77 *Fed. Rep.* 412–20 (1896).

[37] GE testimony, *U.S.* v. *EKCo.*, p. 268.

[38] U.S. Patents #358,848, #370,110, and #370,111.

also differed from these machines by the addition of a moving "hang-up" device for air drying the continuous strip of coated paper.[39] The machine method of coating sensitized printing paper soon demonstrated its superiority to the traditional methods of hand coating, first, in the more even coating of emulsion, and second, in the considerable savings in emulsion, paper, and labor.[40]

A comparison of the Eastman and Anthony company actions illustrates the differing circumstances and strategies of the established house and the young firm. The Anthony company, with a strong position in photographic paper, was unwilling to risk the capital and energy to master the new gelatin technology. It maintained the traditional policy of externalizing the risk and hoping to obtain the reliable sole agent, trade agent, or jobber profits if the external manufacturer were successful. In contrast, Eastman, with no established position and relying upon Anthony for the marketing function, risked his less valuable time and energy seeking to master the new technology. The capital risk was small, but the opportunity was great. However, the new company's technological strategy held promise only if the new developments had a large market potential and could be protected through patents. George Eastman perceived, to some extent, the revolutionary potential of the gelatin emulsion. The Anthony company failed to recognize the significance of this technological change in either dry plates or in their own area of production, print paper.

From 1884 to early 1887 the Eastman company, the sole American producer of bromide paper made by the continuous coating and drying method, promoted the use of bromide paper in the United States and found a large market in England and on the Continent for the Eastman Permanent Bromide paper.[41] When, during this period, Anthony and Eastman severed their exclusive marketing relationship, the Eastman company developed its own marketing capacity in the United States and Europe and expanded its product line to include roll film and roll holders. Concurrent with the development of the marketing department came the creation of the developing and printing service, which later played such an important role in the Kodak system. The developing and printing service addressed two problems. One was the reluctance and inability of the professional photographers to adopt the new gelatin bromide paper. Meanwhile, a new class of serious amateurs was drawn to photography because of the simplifications stimulated by the gelatin revolution. Of special interest to these amateurs was the ability to obtain readily enlarged prints with this gelatin paper. A small camera using only small plates could be used, and yet large prints could be procured. Thus, the Eastman Dry Plate & Film Company provided printing and enlarging service as a means of stimulating the consumption of the new gelatin bromide paper. The second problem that helped to stimulate the service department of the company was

[39] This "hang-up" technique of drying is called festooning; *Eastman* v. *Getz*, 77 *Fed. Rep.* 412–20 (1896).

[40] GE testimony, *U.S.* v. *EKCo.*, p. 268.

[41] GE to Vance & Miller, 3 February 1888 (GEC).

the introduction of the stripping film for roll film photography. This stripping film required complex processing; hence, by providing stripping, developing, and printing service, the Eastman company created a demand for a primary Eastman product and for several ancillary products. This service policy, developed during the middle 1880s to meet the limited demands of the professional and serious amateur markets, later proved an important aspect of the Eastman company's strategy in developing and exploiting the commercially more important mass amateur market. In the meantime, the strategy helped enhance the sales of paper.

Following the break in the marketing relationship between Anthony and Eastman in March of 1885,[42] Anthony lost its jobbing profits on the Eastman products and, relations between the two firms having cooled considerably, began to perceive that its traditional strategies were no longer effective against a firm that threatened its traditional paper market. Relations between the two firms deteriorated further when, early in January 1887, two Eastman employees left Rochester to join the Anthony company: David Cooper, a salesman-demonstrator, and Franklin Millard Cossitt, the operator of the Eastman-Walker continuous coating paper machine.[43] The Anthony company hired Cossitt with the understanding that he was to build and operate for Anthony a machine like the one he operated in Rochester for the production of bromide paper.[44] When George Eastman took legal steps to halt the Anthony-Cossitt activities, Anthony countered by changing the construction of the machine and by inaugurating a suit against the Scovill company, charging that it was selling the Eastman bromide paper, which, it was alleged, infringed the Anthony-owned Roche patent.[45] Eastman stepped to Scovill's defense at once to assure the trade that it need not worry about using the Eastman paper. By early May of 1887, Anthony had its paper on the market and within two months had cut the price 20 percent below that of Eastman.[46] Eastman's strategy was to hold the price and continue legal action, but by the end of the year the Anthony bromide paper had cut so drastically into the Eastman market that Eastman responded by introducing a new, cheaper brand, Eureka, to compete with the Anthony paper.

At first, George Eastman sought to prosecute the case, but from early 1888 to late 1889 efforts were made on both sides to affect a compromise.[47] According to Eastman's perception, while either Anthony or Eastman might inherit complete patent control for bromide paper production, there was the greater likelihood that neither of their patents would hold and the entire business would be thrown

[42] 3 March 1885; "Eastman Companies" (GEN).

[43] GE to George G. Rockwood, 31 May 1886, and GE to Vance & Miller, 3 February 1888 (GEC).

[44] Frank M. Cossitt testimony, *U.S.* v. *EKCo.*, pp. 504 and 509.

[45] GE to W. Irving Adams, 7 January 1887; GE to David Tucker & Co., 21 March 1887; GE to Scovill Mfg. Co., 1 May 1887 (GEC).

[46] GE to Scovill Mfg. Co., 2 June 1887; GE to W. Irving Adams, 4 June 1887; and GE to J. B. Church, 13 July 1887 (GEC).

[47] GE to Moritz B. Philipp, 31 December 1887, and GE to Scovill & Adams Co., 29 January 1889 (GEC).

open to vigorous competition.[48] Negotiations between the two rivals continued, but Anthony faced several reverses in the late 1880s, including the death of Edward Anthony, the senior partner, in December 1888, and substantial loss of business and profits as the failure of the traditional marketing-manufacturing strategy became more evident.[49]

Nevertheless, George Eastman saw his initial sole position in the photosensitive paper market eroded as the market for paper expanded. Even patents and litigation were not sufficient protection of that position. Hence, during the period 1889 to 1893, the number of photo paper producers expanded rapidly. The introduction of several new types of paper, most of them previously developed and introduced in Europe, accompanied this expansion.

The Shift from Unsensitized to Sensitized Papers in the United States, 1889–95

During the first decade of gelatin emulsion production in the United States (1879–89), professional and amateur photographers alike readily adopted gelatin dry plates as negative material, but the revolution in negative material did not extend to printing papers. While gelatin bromide DOP was produced in substantial quantity in the United States during the 1880s, especially for enlargements, professionals only slowly replaced the albumen paper and silver bath with the new paper. During the period from 1889 to 1895 a revolution in printing method parallel to that in negative material a decade earlier occurred. Bromide paper (DOP), however, did not lead the revolution; two classes of chloride papers (POP) stimulated the shift from albumen paper to factory-sensitized papers and, hence, completed the movement from decentralized handicraft to centralized machine production of photosensitive materials.

Phase 1: Growth (1889–92) During the late 1880s Anthony, Allen & Rowell Company, and Buffalo Argentic Paper Company joined Eastman in the production of gelatin bromide developing-out papers, and in the late 1880s and early 1890s collodion and gelatin printing-out papers were introduced and soon seized the market from the albumen and bromide papers. The earliest of the printing-out papers introduced on a commercial scale in the United States was the collodion paper. The collodio-chloride silver emulsion process had been introduced by the English editor of the *Photographic News*, George Wharton Simpson, in 1864. No English manufacturer initiated production, although J. B. Obernetter of Munich did produce some of this paper in Germany in 1868. Only in the late 1880s and the 1890s did this paper, under the

[48] GE to Church & Church, 18 and 22 February 1888; GE to E. & H. T. Anthony & Co., 29 February 1888; GE to W. I. Lincoln Adams, 26 October 1888; GE to Philipp, 28 December 1888, 25 January and 26 April 1889; GE to Scovill & Adams Co., 29 January 1889; GE to Rockwood, 27 December 1888; and GE to Walker, 28 April 1889 (GEC).

[49] GE to Philipp, 14 October 1889 (GEC).

trade name Aristotypie, become popular in Europe.[50] Just as photographers earlier had moved from the direct positive daguerreotype process to the negative-positive collodion process by way of the intermediate collodion direct positive ambryotype, photographers in the early 1890s moved from sensitizing their own printing-out paper to the purchase of factory sensitized developing-out paper by way of factory sensitized collodion printing-out paper.

The requisite technical information for producing collodion chloride printing-out paper in the United States was introduced by Carl Christensen, who had learned the formula from his father in Denmark prior to initiating development experiments in Jamestown, New York. As a consequence of Christensen's success a group of Jamestown businessmen founded in 1889 a new company for the production of collodion paper, American Aristotype.[51] Within this company Charles S. Abbott, an advertising man from New York City, seized the business initiative. The ambitious Abbott gradually revealed an imaginative sensitivity to the interrelated areas of technical innovation and successful marketing. Soon Christensen left the company and went to Berlin, where he operated one of the largest collodion paper companies in Europe;[52] however, before leaving Jamestown he drew contracts with American Aristotype that restricted it from marketing its papers in Europe. Soon Abbott and his partners carefully coordinated their excellent formula, their mechanized methods of production, and their exclusive marketing contract with Anthony and dramatically emerged as a major force in the American photographic paper market.[53] Moreover, shortly thereafter three smaller companies also entered upon collodion printing-out paper production: Kuhn Crystallograph of Springfield, Missouri; Western Collodion Paper of Cedar Rapids, Iowa; and Bradfisch Aristotype of Brooklyn, New York.

The second important new paper introduced in the United States in the early 1890s was the gelatin POP. Captain W. De W. Abney, the well-known English photochemist, experimented in 1882 with gelatin silver chloride emulsions for printing-out paper and published his results in the *Photographic Journal*.[54] Liesegang of Düsseldorf initiated production of the paper in 1886.[55] The firms of Bradfisch & Hopkins and the New York Aristotype Company were two of the earliest American producers of gelatin printing-out papers.

The new entrants into production of printing papers were for the most part single-product manufacturers who had not previously produced photosensitive materials. With the exception of the Eastman company, none of the major dry plate producers was successful in entering the printing paper market. This is understandable from a

[50] Gernsheim and Gernsheim, *History of Photography*, p. 284; and Eder, *History of Photography*, p. 536.

[51] "American Aristotype Company" (GEN).

[52] Wentzel, *Memoirs of a Photochemist*, pp. 65, 69.

[53] "American Aristotype Company" (GEN); and Richard A. Anthony testimony, *U.S. v. EKCo.*, pp. 558–60.

[54] *Photog. J.* 27 (1882): 155.

[55] Gernsheim and Gernsheim, *History of Photography*, pp. 285–86.

Fig. 3.21
Competing types of paper: DOP and POP.

POP	*DOP*
(Printing-Out Paper) Direct printing by sunlight without aid of latent image	(Developing-Out Paper) Printing by quick exposure and development of latent image with chemicals
1. Albumen paper (1870s–1880s) Sensitized by the photographer 2. Collodio-chloride paper (Aristotype) (1890s–) Sensitized in factories 3. Gelatin-chloride paper (1890s–) Sensitized in factories	Silver bromide paper (1884–1890s) Sensitized in factories

technical point of view. Even Eastman moved step-by-step, from dry plates with gelatin bromide emulsions to chloride printing-out paper by way of the technologically intermediate step of producing gelatin bromide developing-out paper. The gelatin chloride or collodion chloride emulsions were quite different from the gelatin bromide emulsions used with dry plates. Several of the leading dry plate companies made efforts to begin printing paper production, but lack of experience with either paper coating or chloride emulsions more than counter-balanced the advantage of having established production and marketing facilities.

During this first phase of the shift from decentralized to centralized production of sensitized paper in the United States, the new entrants into the industry concentrated their marketing endeavor on converting photographers from albumen papers to the "ready" sensitized papers. The initial growth of the printing-out paper market was at the expense of the albumen paper market, which was strongly dominated by European producers. In all cases, the capital and the production of the sensitized paper producers were, at first, quite small. Most producers marketed their paper through the New York jobbers, Anthony, Scovill, Gennert, or Murphy, and thereby avoided the expense of an extensive marketing operation.

The competition of the new printing papers with the albumen papers and to a lesser extent with the bromide paper affected producers in both areas. First, it brought the independence of American photographers from the German albumen paper cartel. Second, it hurt substantially the handful of American albumen paper companies. The Philadelphia firms met the challenge with advertising that attacked the qualities and characteristics of the new papers. Yet, in contrast with the albumen prints, which yellowed with time, the collodion prints were more permanent and possessed sharper definition. The American Albumenizing Company in Rochester, soon after the death of its founder, Douglas Hovey, in 1886, ceased operations.

His partner, A. M. Brown, established in Rochester his own small production of the new printing papers.

The transmission of technology, particularly emulsion formulae and methods of production, between the United States and Europe and among domestic producers of photographic papers played an important role in shaping the photographic paper sector of the industry. The basic forms of printing-out papers were first introduced in Europe, but because of the basic conservatism of the American professional photographers, these papers were adopted considerably later in the United States. Much of the knowledge of emulsion formulae was published in European journals and books. The transmission of emulsion formulae directly from Europe to the United States is documented in the case of Carl Christensen and the American Aristotype Company. Yet, inasmuch as Christensen and American Aristotype made contracts that limited the disclosure of technical information and the marketing of American Aristotype papers in Europe just before Christensen returned to Germany to found his collodion paper company in Berlin, it seems likely that he also carried emulsion and production technology back to Europe.[56]

The channels of transmission of emulsion formulae and production methods among the firms within the United States may be observed by following the movement of key technical personnel. In photographic paper, for example, in the late 1880s Frank M. Cossitt carried production technology from the Eastman company to the Anthony company. In the early 1890s he left Anthony, carrying his experience and knowledge from both the Eastman and Anthony companies to his own business, the New York Aristotype Company. During the 1890s such transmissions became an increasingly important mode of dispersion of product and production technology.

The new papers and new producers, of course, represented a serious challenge to the hegemony of the Eastman company in the production of sensitized paper, an area that Eastman had held exclusively for nearly half a decade. Although overwhelmed with problems in launching the roll film system of photography, Eastman recognized the threats and responded to them in both the collodion POP and gelatin DOP sectors. Eastman was not totally unprepared for the introduction of POP, having experimented with its production prior to 1888; however, the firm did not introduce gelatin POP until the entry of Solio early in 1892. In the DOP sector Eastman's rivals were (1) Blair Camera of Boston, which continued production of bromide paper after acquiring Anthony's facilities in 1891 and Allen & Rowell's facilities in 1892;[57] and (2) Buffalo Dry Plate and Argentic Paper, which entered into production in the late 1880s. The latter, in which Blair also had a part interest,[58] was named by the Eastman Company in a suit that charged the firm with infringement of Eastman patents on the coating process and coating machinery. The suit

[56] "American Aristotype Company" (GEN).
[57] "Allen & Rowell Company" (GEN).
[58] Thomas H. Blair testimony, *U.S. v. EKCo.*, p. 16; *Buffalo City Directory*, 1893–94; Blair Camera Company, "Certificates of Condition," 1892 and 1893, Corporation Records, Boston, Commonwealth of Mass.

lasted until the middle 1890s and concluded in favor of the Buffalo firm. This verdict confirmed Eastman's growing concern with a part of his earlier business strategy, i.e., the utilization of patents as a means of limiting entry to the industry and thereby protecting those markets that the Eastman Company had created. The effectiveness of patents depended, of course, upon effective enforcement against infringement. Therefore, the object of the suits was, first, to pose a threat to small producers who did not have the resources for a long court battle; second, to demonstrate the validity of the patents. Although the failure of this strategy in the specific case of photographic papers led to the development of a new strategy in the second half of the decade, its failure did not vitiate the strategy in general, as illustrated in its immensely successful execution with the roll film system.

Phase 2: "Paper War" (1893–95) In all three sectors of photographic paper manufacture new companies entered and helped to contribute to overcapacity. The new entrants included United States Aristotype of Bloomfield, New Jersey, in collodion POP; Photo Materials of Rochester and Kirkland's Lithium Paper of Denver in gelatin POP; and Nepera Chemical of Yonkers, New York, in bromide DOP. Among these new entrants the technical capacities and the role of technical personnel in management were related to the degree of commerical success attained. The two most successful companies owed their founding at least in part to college trained chemists: Nepera Chemical by the Belgian photochemist and former Anthony employee, Leo H. Baekeland, and by Albert G. C. Hahn, a graduate in chemistry from Cornell and the University of Freiburg; and Photo Materials by two former employees of Eastman, Henry M. Reichenbach and S. Carl Passavant, both with considerable practical experience in the production of photosensitive materials.

By the middle of 1892 the number of new entrants into the field of photographic paper production began to concern the established companies. With the convergence during the next year of the entry of four additional companies into the market, the onset of a general economic depression, and the broad acceptance of sensitized papers, the focus of competition shifted. Instead of competition between sensitized paper producers and albumen paper manufacturers, competition arose among sensitized paper producers climaxing in a vigorous price war.

The leading paper manufacturers developed several strategies in response to the consequent overproduction, overcapacity, and underdemand. One of the more traditional approaches was that of the Eastman Company. George Eastman had assumed from the founding days of his firm that patents could limit entry and, hence, be a control upon oversupply. The strategy of using patents as a means of protecting a given market had worked well in the case of roll holders, roll film, and roll film cameras. Eastman had also seen this strategy work with Hammerschlag in controlling the production and the marketing of paraffin paper.[59] In late 1891 and early 1892 Eastman developed a plan which he, Richard Anthony, and Thomas H. Blair discussed: to have the principal paper and film manufacturers recognize the validity of the Eastman Company's coating and machinery patents, to license each company to use these patents, and in return, to have each licensee informally agree to maintain the traditional price for paper. The plan would have allowed an informal price maintenance agreement among chloride paper manufacturers that could ultimately have been enforced through the licensing agreements.[60] Eastman's plan would have included all chloride paper manufacturers and left the Eastman Company with nearly exclusive production of bromide paper.[61] Apparently Blair, Anthony, and Hopkins were agreeable to the plan, but American Aristotype would not concede the validity of the Eastman patents.[62] Where the pooling agreements popular in many other American industries at the time were legally unenforceable and, therefore, usually effective for only a short time, Eastman's plan would have been a method of effectively establishing a "fair trade" price structure by enforcing it with patents.[63] In this arrangement, the Eastman Company would have retained its dominant position in bromide paper, unmolested by the price cutting of the chloride paper firms. In return, the chloride paper firms would have had their price structure rationalized. The royalties the Eastman Company would have gained by licensing, George Eastman confided to his English partner, William H. Walker, he hoped would pay the litigation expenses and leave a little for profit.[64]

While Eastman promoted his plan, the gelatin paper manufacturers showed considerable concern about the overproduction and price cutting on paper. George Eastman adopted a consistent policy of *not* cutting prices but, if necessary, introducing new brands of paper and maintaining the established price on the quality brand paper.[65] Early in 1892 the Eastman Company, in response to price cutting, introduced "special seconds" of their cheaper brand. Such price cutting techniques stimulated the indignation of other photographic paper companies.[66]

Another response to the increasingly competitive conditions was consolidation through the acquisition or merger of companies. The Blair and Anthony firms had consolidated many of their manufactur-

[59] GE to Strong, 17 October 1894 (GEC).

[60] GE to American Aristotype Co., 25 February 1892, and GE to Blair, 25 February and 8 March 1892 (GEC).

[61] GE to Walker, 11 May 1892 (GEC).

[62] GE to Blair, 25 February and 8 March 1892 (GEC).

[63] This plan was similar to that with which the American Tobacco Company initiated its merger in 1889, based in part on patents held by the Kimball Tobacco Company, a firm located in Rochester. See Patrick G. Porter, "Origins of the American Tobacco Company," *Business History Review* 43 (Spring 1969): 59–76.

[64] GE to Walker, 11 May 1892 (GEC).

[65] GE to J. C. Sommerville, 14 July 1887, and GE to J. W. Queen & Co., 21 July 1887 (GEC).

[66] Bradfisch & Pierce to Eastman Kodak Co. (hereafter EKCo.), 2 February 1892 (GEC).

DOP	POP	Gelatin POP
1. Eastman (Rochester, N.Y.)	1. American Aristotype (Jamestown, N.Y.)	1. Bradfisch & Hopkins (Brooklyn, N.Y.)
2. Nepera Chemical (Yonkers, N.Y.)	2. Kuhn Crystallograph (Springfield, Mo.)	2. New York Aristotype (Bloomfield, N.J.)
	3. Western Collodion Paper (Cedar Rapids, Iowa)	3. Eastman (Solio) (Rochester, N.Y.)
	4. Bradfisch Aristotype (Brooklyn, N.Y.)	4. Blair Camera (Allen & Rowell) (Boston, Mass.)
	5. United States Aristotype (Bloomfield, N.J.)	5. Buffalo Dry Plate and Argentic Paper (Buffalo, N.Y.)
		6. Photo Materials (Rochester, N.Y.)
		7. Kirkland's Lithium Paper (Denver, Colo.)

Fig. 3.22
Photographic paper companies by product sector.

ing facilities in 1891, but with the lack of profitability of the merged facilities they sought throughout 1892 to entice the Eastman Company to gain a controlling interest in the Blair company.[67] While negotiations moved to an advanced stage, the limited assets of the Blair company blocked consummation of consolidation.

In accord with the popular response to overcapacity in nearly all industries at this time, sensitized photographic paper manufacturers formed a pool. C. E. Hopkins, particularly concerned by the threat of the number of small manufacturers both in the New York area and in the West,[68] sought to bring into a pool the five major companies of sensitized photographic paper: the American Aristotype Company, Bradfisch & Pierce, Eastman Company, Hopkins Company, and New York Aristotype Company. While the paper manufacturers did not follow Hopkins's plan in 1893, the major dry plate producers did agree shortly thereafter to an informal pool.[69]

The marketing association of the dry plate manufacturers became the base for an organization including the dry plate and sensitized paper manufacturers and the jobbers. In January 1894, at a meeting in St. Louis, the general terms were drawn for an agreement: first, to place all sales through supply houses, and second, to give an additional 10 percent discount to those houses that maintained list prices on all of a manufacturer's products. Anthony, Scovill, Gennert, and Murphy were the four jobbers who were to assist in the

[67] GE to Walker, 11 May and 9 February 1892; GE to Blair, 8 April, 23 September, and 15 November 1892; GE to D. L. Goff, 18 March 1893; and D. L. Goff to GE, 16 March 1893 (GEC).

[68] C. E. Hopkins to GE, 14 March 1893 (GEC).

[69] GE to George Dickman, 18 March 1893 (GEC).

enforcement of the policy.[70] The paper manufacturers of the New York area, Bradfisch & Pierce, Hopkins, Ilotype, and New York Aristotype, were among the last to join the association.[71] While this association, the Photographic Merchant Board of Trade, provided some stability in pricing for a time, relations deteriorated between two of the largest paper firms: American Aristotype and Eastman. The principal public issue was a survey taken in Chicago among professional photographers to ascertain what paper they were using. American Aristotype reported in its advertising that consumption of its collodion paper far surpassed that of any other, including Eastman's gelatin paper, Solio. In a series of advertisements, charges and countercharges were made, but this minor conflict heralded an era of further intensely competitive conditions in the industry.[72] Complaints soon arose regarding the marketing of "special seconds," a revival of a method of cutting list prices.[73]

The most significant factor in the outbreak of the intensely bitter price war among gelatin paper manufacturers was the entry and serious challenge of Photo Materials Company's Kloro paper. Photo Materials was one of the companies that had been left out of the association; it therefore lacked access through the jobbers to the supply houses. So Photo Materials Company depended upon advertising directed to the professional photographer and selling to him directly at a discount.[74] By early summer of 1894, the success of the Kloro paper challenge was quite clear. The paper manufacturers in the association met in Buffalo, where they agreed that each gelatin paper manufacturer would market a new brand priced to compete with Kloro paper. American Aristotype Company opposed this move, fearing that the cheaper gelatin brands would cut into the collodion market, which it dominated.[75]

With the entry of the new lower-priced gelatin papers in the autumn, the paper war began in earnest. By November, New York Aristotype Company had broken the standard price list, and the executive committee of the Board of Trade soon met and agreed, over the opposition of American Aristotype, to allow unregulated prices in gelatin paper. The collodion manufacturers agreed to maintain their prices and withdrew most of their cheaper brands. The price war finally reached the higher-priced papers. New York Aristotype lowered its price in November, and by the beginning of 1895 Eastman had lowered the price of Solio because Karsak (its

[70] Formal papers in incoming correspondence EKCo., 26 January 1894; final dry plate association agreement in August 1894; and "A Boycott Declared Against the Stanley Dry Plate Company," *Newton Graphic*, 7 September 1894, p. 1.

[71] GE to Strong, 27 April 1894 (GECP).

[72] C. S. Abbott to EKCo., 1 May 1894 (GEC).

[73] Abbott to GE, 7 June 1894; American Aristotype to EKCo., 4 June 1894; Abbott to EKCo., 6 February 1894 (GEC).

[74] GE to Strong, 15 June 1894 (GECP); and Max Brickner testimony, *U.S. v. EKCo.*, p. 438.

[75] GE to H. Q. Sargent, 14 June 1894; and GE to H. Lieber, 22 June 1894 (GEC); and GE to Strong, 15 June 1894 (GECP).

"fighting brand" gelatin POP) paper was gradually accounting for the bulk of gelatin paper sales, at the expense of Solio.[76]

Observing the protected position of the collodion manufacturers, George Eastman decided that his company could no longer await the successful development of its own collodion paper; therefore, in December 1894, Eastman purchased the Western Collodion Paper Company of Cedar Rapids. This company, though small, ranked second among collodion paper producers in quality and sales. Eastman moved the plant and the former owner to Rochester, but the collodion department experienced considerable technical difficulty and never achieved any real success.[77]

The Photo Materials Company strategy of direct sales proved quite effective in obtaining new consumers for Kloro paper. In 1895 Photo Materials had its one year of profitable operation.[78] By late 1895 the price war had ceased, and all manufacturers returned to list prices. The effects of the battle struck the smaller companies hardest, with some going bankrupt. Even Photo Materials Company lacked working capital.[79] The Eastman Company suffered in its paper profits but had the advantage of a diversified product line and benefited from the revival of roll film and roll film camera sales in 1895.

The American Aristotype Company as the dominant producer of collodion paper found itself in a more protected position than earlier. It received new paper formulae and technical information from Carl Christensen in 1893 and 1894 that led to the marketing in 1894 of two new collodion papers. As with its earlier paper, American Aristotype agreed not to market the new papers in Europe, the market preserve of Christensen's new Berlin plant.[80]

By 1896 American Aristotype (collodion chloride) and Eastman (gelatin chloride and bromide) were the dominant firms in their own market sectors. Each of these companies demonstrated unusual sensitivity to technological changes in both production and product. Product innovations, whether of American or European origin, were coordinated and integrated with marketing strategies by entrepreneurial leaders: Charles S. Abbott (American Aristotype) and George Eastman, who were sensitive to the potentialities of technical innovation and surrounded themselves with well-trained technical men. The loss of such leadership in the case of the one potential contender, the Photo Materials Company, may well have made the difference between failure and success, since the other essential ingredients were present. In the photographic industry the entrepreneur had to be prepared to innovate in both the technical and marketing phases of the business. It was the ability to under-

[76] GE to Strong, 12 and 21 November and 20 December 1894; and GE to Walker, 25 February 1895 (GECP); and GE to H. Lieber, 19 November 1894 (GEC).

[77] GE to Walker, 25 February 1895; and GE to Strong, 8 December 1894, 3 January, 18 February, and 11 April 1895 (GECP).

[78] Max Brickner testimony, *U.S. v. EKCo.*, p. 435.

[79] Abram Katz testimony, in ibid., p. 433.

[80] "American Aristotype Company" and "The Vereinigte Fabriken Photo Papiere" (GEN).

stand and integrate both the technical and marketing functions that spelled success for Eastman and for Abbott. Yet, the dry plate and photographic paper sectors did not define the outer boundaries of that success. It was the introduction of the roll film system of photography, which the technological and marketing conditions of those sectors fostered, that opened new and previously unimagined opportunities.

4 / Origins of the Roll Film System, 1884-89

The introduction of gelatin dry plates in Europe in the 1870s stimulated a period of considerable technological experimentation and innovation. The technological mind-set of the collodion period had been broken, and professional and amateur photographers alike considered new technological configurations. One of the important technological issues raised was the problem of the glass plate base. Such plates were heavy, breakable, and bulky. Consequently, they created a substantial burden for the photographer, especially the field photographer. At this time George Eastman and William H. Walker, like many other photographers and some manufacturers, began to explore alternatives to the glass plate. It was their persistence—and perhaps most importantly George Eastman's—in replacing the glass plate that led to the development of the roll film system, a technological-marketing innovation that radically altered the structure and dominant attitudes of the photographic industry in America.

A man of considerable business understanding and experience, Eastman combined business skills and insights with a technological orientation that allowed him to be innovative in both technology and business and sensitive to the potential role of science and technology in his company's overall policy and strategy. He had entered upon the manufacture of gelatin dry plates with little capital and no personal experience in running a business. Yet, he had definite goals and perceptions. His partner, Henry A. Strong, encouraged in Eastman an interest in the goal of growth.[1] Eastman sought to achieve

[1] Henry A. Strong to GE, 14 December 1880 (GEC).

goals of sales and profit growth through economies of scale attained through the use of machinery that had been patented to protect against imitation. His perception—no doubt sharpened and stimulated by his close friend, George Selden, Rochester's leading patent attorney—was that he could obtain and keep a substantial market share in the national market for dry plates, utilizing the sale of foreign patents on his machinery invention to raise capital for his American enterprise. Eastman clearly stated his intention "to engage in the manufacture of Gelatin Plates on a large scale and expect my invention to allow me if necessary, to put the price down to a point which will prevent miscellaneous competition."[2] However, Eastman found that all his calculations and strategies were not fully successful, and his goals of large-scale production, growth, and protection from miscellaneous competition were not to be achieved within the context of the current technological configuration. The plate-coating machinery patents were not defensible protection against potential competitors. By the middle of 1883 a highly competitive situation prevailed in the gelatin dry plate sector, with price cutting the principal method of competition. Still, Eastman did not abandon patents as a method of gaining a "natural monopoly"; instead of relying exclusively on production patents he now sought to develop a system of patents covering both product and production. This strategy, in the long run, was the key to his success.

Eastman could not attain his goals with his method of production of gelatin dry plates because of price competition. As the effects of

[2] GE to Mawson & Swan, 13 October 1879 (GEC).

this competition became evident to him in middle and late 1883,[3] he turned to two strategies: one for the long term and one for the short. The short-term strategy consisted of joining with other leading dry plate manufacturers, such as Carbutt and Cramer, in forming the Dry Plate Manufacturers' Association with the explicit purpose of trying to stabilize dry plate prices. The long-term solution Eastman had in mind was "the prosecution of experiments, having in view the perfection of a system of film photography that would supplant the use of glass dry plates."[4] He conceived of a complete system of products, processes, and machinery for the production of the patented products, which would deliver the entire photographic market into his hands. Therefore, he began in earnest to search for an alternative to the glass plate and, after some success, brought William H. Walker into the firm with the express purpose of working with him on the roll film concept of photography.

William Hall Walker (1846–1917) moved to Rochester from his native Michigan and managed a local camera company (fig. 4.1) there in the early 1880s. During this time, while Walker was engaged in the design and production of a nationally distributed pocket camera for the serious-amateur market, he and Eastman met. Although Walker for a time engaged in the small-scale production of dry plates, his real talent and energies were directed to camera design, where his originality obtained for him several patents in the early 1880s.[5] Walker harbored ambitions but, in contrast to Eastman, betrayed impatience. He preferred to accept a modest success obtained in a short time rather than defer immediate success for a long-term achievement of potentially heroic proportions. This difference in perspective prompted many bitter battles between Eastman and Walker in the late eighties and the nineties; however, Walker's experience and talent for camera design drew him to Eastman and represented a major resource in the new roll film endeavor. As with Strong earlier and numerous technical and business leaders, from Reichenbach to Lovejoy, later, Eastman possessed an ability to recognize people of specific talent who could complement his strengths. Moreover, despite his lack of social confidence, he could attract such

[3] Merchandise Account and Labor Account, Eastman Dry Plate Co., *Ledger*, 1881–1884, Corporation Records, Eastman Kodak Co., Rochester, N.Y. Dollar sales of the Eastman Dry Plate Company for the years 1881 to 1884 rose at a rate of about 100 percent each year until the middle third of 1883, when the rate of sales expansion fell to 75 percent. Then, during the next year, total dollar sales fell absolutely by 3 percent, with substantial drops in sales occurring during the months of September, October, and November 1883 and May and June 1884. Such figures do not, however, tell us precisely what was happening, because price decreases mask true production on a unit basis. Labor costs expanded until the middle third of 1884, which, based upon the assumption that productivity and labor wages did not change markedly, indicates that the unit rate of production actually increased during 1883 and 1884 but that during that period wholesale prices fell by about 60 percent for the Eastman Dry Plate Company.

[4] Affidavit of George Eastman filed 23 May 1891, *Eastman* v. *Blair.*

[5] U.S. Patents #245,180 (28 February 1882), #259,064 (6 June 1882), and #276,311 (24 April 1883); William H. Walker obituary, *Rochester Herald*, 30 November 1917; GE testimony, *Eastman* v. *Blair*, vol. 1, pp. 210–11; and *Brooklyn Advance*, June 1882, p. 9 (advertisement).

Fig. 4.1
William H. Walker and advertisement for his early Photographic outfit. Courtesy of Eastman Kodak Company.

people and draw from them enormously creative achievement. William H. Walker was a notable example.

As Eastman and Walker embarked upon a search for an alternative to the glass plate system of photography, they drew upon knowledge of an earlier roll film system. If technologies are regarded as systems of knowledge, then the Eastman-Walker system may be seen as having grown out of traditions which themselves drew upon bodies of knowledge both external to photography, as manifested in the creation of the roller blind, and internal to photography, as manifested in a series of obscure and commercially unsuccessful continuous strip designs created in the 1850's and in the somewhat more successful Warnerke system of the 1870s (fig. 4.2).[6] While the precise relationship between the Warnerke system and the internal and external bodies of knowledge is not known, it is clear that Eastman and Walker began with a knowledge of the Warnerke system and that that knowledge provided a conceptual foundation and framework for their creative research.

The roller slide system of Leon Warnerke, a Russian immigrant to England, consisted of two principal elements: a roll holder that slipped in place of the plate in the back of a camera and a continu-

[6] Helmut Gernsheim and Alison Gernsheim, *History of Photography: From the Earliest Use of the Camera Obscura in the Eleventh Century up to 1914* (London: Oxford University Press, 1955), p. 291; examples are Scott Archer's collodion-gutta percha stripping film introduced in 1851, British Patent #1914 for 1855, specification: 20 February 1856; Captain H. J. Barr's black calico roll film system, mentioned in the *Journal of the Photographic Society of Bombay*, no. 1, and referred to in *Notes and Queries*, 21 April 1855, pp. 311–12; and Joseph Spencer and Arthur Melhuish's roll film system, British Provisonal Specification #1139, 22 May 1854, referred to in *Journal of the Photographic Society*, 21 April 1856.

Fig. 4.2 Warnerke's roll holder and camera (1877). From Josef M. Eder, Ausführliches Handbuch der Photographie, 3d ed. *(Halle: Knapp, 1905).*

ous roll of collodion tissue. The roll holder contained two rollers on which the sensitive tissue was wound. The holders had a red glass opening that permitted the photographer to observe the numbers on the back of the sensitive tissue. The film consisted of glazed paper with alternate layers of collodion and india rubber and then the photosensitive emulsion. After exposure the emulsion was stripped from the paper and attached permanently to glass plates. In the 1870s Warnerke had covered the stripping film with very insensitive dry collodion emulsions. Although in the early 1880s he sought to employ dry gelatin emulsion, he encountered difficulty in drying his tissue while the gelatin was attached to it and in modifying the measuring and marking mechanisms in the roll holder to compensate for the greater sensitivity of the gelatin emulsions.[7]

Warnerke's roll film system of photography received attention in the British photographic journals during the decade 1875 to 1885, but it never became a commercial success. There were at least five criticisms made of the system. The stripping film itself was very expensive to manufacture, so its cost deterred popularity.[8] The rollers in the roll holders were fixed permanently, so the film had to be put on and taken off the rollers in the camera, an awkward and time-consuming job.[9] The indexing, or measuring, was also troublesome. Even with the collodion tissue, the colored glass window was not perfect protection against entry of light, and the measuring roll method used to mark the gelatin film frequently damaged it. Maintaining tension on the gelatin film as humidity conditions changed its length posed another problem.[10] Finally, Warnerke produced the elements of the system by traditional handicraft methods, making them expensive and unreliable.[11] It was with the Warnerke system and its deficiencies that Walker and Eastman were acquainted when they began to develop their system of roll film photography late in 1883.

As they considered a film system, Walker and Eastman focused on three basic elements they hoped to develop and patent: (1) a film; (2) a film holder mechanism; and (3) film-making machinery.[12] While they worked together in formulating the entire system of photography and production, they did not share fully all the responsibilities of design and development. Eastman, with his extensive experience as an emulsion maker and a commercial producer of bromide paper on a small scale for the Anthony company in 1881 and 1882, focused his attention on the development of a suitable

[7] Leon Warnerke, "A New Departure in Photography," *Brit. J. Photog.*, 18 September 1885, p. 602.

[8] Gernsheim and Gernsheim, *History of Photography*, p. 292.

[9] *Brit. J. Photog.* 34 (1887): 689, 802, and 803.

[10] *Eastman Company* v. *Blair Camera Company*, 62 *Fed. Rep.* 400–403 (1894).

[11] Warnerke, "A New Departure in Photography," p. 603.

[12] The discussion of the mental and experimental process in the development of the roll film system is sketchy because of the lack of primary documentation. Nevertheless, the outline presented here goes beyond previous discussions in reconstructing the sequence of events. See Reese V. Jenkins, "Technology and the Market: George Eastman and the Origins of Mass Amateur Photography," *Technology and Culture* 16 (January 1975): 1–19.

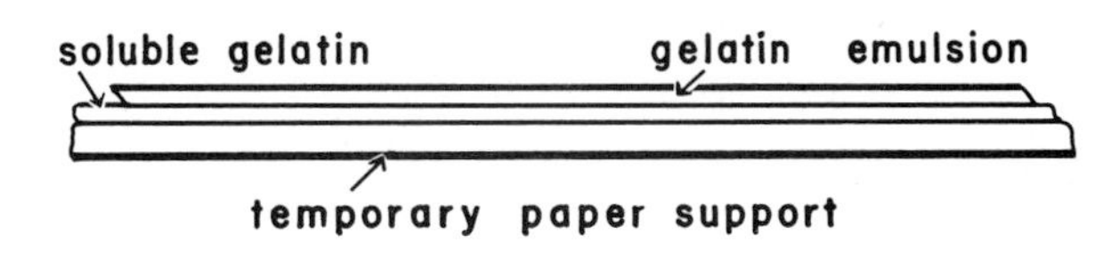

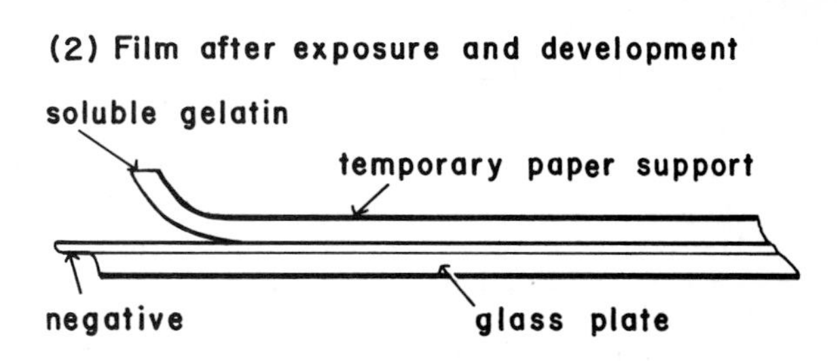

Fig. 4.3
Stripping film.

flexible film for the roll film system. Walker, with his experience in design and manufacture of cameras and camera accessories, took major responsibility for the design and development of the roll holders. Eastman and Walker shared responsibility for the creation of the paper film machinery.[13]

Early in 1884 Eastman began experiments with various materials, seeking a substitute for glass that was flexible, rollable, tough, relatively inert, and transparent—just what the entire field of photographic workers had half-heartedly sought for more than thirty years. Eastman experimented with thicknesses of gelatin and then turned to pyroxyline or collodion, hoping to develop a solution that would, with one application, make a suitable film. When this failed, he tried multiple layers but "found this objectionable."[14] Later, he turned to coating paper first with pyroxyline in alcohol and ether and then with the sensitive emulsion, thus creating stripping film; but he found that neither the collodion nor the pyroxyline gave sufficient support for the emulsion, even with repeated coatings.[15] By early March 1884, Eastman had developed what he regarded then as a workable stripping film that consisted of a base of Rives paper to which was applied a water soluble layer of gelatin and then the photosensitive gelatin layer, made less water soluble by treatment with chrome alum (fig. 4.3[1]). After the paper film had been exposed, it was developed by either the ferrous oxalate or sulphite of soda methods and fixed. Then the film was thoroughly washed and the emulsion detached from the base by floating the film with the sensitive coat downward upon a bath of water and bringing a glass plate from the bath to the sensitive surface. The film was then squeegeed onto the glass and placed in a warm water bath where the soluble gelatin began to dissolve and detach the paper backing from the film surface (fig. 4.3[2]). The final product was a glass

[13] GE testimony, *U.S.* v. *EKCo.*, p. 267.

[14] GE testimony, *Goodwin* v. *EKCo.*, p. 322 (quoted from *Celluloid Manufacturing Company* v. *Eastman Dry Plate & Film Company*).

[15] Ibid., p. 333.

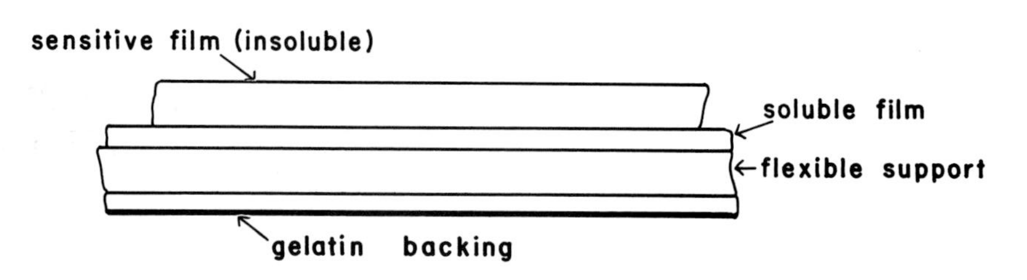

Fig. 4.4 Stripping film with gelatin backing.

plate image from which the print could be made. If the glass was treated with wax or another suitable material, the image film itself could later be detached from the glass and preserved on hardened gelatin for printing.[16] In March 1884 George Eastman filed a patent application on this film, the process of making it, and the method of manipulating it for photographic purposes. The patent was granted in October 1884.[17]

Completion of what appeared to be a successful product did not deter Eastman and Walker from further product tests and efforts to make improvements. From March until early May tests on the film continued, and Walker made the suggestion that a layer of gelatin be added to the back of Eastman's stripping film to keep the complete paper film from curling during periods of expansion and contraction due to major changes in atmospheric humidity (fig. 4.4). Early in May, Eastman and Walker jointly filed a patent application on the film with a gelatin backing, and that patent was also granted in October 1884.[18]

Attempting to overcome the defects of the Warnerke roll holder, Eastman and Walker spent more than six months in research and development before meeting with any success.[19] By the summer of 1884 they had created a holder that to the casual observer appeared similar to the Warnerke holder but differed in important respects. Walker and Eastman's first workable roll holder had two spools and two guide rollers (fig. 4.5). Being concerned with the effect of changes in atmospheric humidity on the gelatin film and on the tension with which it was held in the roll holder (failure of tension due to contraction could pull the film out of the focal plane), Walker and Eastman provided one spool with a friction device, or brake, and the other spool with a spring, so that the film tension could be maintained at all times. This represented a major improvement over the Warnerke roll holder. In addition, Walker and Eastman made the spools in their roll holder removable. A measuring roll was provided, but in the Walker-Eastman device there was no external indicator, just a clicking device that provided an audible marking of the turning of the film. A small perforating pin on the measuring roll marked the film for later cutting. Thus, the Warnerke

[16] U.S. Patent #306,594 (14 October 1884), GE, assigned to Eastman Dry Plate Company.

[17] Ibid.

[18] U.S. Patent #306,470 (14 October 1884), George Eastman and William H. Walker, assigned to Eastman Dry Plate Company.

[19] *Eastman* v. *Blair*, vol. 1, pp. 209, 211, and 227.

Fig. 4.5
Walker-Eastman roll holder and camera.

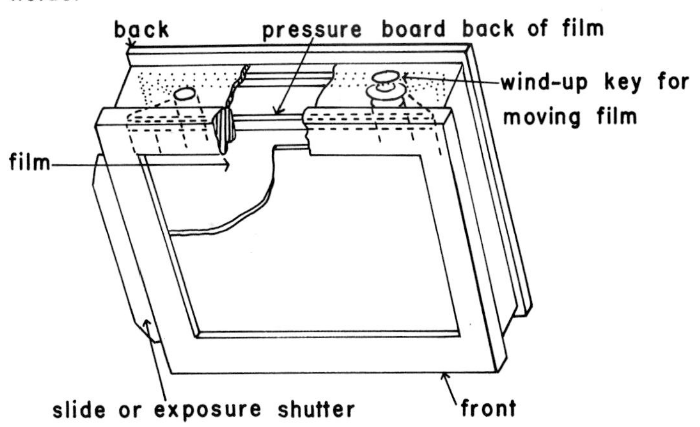

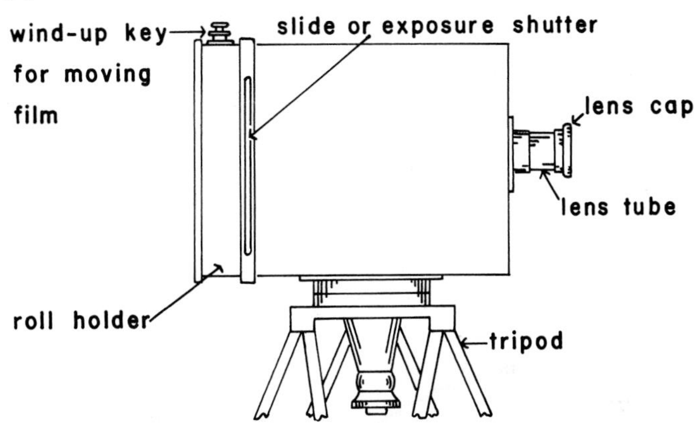

holder had been improved by (1) creating tension on the film; (2) marking the film with the pin on the measuring spool; and (3) providing removable spools.

Walker and Eastman completed a second model of their roll holder and filed an application for a patent on it by early August. As they had done with the film, however, they continued to work on the roll holder, seeking to make improvements that would aid in its operation and simplify its manufacture.[20] At the same time that they applied for the first patent on the roll holder, Eastman and Walker also applied for a patent on a plate holder for sheets of photosensitive paper (fig. 4.6). Therefore, Eastman had apparently decided to produce not only a roll film substitute for glass plates but sheet paper plates as well, perhaps seeing this as a transition between the glass plate and the roll film approaches.[21]

The film and film holder having been designed and created, the third element of the system remained: the production machinery and the process for making the film. Aware from his experience with

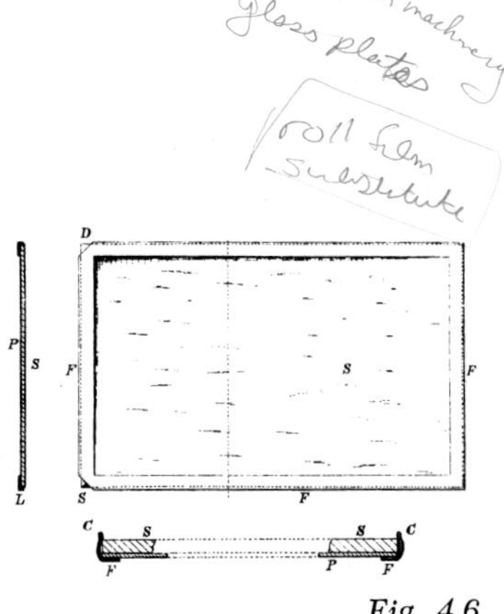

Fig. 4.6
Walker-Eastman plate holder for sheets. From U.S. Patent # 316,952.

[20] Ibid., pp. 209–11; and U.S. Patent #317,049 (5 May 1885), W. H. Walker and George Eastman.

[21] U.S. Patent #316,952 (5 May 1885), George Eastman and William H. Walker, application filed 8 August 1884.

Fig. 4.7
"Hang up": drying section of paper coating machine. Courtesy of Eastman Kodak Company.

dry plates that a patent on the production machinery was not in itself adequate protection, Eastman decided to patent each phase of the film system, including the process for making the film and the necessary production machinery. In the late summer and early fall of 1884, Eastman and Walker, drawing upon their own ingenuity and background information about coating machinery, worked closely together to develop the vital third element of their system. The resulting coating machine was somewhat similar to an English machine Eastman had seen at Anthony's and to Allen and Rowell's coating machine, which had been viewed by Walker. The Allen and Rowell machine and the Anthony machine were similar in design, consisting of roller, trough, and heating device, but Eastman and Walker added a hang-up drying mechanism that allowed continuous coating and drying of paper (fig. 4.7).[22] Eastman described the

[22] The "hang-up" design is called "festooning." *Eastman Company* v. *Getz*, 77 *Fed. Rep.* 412–20 (1896) and 84 *Fed. Rep.* 458–63 (1898); *EKCo.* v. *Blackmore*, vol. 2, pp. 1001–2, 1005, and 1034.

Eastman and Walker machine as consisting of "a trough in which there was a partially submerged roller. The roll of raw paper was hung in bearings behind that trough, and the paper led over an idler and under the roll into the emulsion, then over other rolls through a cooling device to set the gelatin. The gelatin will [*sic*] not stop running until it is set by the cold. Then it was hung on what is called a hang-up machine, a machine having movable slats, which took it as it came and hung it in loops in the drying chamber." Eastman further stated nearly thirty-five years later that the machine was "so satisfactory that the machinery has never been changed except in detail, since."[23]

By the early fall of 1884, Eastman had fulfilled the first phase of his plan. The three basic elements of the system—film, film making, and roll holder—had been developed and patents applied for. Eastman, Walker, and other company employees continued to improve the system, but Eastman's attention began to turn to the organizational modifications necessary to exploit the technical achievements of the first phase of the innovative endeavor.

George Eastman and Henry A. Strong had organized in January 1881 the Eastman Dry Plate Company as a partnership, with Eastman as the technical man, superintendent, and business manager and with Strong as the source of capital and the business adviser. As the film system developed during the early and middle part of 1884, it became clear to Strong and Eastman that additional capital beyond their means would be needed to finance the production of the roll holders and the film and to promote this new system of photography. In the short period from January 1881 to September 1884, the company had made gross profits in excess of $50,000, and Strong and Eastman had reinvested nearly all profits in the real estate, machinery, and inventory of the business. The expenses for developing the roll film system, including Walker's salary, were in excess of $2,000. In spite of the remarkable growth of the company in the first four years, it could not be expected to provide the necessary capital for engaging in this new line of business, certainly not with the profit trend in the dry plate industry.[24]

As a move to increase the operating capital of the firm and to indicate the new direction, Strong and Eastman dissolved their partnership and incorporated the company as the Eastman Dry Plate & Film Company on 1 October 1884. The new company was capitalized at $200,000, purchasing the Eastman Dry Plate Company for $162,000 in stock and selling 380 shares of stock at $100 per share. The stock in the old company was distributed largely to Strong and Eastman, but Walker figured prominently in the settlement of the patent account. The company patents were evaluated at $100,000, with Eastman and Strong receiving $40,000 each and Walker $20,000. Of the remaining capital, Eastman received $25,000 and Strong, $35,000.[25] The new capital in the Eastman Dry Plate &

[23] *Blackmore* v. *EKCo.*, vol. 2, p. 1004.
[24] Eastman Dry Plate Co., *Ledger*, 1881–84; "Eastman Companies" (GEN).
[25] Eastman Dry Plate Co., *Ledger*, 1881–84.

Fig. 4.8
Eastman Dry Plate & Film Company building, State Street, Rochester, New York. Courtesy of Eastman Kodak Company.

Fig. 4.9
John H. Kent. Courtesy of Eastman Kodak Company.

Film Company came from local Rochester businessmen who were acquainted with both Strong and Eastman.[26]

Reflecting the ownership in the new company, the list of officers consisted of Strong as president, Kent as vice-president (fig. 4.9), Eastman as treasurer, and Walker as secretary. These four men were joined by Sage to compose the first Board of Directors. It is clear, however, that the business continued to be run by the original partnership for the next thirty to thirty-five years, with Eastman in general control. As treasurer, Eastman knew every detail of the business, and despite his junior years, he had gained the complete loyalty and confidence of his senior partner, Strong. Eastman also knew, for at least the next decade, more about the overall technical side of the business than any other stockholder. Although he received sound counsel from his Board of Directors regarding general policy, both his technical knowledge and business acumen gave him in counsel with Strong an authority in the conduct of the company's affairs that went unchallenged. As he once said, when it came to new technical or business matters, "I made it my business to keep thoroughly posted. . . ."[27] And he did.

The new company began to deal with the problems of producing the new system in the fall of 1884. Machinery for production of the film was ordered. In order to avoid a major investment in metal and woodworking production facilities, the company subcontracted to a local metalworking firm production of the roll holder mounts (frames) and to a local cabinet maker and manufacturer of studio cameras, Frank A. Brownell, the assembly of the roll holders.

The disposition of the foreign patent rights of the old Eastman Dry Plate Company provides a clue to the internal ambivalence regarding the geographical boundaries of the company's market and the company's general marketing conception and strategy. Early in November 1884, a Board of Directors resolution allowed Eastman and Walker to retain rights to their foreign patents. Possibly Eastman and Walker wanted to move into European markets, and Strong was reluctant to move so quickly, but by early spring 1885, any differences regarding marketing strategy had been resolved and a whole new plan embarked upon. First, the exclusive agency with the Anthony company was canceled in early March. The reason for the break between the two firms is not clear, but the entry of Eastman into paper manufacture may have posed a threat to Anthony, and the jobber's profits may have looked enticing to Eastman, and Anthony may have wanted nothing to do with flexible films. With an entirely new line of products and a new system of photography, the Eastman firm may have been unable to negotiate the kind of support and terms with Anthony that it thought it should have. Clearly, the Eastman Dry Plate & Film Company confidently started its own marketing department complete with demonstrators and salesmen on the road.

[26] E. O. Sage ($10,000), J. H. Kent ($10,000), B. H. Clark ($9,000), and G. H. Clark ($1,000) were the major outside stockholders.

[27] GE testimony, *Goodwin* v. *EKCo.*, p. 322.

The second part of the new plan was a reversal for the company with regard to the foreign patent rights. The company arranged to buy all the foreign patents of Strong, Eastman, and Walker, with an accompanying increase in capitalization. The capitalization was increased by $100,000, with Strong and Eastman receiving $32,000 worth of stock each and Walker $16,000 worth. The remaining $20,000 worth of stock was sold to the other shareholders on a pro-rata basis with the proceeds going to Strong, Eastman, and Walker. This committed the company to an interest in the foreign patents.[28]

As the third part of the plan, the company opened a wholesale outlet on Soho Square in London with William Walker as manager. Thus, in just the few months between November 1884 and April 1885, the company made the significant move from a conservative policy of operating exclusively within the domestic market to a policy of involvement in international marketing. The Eastman store in London featured not just Eastman goods but other American photographic products as well, particularly the apparatus of the Scovill Manufacturing Company.[29] Relations with Scovill became fairly close after the ties with Anthony had been broken, and Eastman became a dealer in Scovill cameras, adding roll holders to them at customers' requests. Scovill benefited from having a full-time representative and outlet in the center of the world trade in photography. Eastman, however, ambitious to expand further, directed Walker to begin explorations for outlets on the Continent. While representatives were soon established in most capital cities of Western Europe, the expansion proceeded carefully to avoid possible patent difficulties in the introduction of the new film system products.[30]

The understanding and responsibility for attending to the complications of such patent matters were left to George Eastman's personal technical and diplomatic representative in Europe, Joseph Thatcher Clarke. Interest in a camera that the Bostonian had developed brought Eastman and Clarke together, and in 1886 Eastman employed Clarke as his correspondent in Europe, where he gathered and conveyed patent and technical information to Eastman and acted as the company's legal representative and negotiator. Employment of Clarke indicated not only that Eastman had global ambitions as early as 1886 but also that he respected the scientific and technical developments in Europe. Sensitive to the importance of innovation in the photographic business, he recognized that failure to understand and control new developments could destroy an otherwise strong firm. Eastman charged Clarke with the responsibilities of linking the United States (through George Eastman directly) with Europe in knowledge of photographic developments and of protecting American innovations in Europe.[31]

28 "Eastman Companies" (GEN).
29 GE to Scovill Mfg. Co., 20 and 30 May 1885 (GEC).
30 GE to Prof. E. Stebbing, 5 October 1885 (GEC).
31 Interview by Linda Allardt, 8 June 1953, with Hans T. Clarke, transcript of interview at Office of Corporate Information, EKCo., Rochester, N.Y.

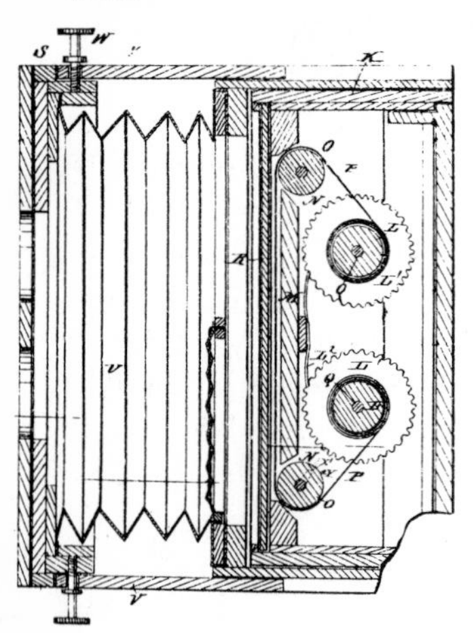

Fig. 4.10 David H. Houston camera design. From U.S. Patent # 248,179.

Early in 1885 George Selden, Eastman's patent attorney, learned that a feature of the Walker-Eastman roll holder infringed on a roll holder camera patent held by a farmer from Wisconsin and the Dakota territory, David H. Houston.[32] Houston had obtained in October 1881 a patent for a roll holder similar to Warnerke's but differing in its use of a sharp point or stud to perforate the edge of the film (fig. 4.10). It was this novel feature that the Walker-Eastman holder infringed.[33] Needless to say, Strong, Eastman, and Walker sought to make some arrangement with Houston whereby this potential threat to their "secure" protection would be eliminated. Finding the price of the patent too high, Walker secured instead an exclusive shop license for Monroe County, New York, for $700.[34]

The early months of 1885 were a period of considerable difficulty in initiating production of the new system. Finally, in the spring of 1885, the company marketed the roll holders.[35] They were warmly received by a number of American and British photographic journals,[36] attracting accolades such as "ingenious mechanism" and "one of the most perfect pieces of mechanism yet introduced into photography."[37] In reviewing the Walker-Eastman roll holder, Leon Warnerke noted that it was "of different construction" from his own, and he went on to observe that "details of the mechanism of this apparatus elicit general admiration. It is made on the interchange system, so useful when large numbers of the apparatus of the uniform size is [*sic*] to be produced; it is devised and made like Americans are in the habit of doing to special machinery. . . ."[38] Of particular interest is Warnerke's reference to the "interchange system," the system of standardization and machine production utilizing interchangeable parts. Eastman's orientation prior to entry into the dry plate business was to large-scale, machine production; clearly, production of the roll holder system followed a similar philosophy. The admiration of the design and construction of this roll film system with interchangeable parts by one of Eastman's precursors was a tribute to Walker and Eastman, to their originality, and to their ability to produce a piece of mechanism.

The film for the system proved more obstinate. Initially, the stripping film could not be successfully produced because of blistering of

[32] *Eastman* v. *Blair*, vol. 1, pp. 229 and 234.

[33] Ibid., p. 234; and U.S. Patent #248,179 (11 October 1881), David H. Houston, filed 21 June 1881.

[34] *Eastman* v. *Blair*, vol. 1, pp. 229 and 252–54. Purchase of Houston patent, 14 May 1889. This shop license provided the Eastman company with exclusive right to produce the patented item in Monroe County.

[35] *Eastman* v. *Blair*, GE affidavit filed 23 May 1891; roll holders with paper film produced in April of 1885, see GE to B. C. Forbes, 27 December 1916 (GEC).

[36] *Times* (London), 11 August 1885; *Anthony Photog. Bull.* 16 (1885); *Philadelphia Photographer*, (September 1885); *Brit. J. Photog.*, 28 August 1885; *Photog. N.* 29 (1885); and *London Amateur Photographer*, 11 September 1885.

[37] *Brit. J. Photog.*, 28 August 1885, pp. 547–48.

[38] Ibid., 18 September 1885, p. 603.

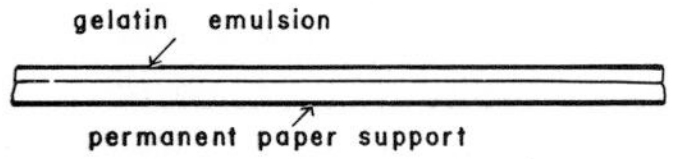

gelatin emulsion

permanent paper support

Fig. 4.11
Paper film.

the coating.[39] In order to supply a film for the roll holders, a paper strip was coated with regular dry plate emulsion (fig. 4.11). This negative paper film was not nearly as satisfactory as a transparent film because the printing had to be done through the grain of the paper, giving the prints a washed-out, faded, or even grainy appearance. Treatment of the paper with light oils prior to printing helped to increase the transparency, but the paper film was never popular. The introduction of stipping film, or American Film, was delayed until late in 1885.[40] The photographer's operation with the stripping film was so complicated and delicate—development, soaking, separation, squeegeeing, and varnishing—that even during 1886 American Film was not promoted. Furthermore, its production required three times as long as that of negative paper. The company policy was to persuade photographers to purchase the roll holder and acquaint themselves with its operation using negative paper and then hope to introduce them to American Film later, when they already knew how to use the roll holder.[41] However, when further improvements in the American Film allowed it to be independent of the glass base, the company began to promote stripping film over negative paper.[42] But by late 1887 the company had openly admitted the failure of the roll film system to replace the glass plate. Studio photographers had not accepted American Film because of the complications of processing and because of the low quality of the negative produced. The Eastman Company continued to promote stripping film but emphasized its merits for landscape photographers.[43]

While the film system was not truly successful, one of its by-products did open a new and prosperous line of business. The development of coating machines for the film stimulated the introduction of machine-made continuous-roll presensitized paper. Eastman's Permanent Bromide became a major sales item as the company developed American markets and utilized its new London branch to promote introduction of the new paper in Europe. By late 1885 the company was producing American Film, negative paper, and three grades of permanent bromide paper.[44] Furthermore, the production of this paper, especially the bromide developing-out printing paper, stimulated the opening in the early spring of 1886 of the service sector of the Eastman operation: the printing and enlarging service. A natural outgrowth of a marketing technique to stimulate the use of bromide paper for enlargements, the service soon became a sizable operation as demonstrators promoted not only the sale of dry

[39] GE to George Davison, 5 December 1902 (GEC).

[40] Eastman Dry Plate & Film Co. (hereafter EDPF) to W. E. Lewis, 11 December 1885; EDPF to Henry N. Sweet, 18 December 1885; EDPF to Douglas S. Thompson & Co., 30 December 1885 (GEC).

[41] GE to Andrew Pringle, 2 April 1886 (GEC).

[42] GE to George Bullock, 4 December 1886, and GE to H. L. Aldrich, 13 December 1886 (GEC).

[43] GE to A. H. Pitkin, 19 November 1887 (GEC).

[44] EDPF to Douglas S. Thompson & Co., 30 December 1885 (GEC).

plates, the roll film line, and bromide paper but also orders for enlarging.[45] By late 1886, despite use of a large automatic printing press with a capacity to produce from 5,000 to 10,000 prints in a ten-hour day,[46] the Enlarging and Printing Department was working day and evening to keep up with the orders. The machine utilized electric arc lamps, thereby freeing the enlarging and printing processes from dependence on solar illumination.[47] This service provided a market for the bromide paper produced by the company and, therefore, represented a step toward forward vertical integration.

Technical failures continued to plague the company. Beginning in 1883 the company had difficulty in matching the quality of some of its dry plate competitors. Poor and spotty emulsion was a continual problem.[48] Eastman, who served as the head emulsion maker and chemist, came to recognize that leadership of the business did not allow even such an avid reader as himself enough time to keep adequately informed. As dry plate sales fell in 1885 and 1886, he sought technical assistance.[49] Professor Lattimore, chairman of the Department of Chemistry at the University of Rochester, recommended that Eastman hire Henry M. Reichenbach, an undergraduate science major who had assisted Lattimore in the instructional laboratory. Not completing the requirements for his degree, Reichenbach joined the Eastman firm in August 1886 and undertook improvement of light-sensitive emulsion.[50] Explaining this addition to the staff, Eastman wrote to Walker: "We have a young chemist who devotes his time entirely to experiments and we hope he will strike the right emulsion sooner or later, but it may be a long job. . . . He knows nothing about photography which was all the better. I told him what was wanted and that it might take a day, a week, a month or a year to get it or perhaps longer but that it was a dead sure thing in the end. . . ."[51] While Reichenbach devoted much time to the emulsion, he also performed some analytical work for the company. His presence marked the first step of George Eastman's movement away from direct activity in the technical matters of the company; it also showed how desperately Eastman wanted to save the foundation of the company's business, the gelatin dry plate.

Yet, the company survived because of its diversification. During 1886, nearly half of all sales were in printing paper and two-thirds in either printing or negative paper.[52] The dividends paid by the com-

[45] EDPF to David Cooper, 2 April 1886 (GEC).

[46] GE to Matthews Northrup & Co., 18 December 1886 (GEC).

[47] GE to E. W. Mealy, 18 March 1887 (GEC).

[48] GE to D. Cooper, 19 August 1885; GE to E. & H. T. Anthony & Co., 28 August 1885; GE to Scovill Mfg. Co., 29 September 1885; GE to C. Gentile, 5 November 1885; and GE to Walker, 30 April 1887 (GEC).

[49] GE to D. Cooper, 29 May 1886 (GEC).

[50] *Rochester Union and Advertiser*, 26 September 1879, p. 2; student files at Archives, University of Rochester; Carl W. Ackerman, *George Eastman* (London: Constable, 1930), p. 56; and *Goodwin* v. *EKCo.*, vol. 1, p. 349.

[51] GE to Walker, 7 November 1886 and 30 April 1887 (GEC).

[52] Paper: positive 46 percent; negative 20 percent; apparatus 23 percent; and dry plates 11 percent. ("Eastman Companies" [GEN]).

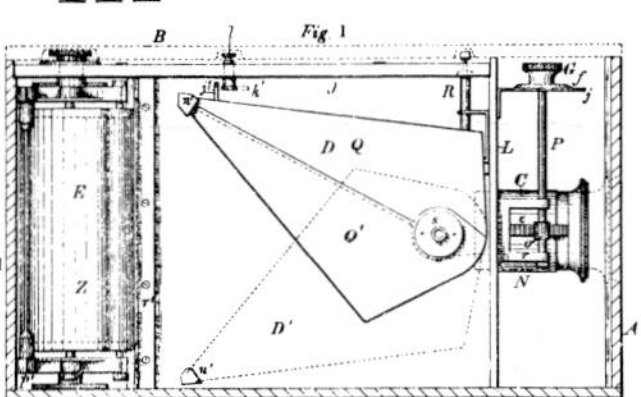

Fig. 4.12 Eastman and Cossitt detective camera. From U.S. Patent # 353,545.

pany during this period were due largely to the sales of photographic paper. It is also likely that there was a greater profit per dollar of sales on paper than on plates because of the difference in competition in the two areas. Not only was the dry plate sector of the business declining, but the roll holder sector was not growing. A drop in sales of roll holders during the first three years of production signaled the failure of the newly introduced roll film system.[53]

In the face of these problems, George Eastman and his associates did not despair but continued to develop new products in the hope of creating a large market that would be protected from competition by patents. Eastman joined forces with a new employee of the firm, Franklin M. Cossitt, formerly of Rochester Optical Company, in an attempt to produce a detective camera, a new class of popular hand camera.[54] Although Eastman and Walker had conceived of the detective camera while they were developing the roll film system, work on it was postponed until the fall of 1885.[55] As initially conceived, the camera was to have been six inches square and ten inches long and equipped only for films (fig. 4.12).[56] While the company had experience with the assembly of the roll holders, the entry into production of a complete camera proved to be long, difficult, and most important, more costly than expected. Initially, Eastman believed the camera might be "put on the market at such a low price that it would be a leading card for us and defy competition from other makers."[57] However, the failure of the negative paper and stripping films to gain popularity prompted the decision, in June 1886, to include a plate holder in the new camera. Production difficulties delayed its appearance until late 1887 or early 1888 (and Eastman had for two years to write letters of explanation), and the price was then set at fifty dollars apiece. As it had done for roll holders, the Eastman company turned to local Rochester firms for materials and for assembly of the apparatus. Metal parts were provided by Yawman & Erbe Company, lenses by Bausch & Lomb Optical Company,[58] and Frank Brownell accepted responsibility for assembly. A further indication of Eastman's disillusionment over the roll film system was the introduction early in 1887 of a view camera, the Genesee. It was a good quality camera adapted to take either roll holders or plates and was priced from thirty-six to sixty dollars.[59] Neither of these cameras was commercially successful.

By 1887 the Eastman company had largely failed in the areas of intended endeavor—dry plates and the roll film system—while in

[53] Sales of roll holders: 1885—1,334; 1886—578; and 1887—568 (*Eastman v. Blair*, GE affidavit filed 23 May 1891).

[54] "W. H. Walker & Co. (Rochester Optical Co.)," D & B, N.Y., vol. 167, p. 338.

[55] GE to H. L. Ensign, 11 September 1885 (GEC); and U.S. Patent #353,545 (30 November 1886) filed 1 March 1886.

[56] EDPF to E. S. Osborne, 18 and 22 December 1885 (GEC).

[57] GE to Ensign, 11 September 1885; GE to W. H. Walmsley, 27 January 1888; EDPF to J. W. Buel, 25 June 1886 (GEC).

[58] Establishing a long and close relationship with that leader of American optics.

[59] GE to A. Schefler, 5 January 1887 (GEC).

areas that had been developed as the opportunity presented itself—photographic paper and the enlarging service—it was doing very well. But even in the area of photographic paper, competition was increasing, as Cossitt and Cooper's defection to the Anthony company with the Eastman process and machinery plans indicated. With regard to the past and the future of the roll film system, George Eastman told the story of failure and success quite succinctly: "When we started out with our scheme of film photography, we expected that everybody that used glass plates would take up films, but we found that the number that did this was relatively small and that in order to make a large business we would have to reach the general public and create a new class of patrons."[60] It was this shift in market conception and goal that constituted the most revolutionary moment in the history of photography.

The market for photographic goods expanded during the period from 1870 to 1900, and the dynamic forces behind the expansion were technological. In 1870, as in the thirty years preceding, the market for photographic materials and apparatus was largely composed of professional studio photographers and a handful of very serious amateurs. With the advent in the early 1880s of the gelatin dry plate and the detective camera, the number of serious amateurs increased substantially, because the delicate and very technical operation of producing the photosensitive material had been removed from the hands of the photographer and placed in the hands of production experts in factories. Still, the vast majority of persons never entertained the thought of taking a photograph, let alone pursuing the complicated operations of developing and printing that were required following exposure. Yet, it was to this vast majority that George Eastman now turned. With a well-protected system of photography that had been largely rejected by both professional and serious amateur photographers because of the poor quality and complicated operation of the film material itself, Eastman reconceptualized and combined in a new way his major assets: roll holders, the American Film, the Enlarging and Printing Department, and a new potential market. If a camera could be produced that was both simple to operate and relatively inexpensive, the novice photographer could take his own pictures and let the Eastman Service Department do all the complicated developing and printing for him (fig. 4.13). As Eastman advertised throughout the world, *You press the button and we do the rest.* The gelatin revolution had simplified photography by introducing factory-sensitized photographic materials; stimulated by the original failure of his roll film system, Eastman conceived of the crucial second step, factory processing of the finished product. The refusal of the professional to adopt a new development led to the circumvention of the professional and the creation of a vast new market the like of which was undreamed of by Daguerre, Maddox, or even Eastman. Eastman's failure to realize his first goal (replacement of dry plates) propelled his creative tal-

[60] *Goodwin* v. *EKCo.*, vol. 1, p. 353.

Fig. 4.13
Photofinishing, c. 1889.
Courtesy of Eastman Kodak Company.

ents toward a goal that was a quantum jump greater than the first. But without the experience with the service department during 1886 and 1887, the feasibility of such an undertaking might never have occurred to Eastman or anyone else in the Rochester organization.

George Eastman began working on the new, simple-to-operate camera for the mass market in the summer of 1887. It is likely that he had assistance from some of his workers and possibly from Frank Brownell, but he was ultimately responsible for the conception and execution of the basic design. In one of the earliest references to this camera in a letter to W. J. Stillman at Scovill Manufacturing Company late in October 1887, Eastman wrote, "I believe that I have got the little roll holder breast camera perfected." But secrecy was the word, for as he stated in the same letter, "I will never be caught advertising anything again until I have it in stock."[61] He had learned a lesson with the detective camera.

Eastman also learned from his experience with the detective camera that a camera must be designed not only for the convenience of the user but also for simplicity and efficiency of manufacture. As he himself said, "I think the experience we have had in getting out the detective camera will enable us to avoid most of the difficulties in

[61] GE to W. J. Stillman, Scovill Mfg. Co., 22 October 1887 (GEC).

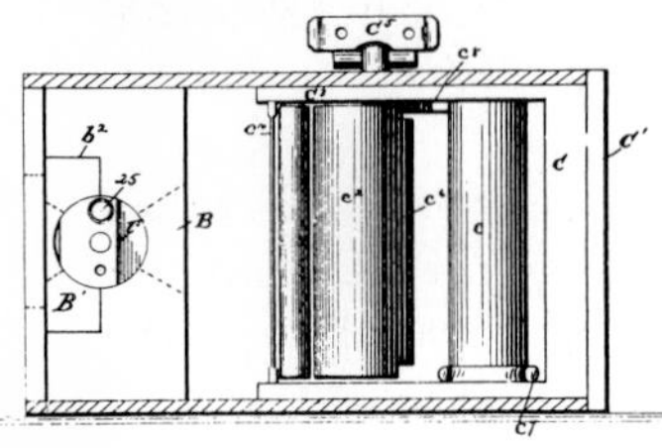

Fig. 4.14
Basic design of first Kodak camera.
A. *External view. Courtesy of Don Ryon, Eastman Kodak Company.*
B. *Roll holder section with film in focal plane. From* Scientific American, *15 September 1888.*
C. *Side view showing roll holder and cross-section of cylindrical lens-shutter. From U.S. Patent # 388,850.*

manufacture. The trouble with the detective is that no matter how successfully it works, it will always be hard to make."[62] Even in the patent Eastman obtained on the breast camera, mention is made of certain design steps having been taken in consideration of "convenience of manufacture and simplicity of construction."[63]

George Eastman's "little roll holder breast camera" needed a name. The word *Kodak* was first used in December 1887, after Eastman had contemplated an appropriate name for the new system, which he did not want to be mistaken for just another detective camera. Fond of the letter *k* (his mother's maiden name was Kilbourn), he created a word that could not be mispronounced and that was unlike any other word, thereby meeting the most stringent requirements for trademark registration, those that obtained in England.[64]

The unique Kodak camera contained several new elements. It was a fully enclosed camera in its own case, a leather-covered box 6½″ x 3¼″ x 3¾″ (fig. 4.14[A]). Hence, it was one of the class of the newly popular detective camera, except that it used roll film rather than glass plates. The roll holder section constituted the rear of the camera case, sliding into the back portion of the camera (fig. 4.14 [B] and [C]). The roll holder mechanism was on a wooden frame that made up the sides of the back insert (fig. 4.14 [B]). The basic arrangement of spools and rollers was similar to that of the earlier Eastman-Cossitt detective camera (fig. 4.12). An external indicator extending from the internal measuring spool showed when the spool had been turned sufficiently to give a new exposure. There was no consecutive indication of exposure number.[65]

One of the unique elements of the first Kodak camera was the new arrangement of lens and shutter (fig. 4.15). Eastman placed the lens in a cylinder with cut-out sides, and the rotation of the cylinder, with the lens fixed, acted as a shutter (fig. 4.16). A spring, tensed after each exposure by the photographer's pulling on a short string, drove the shutter, which was released by a small button on the side of the camera. Furthermore, the shutter was self-capping. Previously, after the shutter had been released, the lens had to be capped while the shutter was cocked again. Without this capping, the film would be exposed again during the cocking procedure. With the Kodak, the shutter operated on a continuous basis; therefore, the lens did not have to be capped between exposures.[66] The lens, produced by Bausch & Lomb Optical Company of Rochester, was a simple symmetrical doublet used as a fixed-focus lens. Its chromatic aberration and curvature of field were alleviated by masking the image down at the focal plane, producing circular negatives 2½

[62] Ibid.
[63] U.S. Patent #388,850 (4 September 1888), p. 2.
[64] The statement in the issue of trademark indicated its use since December 1887. See U.S. Patent Office, *Official Gazette*, 4 September 1888, p. 1072.
[65] Beaumont Newhall, "The Photographic Inventions of George Eastman," *Journal of Photographic Science* 3 (March–April 1955): 33–40; and U.S. Patent #388,850 (4 September 1888).
[66] GE testimony, *U.S.* v. *EKCo.*, p. 204.

Fig. 4.15
Front view of first Kodak camera with cylindrical lens-shutter. From Scientific American, *15 September 1888.*

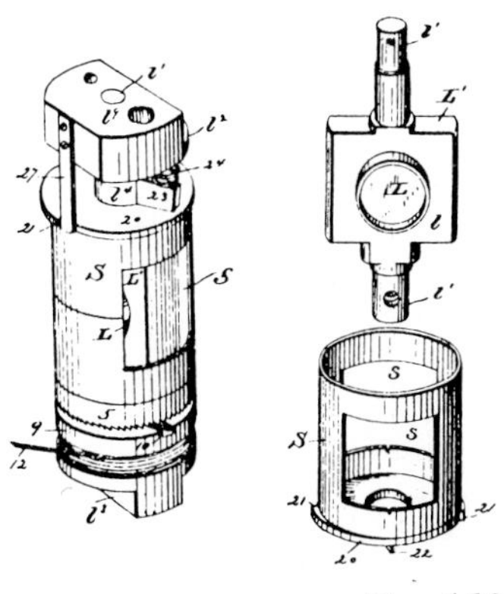

Fig. 4.16
First Kodak camera lens and shutter assembly. From U.S. Patent # 388,850.

inches in diameter. The lens-shutter combination was remarkable for its simplicity, economy, and reliability.

Eastman had designed a simple-to-operate camera, but that was only part of the simplification that the Kodak camera system brought to amateur photography. The Kodak camera was sold to the public for twenty-five dollars, already loaded with a one-hundred-exposure roll of American film. The novice photographer had only to turn the key (to bring fresh film into the focal plane), pull the cord (to cock the shutter), and press the button. After the film had been exposed (when it could be wound no more), the entire camera was returned to the Eastman factory, where the exposed film was unloaded, developed, and printed. The developed film (or negatives), prints, and the reloaded camera were returned to the novice photographer. The basic elements of the photographic process had been isolated and technical specialists trained to perform all the functions except aiming the camera and pressing the button. That separation of functions, fully realized by the Kodak camera system, revolutionized photography, for how the novices liked to press the button![67] The Kodak camera system also solved the basic problem with American Film. Too complicated for professional photographers to operate conveniently, the film could now be fully processed by the Eastman Company, and the novices were quite willing to pay for that service.

By the end of January, several models of the new camera had been constructed and sent to selected persons, for example, the patent attorneys in Washington who were preparing the patent application, and Kilbourne Tompkins, a relative of Eastman who was an advertising writer in New York City. Tompkins was commissioned to write the manual of instructions for the Kodak.

By late April 1888, development of production facilities was under way, the application for the patent had been filed, and Eastman was struggling with Tompkins over the writing of the instruction manual. Eastman argued: "The elimination of the word 'camera' is a good thing and comes back to my original idea, to make a new word to express the whole thing. . . ."[68] His success with his "new word" later required, much to his chagrin, considerable advertising expense to retain rights to the word *Kodak* as a trademark. Dissatisfied with Tompkins's work, Eastman eventually rewrote the manual himself.

In February 1888 preparations for the production of the new camera—in time for spring and summer, the busy seasons in the industry—were interrupted by a fire that did extensive damage to the operating plant. For about two months the company focused most of its attention on getting back into production with the paper and film. By late June, however, the first Kodak cameras were in production and ready for the market. George Eastman personally attended the annual photographers' convention held in early July in 1888 in Minneapolis. At the convention, where he exhibited the

[67] Newhall, "The Photographic Inventions of George Eastman," pp. 33–40.
[68] GE to K. Tompkins, 19 April 1888 (GEC).

Kodak camera, the panel of judges awarded it the medal as *the* photographic invention of the year.[69]

By early August Eastman realized that something spectacular had happened. Late in July he observed: "From present indications it will be the most popular thing of the kind ever introduced. . . ."[70] Shortly thereafter he noted:

The few dealers who have had them have made remarkable sales considering that they have not been pushed, in fact, wherever one of them is seen, it had secured the sale of several others. A feature which makes it a great favorite is our proposal to develop the 100 negatives and make one print from each and reload the camera, at a cost of $10. We have already had a good many returned for this purpose, and . . . many of them are made by people who never handled a camera before. . . .[71]

The experience of the major owner of the Eastman company likely epitomized the first experience with a Kodak of millions of people all over the world. Early in July of 1888, George Eastman related this story:

I gave one of these cameras to Mr. Strong who took it with him on a trip to Tacoma on Puget Sound a few weeks ago. It was the first time he had ever carried a camera, and he was tickled with it as a boy over a new top. I never saw anybody so pleased over a lot of pictures before. He apparently had never realized that it was a possible thing to take pictures himself.[72]

After more than eight years together, Eastman had finally made a photographer out of his old partner, Henry Strong, and out of millions of others as well, no doubt much to the delight of Strong and all the other Eastman company stockholders. The public reaction was immediate and most favorable, and soon the production facilities were taxed.

With a new system and a new market, many changes in the style of operation and the organization of the company became necessary. With the major shift in market, the practice followed by the company since the early 1880s of advertising extensively and almost exclusively in the photographic trade journals was no longer valid. While the company did not abandon the products that were of interest to the professional or its advertising in the traditional places, it established an advertising department that began to buy advertising space in popular national circulation magazines, including *Scribners, Century, Harpers, Popular Science, Outing, Scientific American, Frank Leslie, Puck, Judge, Life, Time,* and *Truth* (fig. 4.17).[73] Also, with the new market, new channels of distribution

[69] EDPF to Minn. Tribune Company, 23 July 1888 (GEC).
[70] Ibid.
[71] GE to Messrs W. S. Bell & Co., 9 August 1888 (GEC).
[72] GE to Stillman, 6 July 1888 (GEC).
[73] GE to J. Walter Thompson, 2 August 1888 (GEC); and Ackerman, *George Eastman*, p. 80.

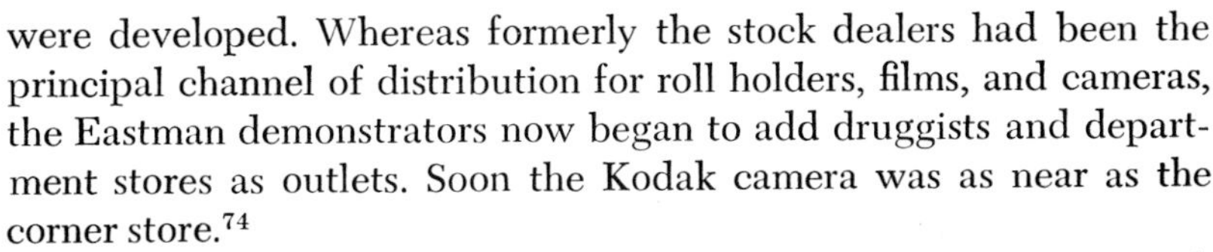

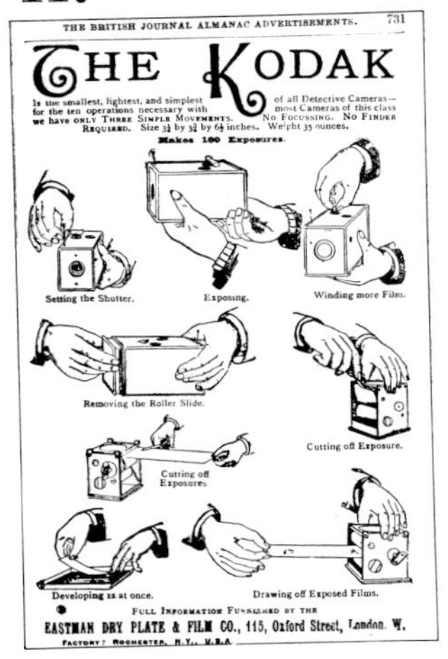

Fig. 4.17
Early advertising of Kodak camera. From British Journal of Photography Almanac, *1889.*

were developed. Whereas formerly the stock dealers had been the principal channel of distribution for roll holders, films, and cameras, the Eastman demonstrators now began to add druggists and department stores as outlets. Soon the Kodak camera was as near as the corner store.[74]

The change in England symbolized even more markedly this shift in distribution. The London store moved to larger quarters and sold both wholesale and retail. Retail sales were introduced in England because stock dealers had shifted many prospective customers from Eastman goods to those of other manufacturers—and also in preparation for addressing a new market for the first Kodak camera.[75]

From the time Walker opened the London branch, he worked diligently to expand the sphere of operations of the company throughout Europe, Asia, and Australia. By 1889 the London branch was supervising Eastman agencies in Paris, Berlin, Milan, St. Petersburg, Constantinople, Melbourne, Sydney, Shanghai, and Tokyo. With the growth of the foreign markets the London branch began considering manufacture of the company products. When the issue was raised at a Board of Directors' meeting in Rochester early in 1889, the question centered around either selling the foreign branch with the patents and licenses or developing a strong manufacturing and marketing capability abroad. By September, Strong, Eastman, and Walker had negotiated an acceptable plan, consisting of the establishment of an independent company in England for the operation of the company outside the Western Hemisphere. The parent company in Rochester held 78 percent of the stock. More than £33,000 worth of stock was sold for cash in London, with a quarter of that amount going to Rochester in partial payment for the assets of the foreign business. The plan succeeded in raising capital for both the parent and foreign portions of the business and yet left the parent company in clear control. In order to accomplish the sale of new stock, five English directors were included on the board along with two American directors.[76] The new British company was established, in late November 1889, as the Eastman Photo Materials Company, Limited. Eastman learned much from this reorganization that allowed him to exert strong leadership in future reorganizations.

The American company underwent reorganization at about the same time. The expansion of production facilities was creating a need for additional capital, and the growth and success of the company since the introduction of the Kodak camera necessitated revaluing the stock of the company. Late in December 1889, each share of stock in the Eastman Dry Plate & Film Company was exchanged for three shares of the Eastman Company stock; with 1000 shares left to sell, the capitalization of the new company was $1 million. The £8,350 cash from the creation of the Eastman Photo

[74] EDPF to Sam C. Partridge, 5 December 1888 (GEC).

[75] "Eastman Companies" (GEN).

[76] Colonel J. T. Griffin, Colonel J. N. Allix, George Davison, Andrew Pringle, and H. W. Bellsmith. American Directors: Strong and Walker (ibid).

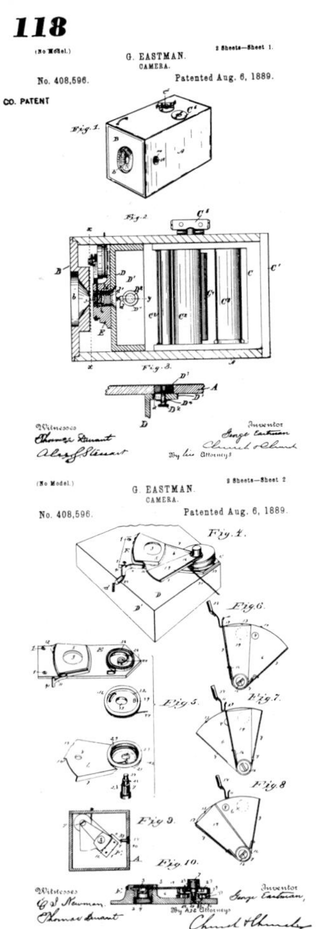

Fig. 4.18
Patent design for No. 1 Kodak camera. From U.S. Patent # 408,596.

Materials Company, the $100,000 from sale of Eastman Company stock, and the retained profits were to provide the capital base for the rapid expansion of the next three years.

This reorganization and refinancing of the company was characteristic of the conservative Strong-Eastman financial philosophy, which the Eastman companies followed throughout the years of their influence. The major sources of new capital for the ever-expanding firm were retained profits and limited sale of stock. Unlike Eastman and Strong, Walker wanted to "boom it" and sell out; but Strong and Eastman were building for the future and avoided quick-profit schemes and diluted ownership. Although within the next decade the ownership of stock became more widespread, the major block of stock remained in the hands of the original partnership. In spite of this conservative financial policy, the profits of the company were great enough to provide both generous dividends and capital for rapid expansion.

While the market and the organization of the company changed in the late 1880s, the basic philosophy of technical innovation did not change. During the late 1880s and early 1890s, George Eastman continued to play an important personal role in the innovative endeavor, although after this period his own personal technical activities ceased. Late in summer of 1888, as the Kodak camera gained in popularity, Eastman began to think of creating a new and larger model of the Kodak camera. Realizing that the current production facilities were already taxed, he aimed for introduction of a new model late in the next season. In December 1888 it was decided to substitute a new shutter for the "old" one in the first Kodak camera, the principal consideration being that the novice would find the operation of the two shutters similar, but the new shutter cost considerably less to manufacture (fig. 4.18). Confusingly, this new style Kodak camera was called the No. 1 Kodak camera (fig. 4.19). Eastman and others in the camera department were working on three different Kodak camera models: one producing 3½″-round negatives, one producing negatives 3¼″ X 4¼″, and one producing negatives 4″ X 5″. By early March the physical models had been completed, and plans were under way for setting up production and for promoting the new cameras. The 3½″ model became the No. 2 Kodak camera, which was introduced in October, loaded with film for sixty exposures, although the capacity was one hundred exposures.[77]

The key to Eastman's business policy during these early years continued to be his innovation-patent policy. He was determined to protect his successful product by personally stimulating continued innovation, by carefully patenting all the company's own inventions,

[77] Donald C. Ryon, "Development of the No. 1 Kodak Camera," in Photographic Historical Society, *Symposium, 19–20 September 1970* (Rochester, N.Y., 1970); see also: GE to C. W. Hunt, 6 September 1888; GE to E. Waters & Son, 31 December 1888 (GEC); GE to Walker, 3 January, 3 March, and 27 October 1889 (GECP).

Fig. 4.19
Left, *First Kodak camera with barrel shutter;* right, *No. 1 Kodak camera. Courtesy of Don Ryon, Eastman Kodak Company.*

and by trying to acquire all patents that related to the company's principal products. In pursuit of that policy, Eastman purchased the Houston patent for $5,000.[78] The marking pin was the particular feature of that patent that pertained to the Kodak camera. The patent as a whole was important in the Eastman policy because George Eastman wanted not just to protect the Eastman Company from infringement suits but to use it as a threat against possible infringers of Eastman patents. Eastman explained his policy to his English partner, Walker, with regard to roll holder patents:

Would it not be well for you to secure options on any English patents not owned by us. I am endeavoring to secure the sole right to the Houston patent here, and it will be our endeavor to fortify ourselves as much as possible by controlling all the Roll Holder patents that we can.[79]

I do not quite agree with you upon the fraility of our English patents. . . . I believe that $25,000 would put our patents in England on a foundation that would be unassailable. We have got so many patents that if we get beaten on one we could try another and it would take our competitors ten or fifteen years to break them all down. . . .[80]

In pursuit of his policy, Eastman obtained a list of all photographic patents ever issued, read the list of new patents each week in the *Official Gazette* of the Patent Office, and kept attuned to informal sources of information on new inventions and patent applications. He acquired roll holder and accessory patents on a system-

[78] In pursuit of this policy, Eastman began negotiations with Houston in the fall of 1888, inquiring of him his price for the 1881 patent. Houston's response was $30,000. While Eastman dropped the matter at that time, by the spring of 1889 Houston and Eastman were in negotiations again, which culminated on 14 May with the purchase of the 1881 patent for $5,000 and another patent for $50 ("Eastman Companies" [GEN]).

[79] GE to Walker, 3 March 1889 (GECP).

[80] Ibid., 23 October 1890.

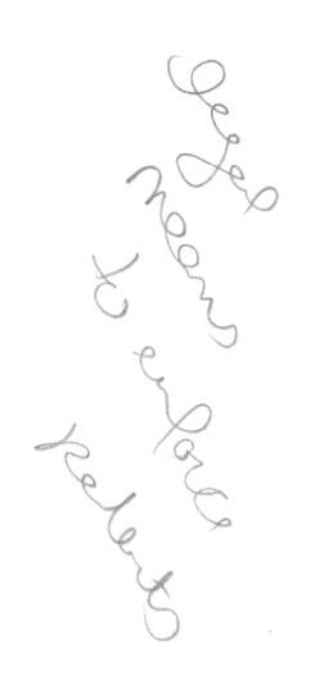

atic basis. Furthermore, he was prepared to use legal means to enforce his patents, as in the machinery suit against the Anthony company. Thus, the patents were part of a protective barrier to entry of possible competitors in the entire area of roll film photography. This policy was pursued with considerable success and vindicated Eastman's prediction that "success means millions."[81] Yet, Eastman did not just depend upon the acquisition of patents for technical innovation but also continued to emphasize technological research and development. The significance of this policy to the continued success and growth of the company was well demonstrated in the commercial development of the vitally important celluloid roll film base.

[81] Ibid.

Fig. 4.20
Women assembling Kodak cameras, c. 1889. Courtesy of Eastman Kodak Company.

5 / The Development of Celluloid Roll Film, 1887–95

The development of a satisfactory roll film depended on one of the most important synthetics developed in the nineteenth century: celluloid, a nitrocellulose compound. We have already seen the importance of collodion, a nitrocellulose material, in the shift from daguerreotypy to glass plate–paper print photography at about mid-century. From the early 1850s photographers sought a substitute for the glass plate—a light-weight, flexible, tough, transparent material that was inert to the photographic chemicals. Obtaining all of these characteristics in one film made from either collodion or gelatin proved extremely difficult. Yet, nitrocellulose materials eventually came to play a major role as a base for photosensitive materials in photography, but only after appropriate mixtures of certain solvents for pyroxylin had been discovered.

Celluloid had its origins in Europe. Alexander Parkes, an Englishman, developed a mixture of oils and gums as a solvent for nitrocellulose, producing a tough, solid material known as Parkesine. He established a company for its production in 1864, but owing to careless methods of production and pricing, his company failed in 1867. An associate of Parkes, Daniel Spill, developed another solvent for nitrocellulose that consisted of camphor and alcohol. In 1875 Spill received a British patent for it under the name *Xylonite*. He formed a company which in 1877 became the British Xylonite Company Ltd., a firm that met with success.[1]

[1] Williams Haynes, *Cellulose: The Chemical That Grows* (New York: Doubleday & Co., 1953), pp. 222–24.

In the United States at the time of the Civil War the Hyatt brothers of Albany, New York, introduced nitrocellulose materials for the production of checkers, dominoes, and billiard balls. John and Isaiah Hyatt discovered that camphor under heat and pressure acts as a solvent for certain nitrocellulose compounds. Isaiah called the new material celluloid.[2] After applying for patents, the Hyatt brothers obtained financial support, and in the early 1870s they founded the Celluloid Manufacturing Company. The company, which became America's largest manufacturer of plastics and synthetic materials during the next quarter century, soon moved to Newark, New Jersey.[3]

Development of new nitrocellulose materials, production processes, production machinery, and new uses for celluloid commanded the attention of the founders and managers of the Celluloid Manufacturing Company. They sought to patent every conceivable use of celluloid and every new development in the field. Among the hundreds of patents acquired by the company were three of some importance to photography at this time: (1) a method of producing thick clear blocks of celluloid; (2) a slicing mechanism that could cut sheets of celluloid as thin as ten one-thousandths of an inch; and (3) a new solution of nitrocellulose in amyl acetate. Amyl acetate combined with other solvents made possible the production of thin films of celluloid, which became important in the production of lacquers, airplane dope, artificial leathers, and photographic film.

[2] U.S. Patent #156,353 (27 October 1874).
[3] Haynes, *Cellulose*, pp. 224–29.

During the 1880s the Celluloid Manufacturing Company formed several subsidiary companies, including the Krystalline Company and the Celluloid Varnish Company, to exploit the products formed with the amyl acetate solvent.[4] Initially Charles Cooper and Company of Newark, New Jersey, an old established collodion producer, supplied the chemical materials to the Celluloid company, but during the 1880s both companies integrated vertically and thereby became competitors. The new solvent, amyl acetate, proved to be a major force in the celluloid industry in the middle and late 1880s, and Newark was the scene of the greatest activity.[5]

As the celluloid material became known in England and the United States in the 1870s, photographers became increasingly interested in its possibilities. One of the most avid and persistent experimenters, the French photographer David, in the early 1880s demonstrated and published the results of his work. British photographic periodicals kept a close account of his progress, realizing the potential if he met with success. David, like many other experimenters, worked assiduously but with little success because of all the requirements (transparency, toughness, flexibility, inertness to photographic chemicals) placed upon a substitute for glass in photography.[6]

The first commercial photographic use in America of celluloid as a substitute for glass came as a consequence of the developmental work of the Hyatt brothers and the recognition of its usefulness by the pioneer dry plate manufacturer of Philadelphia, John Carbutt. Carbutt experimented with celluloid in photography from 1884 to 1888 and in 1888 commercially introduced celluloid sheet film, which was used in place of the glass plates in plate cameras. Carbutt purchased the shaved sheets, made according to the Hyatt method of slicing from blocks of compressed celluloid, from the Celluloid Manufacturing Company and then coated the sheets with his dry plate emulsion.[7] Other American photographic manufacturers followed Carbutt's lead and during the late 1880s and early 1890s manufactured and sold cut sheet film. These firms included the Allen & Rowell Company of Boston, the Seed Dry Plate Company of St. Louis, and the E. & H. T. Anthony & Company. The Anthony cut films were produced from 1888, their improvement being one of the major responsibilities of Leo Baekeland when he entered the employ of the Anthonys in 1890. Typical of the Anthonys' policy, the firm not only produced and sold its own sheet film but also sold the film of Carbutt, Seed, and Allen & Rowell.[8]

The cut film, or sheet film, posed numerous technical problems to the manufacturers, who wanted to produce a reliable, high quality

[4] Ibid., pp. 98–100.

[5] Ibid., p. 236.

[6] *Les Mondes* (Paris) 61 (1883): 624–26; *Brit. J. Photog.*, 9 June 1882, p. 336; *Photog. N.* 26 (1882): 139; and *Bulletin de la Société Française de Photographie*, March 1882, pp. 64–65; June 1882, pp. 151–53; May 1883, p. 120; and July 1883, pp. 169 and 184–86.

[7] Marshall C. Lefferts testimony, *Goodwin* v. *EKCo.*, vol. 3: 2442–43.

[8] F. A. Anthony testimony, *Goodwin* v. *EKCo.*, vol. 1: 380, 386–87, 424, and 428.

Fig. 5.1 The Reverend Hannibal Goodwin. From Fritz Wentzel, Memoirs of a Photochemist *(Philadelphia: American Museum of Photography, 1960). Courtesy of Minnesota Mining and Manufacturing Company.*

substitute for glass. Soon complaints about an alleged deleterious effect of the camphor on the emulsion were heard. This allegation was later hotly contested in patent litigation. During the late nineteenth and early twentieth century the films did not attain popularity.

The celluloid material, as it was produced for sheet film, was too thick to be rolled for roll holders and, therefore, provided no direct technological link with the development of the celluloid roll film. Nonetheless, the idea of replacing glass with nitrocellulose materials was part of the general technological climate at the time, especially in Newark, and the desire for such a substitution was an old one in photographic circles. It is not surprising, then, that the first successful use of celluloid for roll film occurred in Newark, New Jersey.

One of the men most heralded for the early development and patenting of a nitrocellulose roll film was the Reverend Hannibal Williston Goodwin, a Newark clergyman (fig. 5.1). His use of illustrated lecture materials prompted in him an interest in photography and, from that, an eventual interest in finding a substitute for the heavy photographic plates. During the late 1870s and early 1880s, Goodwin took a considerable interest in photomechanical printing methods, securing patents upon his inventions and establishing a small enterprise, the Hagotype Company, which produced halftone screens.[9]

By 1886 two significant developments were evident to him. First, the growth of the Celluloid Manufacturing Company and Charles Cooper & Company indicated that celluloid was commercially significant as a substitute material for numerous products and processes. Second, the commerical introduction of the Walker-Eastman roll holder, initially hailed by the photographic journals, made evident the need for a substitute base for flexible roll films. Aided by a paid assistant Goodwin conducted experiments with various nitrocellulose materials in his home laboratory.[10] He obtained materials and probably also knowledge of nitrocellulose materials from Charles Cooper & Company.[11] Acquainting himself with practical literature Goodwin directed months of experimentation, producing a large number of sheets and pellicles, some quite thin and some fairly thick.[12]

[9] Goodwin's patents included one English patent and fourteen U.S. patents; *Dictionary of American Biography,* s.v. "Hannibal W. Goodwin"; Louis Walton Sipley, *Photography's Great Inventors* (Philadelphia: American Museum of Photography, 1965), pp. 117–18; Edward C. Worden, *Nitrocellulose Industry . . . ,* 2 vols. (New York: D. Van Nostrand, 1911), 2: 846; and Robert Taft, *Photography and the American Scene: A Social History, 1839–1889* (New York: Macmillan Co., 1938), pp. 391–403.

[10] Statement of James L. Crockett in "Goodwin File Wrapper and Contents," U.S. Patent Office, reproduced in *Goodwin* v. *EKCo.,* pp. 3751–52.

[11] Haynes, *Cellulose,* pp. 101–2.

[12] The earlier part of this experimentation was conducted by Goodwin with Crockett's personal help, but even the coating of these supports, which Goodwin performed, was accomplished in the presence of Crockett. As late as 1896 Crockett still had in his possession photographic films produced by Goodwin during these experiments. See Crockett, in *Goodwin* v. *EKCo.,* pp. 3751–52.

During the early spring of 1887 Goodwin went to his patent attorneys and initiated the preparation of an application that was filed with the U.S. Patent Office on 2 May 1887. From his original specification, it is clear that Goodwin's intentions were to produce a synthetic substitute for glass plates in photography, but he also recognized the similar need for a substitute for paper film for roll cameras.[13] He conceived of his development as radically new and, therefore, laid claim to virtually any substitute for glass or paper that was produced by flowing and evaporating and also to the process of such production—a rather audacious claim in light of the European photographic and chemical literature of the previous generation.[14]

From the summer of 1887 until the summer of 1889, the patent application lay in the Patent Office unissued. During the first year the claims were rejected, amended, and rejected seven times.[15] Goodwin and his attorneys sought to narrow carefully the scope of the claims. Initially, he excluded the best known nitrocellulose material, collodion, and one of the solvents requisite in its production, ether, arguing that collodion could not produce a suitable photographic support. At the same time, Goodwin, who apparently had only a superficial acquaintance with chemistry, described the product of flowage and evaporation as celluloid, thinking that celluloid was a form of nitrocellulose that had been dissolved in any one or more of many solvents. He was not aware at this time that camphor was typically a major component of celluloid. By the end of the first year, Goodwin conceived of his process and his product as a thin celluloid support made by flowing and evaporating liquid celluloid. He continued to claim every solvent of nitrocellulose except ether. The Patent Office examiner still held that the claims were anticipated and untenable.[16]

The original specification and the subsequent amendments indicate the limited chemical knowledge of Goodwin and the combined inarticulateness of Goodwin and his attorneys in framing the specification and amendments. Furthermore, Goodwin's attorneys ignored several of the examiner's criticisms, responding with neither modified claims nor arguments in support of the originals. At one point during the first year in the Patent Office, the original claims for both the product and the method of production were amended to drop the production claim and then amended again to drop the product claim and add the production claim. The preparation and prosecution of the application failed to indicate the level of professional competency that most patents in related areas possessed at this time.

Following the activity during the first year in the Patent Office,

[13] Original specification of Goodwin (Application #236,780), "Goodwin File Wrapper and Contents," pp. 3577–78.

[14] Ibid., pp. 3578–83.

[15] Dates of filing of Goodwin amendments: 18 July 1887; 21 January, 3 February, 8 March, 6 April, 14 May, and 9 June 1888. See ibid., pp. 3824–30.

[16] Ibid., pp. 3576–3616. See also William Main testimony (chemical expert witness for EKCo.), Hannibal Goodwin statement (8 October 1896), and "Brief for Complainant-Appellee" (Goodwin), in ibid., pp. 1897–1935, 3717–49, and 9–45, respectively.

the application lay dormant for nearly sixteen months and was not amended again until the office brought possible competing claims to Goodwin's attention. Why Goodwin ignored the application during this period is not clear. Even when, during the second half of 1888, both the Kodak camera and Carbutt's sheet celluloid film were introduced and met with initial public enthusiasm, the initial popular response did not stimulate Goodwin to proceed further. It is possible that Goodwin, who complained at a later point of being unable to pursue the application because of the mounting cost of counsel, may have found himself in a similar position following the number of revisions during the first year.

Meanwhile, other persons also actively sought to produce a suitable pyroxylin-based film. In 1885 Edgar E. Ellis of Rochester, New York, developed a pyroxylin-based film utilizing a gum camphor, Canada balsam, venetian turpentine, and dissolved rubber and commercially produced and marketed such films for a short time under the brand Sunart.[17] Many other Americans were thinking in terms of a celluloid base. For example, W. N. Jennings of Philadelphia suggested the use of celluloid to George Eastman in 1887. Jennings experimented with celluloid for at least another year, keeping Eastman informed in a general way of his progress.[18]

From 1884 Eastman himself made experiments seeking to find a suitable flexible, transparent base. He worked with solutions of pyroxylin, Irish moss, Japanese isinglass, and seaweed, among other substances, without success.[19] Early in March 1888, Eastman began seriously to consider celluloid as a substitute for paper film. He purchased various thicknesses of sheet celluloid from the Celluloid Manufacturing Company and experimented throughout the summer.[20] He read avidly on the subject of pyroxylin and celluloid, hoping to learn how to make a solution of sufficient thickness to possess, when dry, the appropriate properties.[21] At the same time Frederick Crane of the Crane Chemical Company in New Jersey wrote to William H. Walker in London suggesting joint experiments aimed at production of a new film support.[22] No such agreement was forthcoming, but the overtures indicate the extent to which a substitute, especially celluloid, was a part of the technological climate of the time. The Celluloid Varnish Company, the Celluloid Manufacturing Company subsidiary, supplied the celluloid varnishes that the Eastman company used with the stripping film. Late in the summer, the company sent samples of a new celluloid varnish that suggested to Eastman its potential for producing a new film base.[23] Eastman then set his chemist, Henry Reichenbach, to work

[17] Ibid., pp. 1836–41, 3307–11, and 3428; and GE to J. J. Kennedy, 23 August 1905 (GEC).

[18] GE to W. N. Jennings, 19 March 1887 and 24 April 1888 (GEC).

[19] GE Patent Office affidavit, Interference #15,529, *Goodwin* v. *Reichenbach*, contained in *Goodwin* v. *EKCo.*, pp. 2654–63; see esp. p. 2657.

[20] GE to Celluloid Mfg. Co., 3 March 1888; GE to Jennings, 24 April 1888; and GE to Lt. Col. Pennington, 27 June 1888 (GEC).

[21] GE to Moritz B. Philipp, 18 October 1889 (GECP).

[22] GE to Frederick Crane, 27 June 1888 (GEC).

[23] GE to Celluloid Varnish Co., 19, 25, and 29 September 1888, (GEC); last letter quoted in part in *Goodwin* v. *EKCo.*, p. 2658.

experimenting with the new varnish to see if a suitable flowed-film could be produced from it utilizing other solvents. Reichenbach's experiments continued during the fall, and in December he began to meet with some success.[24] In January, Eastman began to make inquiry regarding the cost of solvents in large quantities and the appropriate sizes and quality of glass for coating-tables. Late in January he requested his Washington patent attorneys to obtain copies of all the celluloid patents and to make plans to meet him to prepare applications for the production of thin celluloid films.[25]

Eastman's role in this development should be clarified. George Eastman initiated and, with the exception of the month of November when he was seriously ill, worked closely with his chemist, Reichenbach, on the development of the celluloid film. Although Reichenbach brought to this work a chemical knowledge and expertise that surpassed Eastman's, Eastman made every effort to learn the state of the art in celluloid, just as he had with gelatin photography nearly a decade earlier. When the patents were finally issued, Reichenbach's name appeared alone on the film patent and Eastman's alone on the machinery patent. The sole name of Reichenbach on the film patent does not fully reflect the Eastman input into the new development. Eastman wanted Reichenbach to receive recognition for solving the chemical problem that Eastman had begun to solve alone, so two separate patent applications were made, one in the name of Reichenbach for "manufacture of flexible photographic films" and one in the name of Eastman for "improvement in flexible photographic films." Upon examining the two applications prepared by his patent attorneys, Eastman asked that certain claims in his application be deleted, forthrightly declaring: "I can lay no claim to the chemical part of the process. The mechanical part only is mine and the chemical part Henry M. Reichenbach's."[26] Eastman wanted the patent issued that way as a special indication of appreciation to Reichenbach.

Undoubtedly Strong had been regularly informed of the progress of the developmental work of Eastman and Reichenbach, but Eastman did not rely on the support of the major stockholders only; he wanted to gain the support of all the major financial interests in the firm. Before Eastman took the new product to the Board of Directors, he carefully prepared his data for presentation. Not only did he have the product and the production method completely designed, but he had cost data on the raw materials and the essential elements of the necessary production machinery. As part of his exhibit, he included samples of the new celluloid film itself. The presentation was made on 23 February 1889, and the board unanimously approved the initiation of manufacture of such film as soon as possi-

[24] GE to Philipp, 18 October 1889 (GECP); and GE testimony, *Goodwin v. EKCo.*, p. 360.

[25] EDPF to Buffalo Alcolene Co., 24 December 1888; GE to Philipp, 23 January 1889; and GE to F. B. Church, 24 January 1889 (GEC).

[26] GE to J. B. Church, 3 March 1889 (GEC); Eastman Application #306,284; Reichenbach Patent #417,202 (10 December 1889); and see also Reichenbach Patent #479,305 (19 July 1892).

ble.[27] The success at this time of the first Kodak cameras no doubt helped the board members to place further confidence in Eastman's technical and business ability and judgment. The manufacture of the new film was no small undertaking for the infant Rochester enterprise. A building was leased from Strong for the film factory, and about $15,000 was spent on machinery and patents.[28] A watchful eye was kept on the development of the production capacity and the progress of the patent application, both essential elements in the successful exploitation of the new product.

The patent applications filed in April 1889 by Reichenbach and Eastman soon interfered with each other and with the application of Hannibal Goodwin, setting in motion a legal battle that spanned nearly a quarter of a century. Reichenbach's application claimed the use of various combinations of fusel oil, amyl acetate, and methyl alcohol with fluid solutions of nitrocellulose and camphor for coating a rigid surface, drying, and removing the resulting thin film.[29] In September, the examiner who was handling all three applications declared an interference among the Reichenbach, Eastman, and Goodwin applications.[30] The interference concerned the similarity of the specific claims made by Reichenbach and those made by Eastman and the inclusion of those claims under Goodwin's broad claim covering any process of making flexible photographic films by flowing liquid celluloid.[31]

Goodwin, even though alerted in advance that an interference was to be declared, that part of his claim might be patentable, and that he should prepare to amend his application, appears to have done nothing during the fall of 1889 when the interference was under consideration. In contrast, the Eastman company, which had already begun late in the summer of 1889 to market the celluloid film, seized the initiative by dropping some claims from the Eastman application which interfered with those in the Reichenbach application and by removing some claims from the Reichenbach application which might have been subsumed under Goodwin claims. In early December the examiner discovered a new anticipation of Goodwin's principal claim, so he rejected the Goodwin application again and suspended the interference between the claims of Goodwin and those of Reichenbach and Eastman. Soon the examiner decided to issue the Reichenbach patent, which contained the claim essentially as originally submitted but, in contrast to Goodwin's claim, recommended specific proportions of the nitrocellulose, camphor, and solvents.

27 "Eastman Companies" (GEN).

28 Ibid.; and GE to Harris Hayden, 13 June 1889 (GEC).

29 Original Specification, "Reichenbach File Wrapper and Contents," U.S. Patent Office, Patent #417,202, in *Goodwin* v. *EKCo.*, p. 3358.

30 This is contrary to the explanation provided by Haynes (*Cellulose*, p. 102), who claims that Reichenbach's patent was issued and Goodwin's was not because Reichenbach's patent fell "into the hands of a more liberal-minded examiner." William Burke was the first examiner of both applications. This is representative of numerous other minor errors in Haynes's book with regard to the Goodwin-Eastman-Reichenbach inventive and legal activities.

31 Declaration of Interference, Goodwin-Eastman-Reichenbach patent applications, in *Goodwin* v. *EKCo.*, pp. 3364–65.

Shortly after the issuing of the Reichenbach patent the disappointed Goodwin went to the Patent Office in Washington and consulted with the examiner.[32] During the period from December 1889 until September 1890, the aging inventor submitted five new amendments. In the fall of 1890, just as the examiner was about to declare an interference between the amended Goodwin application and the Reichenbach patent, Goodwin requested a delay that was continued for another year.[33] Delays and periods of inactivity, during the fall of 1889 and the second year of the application in the Patent Office, may reflect Goodwin's inability to finance the legal costs of further amendments and legal counsel in proceedings such as interferences.

The Eastman Company had some advantage over Goodwin in the Patent Office. First, thanks to George Selden, George Eastman was experienced at preparing patents, in contrast to the unsophisticated Hannibal Goodwin. Second, it seems clear from the applications and amendments that the Eastman legal counsel was far superior to that of Goodwin. Third, the Eastman Company had far superior financial resources to those of Goodwin and, therefore, could promptly and fully protect its legal interest. By 1890 the Eastman Company possessed a patent that protected its mode of producing celluloid film support for its flexible film.

The three principal figures in the development or introduction of nitrocellulose film, Goodwin, Eastman, and Reichenbach, did their work following essentially trial-and-error methods. Reichenbach had the best training in chemistry, but there is no indication that any theoretical principles guided his research work, even though theoretical principles applicable to nitrocellulose chemistry did exist in the literature. Nonetheless, the trial-and-error methods served Reichenbach and Eastman well, just as they had other pioneers in the development and use of celluloid.

While the legal battles continued in and out of the U.S. Patent Office, the battle of production implementation raged in Rochester. No new product is translated from the drawing board to the merchant's counter without much long, tedious, and frustrating effort. George Eastman knew well such difficulties. Expecting delays, he cloaked the new film and its production in secrecy. Orders for raw materials were placed with people Eastman could trust. In order to preserve the secrecy of the precise quality, quantity, and specific ingredients of the new film, he established a number code system with Charles Cooper & Company, a practice that continued for some time. Likewise, the out-of-town stockholders were informed confidentially during the spring of 1889 that a new transparent, flexible, nonstripping film was being readied for production and the market.[34]

As the new film plant was secretly being prepared on Court Street

[32] Goodwin statement of 8 October 1896, "Goodwin File Wrapper and Contents," *Goodwin* v. *EKCo.*, pp. 3719–20.

[33] Ibid., pp. 3630–54.

[34] GE to Charles Brewster, 26 April 1889; GE to Charles Cooper, 29 April 1889; GE to W. I. Adams, 17 May 1889; and GE to R. A. Adams, 1 June 1889 (GECP).

Fig. 5.2
Court Street film plant, c. 1891. Courtesy of Eastman Kodak Company.

in Rochester, George Eastman felt triumphant (fig. 5.2). With Walker in London, Eastman had someone away from the action with whom he could safely share his enthusiasm:

The new film is the "slickest" product that we ever tried to make and its method of manufacture will eliminate all of the defects hitherto experienced in film manufacture. . . . The field for it is immense and no estimates so far made are based on a more exclusive control of it than we have of Bromide Paper. If we can fully *control it I would not trade it for the telephone. There is more millions in it than anything else because the patents are young and the field won't require 8 or 10 years to develop it & introduce it.*[35]

This enthusiastic statement epitomizes Eastman's goal at this time: the exclusive control of the market through the use of patents. And Eastman's patent strategy was no haphazard affair, but a consistent, systematic approach of introduction to the market of new products with the product and production systematically protected by patents. By now, Eastman was beginning to recognize the full potential of the creation of the Eastman-Walker roll film system, but the hard work of implementation remained before the dream could be concretized.

The hard work turned out to be greater than even Eastman had expected. He anticipated that the factory would be running by the first of May, but the machinery and glass tables were not ready until late in May. Large numbers of tables of film were made before any salable film was produced because the operation of making film by the evaporation process proved to be much more difficult than anyone expected. Bubbles in the celluloid and black spots in the emulsion plagued the operation during the summer, but finally, on 27 August 1889, the first film for public sale was produced. At about the same time Eastman filed application for a patent on the coating machinery, a product of the recent development.[36]

The basic mode of production was cumbersome and slow. Tables three and one-half feet by fifty feet with plate glass tops were prepared for the nitrocellulose solution (fig. 5.3). The solution was spread on the tables by means of a hopper that had valves to control the flow of solution. A spreader controlled the thickness of the solution on the surface of the glass. The character of the solution was critical, for if it was not sufficiently viscous it would run off the table, but if it was too viscous it would not spread evenly. The coating on the tables was then allowed to dry by evaporation. Heaters and fans assisted in this process, which generally was conducted at night. The next morning the dried celluloid film was prepared for the emulsion coating, which was applied while the film remained on the tables. The solution consisted of nitrocellulose dissolved in wood alcohol plus additional solvents such as acetate of amyl, fusel oil, and camphor. The basic proportions used initially

[35] GE to William H. Walker, 3 March 1889, and GE to Walker, 5 May 1889 (GECP).

[36] GE to Walker, 7 March 1889 and 23 June 1889; GE to Charles Cooper and Co., 27 May 1889; GE to Hayden, 23 June 1889 (GECP); GE testimony, *Goodwin* v. *EKCo.*, p. 325; U.S. Patent #471,469 (22 March 1892), application filed 3 August 1889 ("Machine for forming flexible photographic films").

Fig. 5.3
Glass coating tables for celluloid film. Courtesy of Eastman Kodak Company.

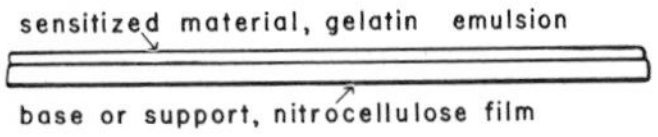

sensitized material, gelatin emulsion

base or support, nitrocellulose film

Fig. 5.4
Eastman Kodak Company celluloid roll film.

followed closely those of Reichenbach, especially on the critical proportion of nitrocellulose to camphor (twenty-two to fourteen parts).[37] After the emulsion had dried, the film was stripped from the table, cut, and spooled for roll film cameras.

The worldwide reception of the new film was most enthusiastic, and the demand for film for Kodak cameras soared. By 1890 the Eastman Company, unable to meet the demand, initiated plans for vast expansion of production facilities in the United States and the erection of production facilities for the English company at Harrow, near London.[38] However, there remained production problems such as dark spots in the emulsion and streaks due to electrical discharges, the solutions to which Eastman himself pursued. Eastman's report to Walker on the resolution of the problem of static discharges from the film in cold weather reflected the character of the technical thinking and teamwork between Eastman and Reichenbach:

One day, reflecting upon the theory that the discharge was caused by two surfaces, one of which was positive and the other negative, it occurred to me that if one of the surfaces was metallic there could be no generation. The idea of making one of the surfaces metallic naturally followed. A little further reflection, however, staggered me, because it seemed that the emulsion must be metallic, but I knew that it would spark as badly almost as the dope. I finally decided that every metallic particle in the emulsion must be insulated by the surrounding gelatine.

Thinking about this matter convinced me that if the gelatine substratum which we were then experimenting with, could be rendered a conductor at all it would not be by the use of any insoluble matter. I then naturally thought of the soluble salts and knowing

[37] D. de Lancey testimony, *Goodwin* v. *EKCo.*, p. 2341; and GE testimony, in ibid., pp. 325–27 and 354–55.

[38] GE testimony, *Celluloid Manufacturing Co.* v. *Eastman Dry Plate & Film Company*, p. 77.

that nitrates would not interfere with the emulsion, I decided to try them first. I directed Reichenbach to try the first experiment with Ammonium Nitrate, but he tried it with Potassium Nitrate, and found it worked perfectly.[39]

Eastman clearly had a command of the nature of the technical problems, addressed them, and with Reichenbach's assistance, often solved them.

With the nitrocellulose film in production, a decade of fairly radical technological innovation by George Eastman, William Walker, and Henry Reichenbach came to a close. However, even this decade was not one of constant change or discontinuity; there were periods of discontinuity, characterized by radical innovation, but longer periods of continuity, characterized by less notable but continuous innovation. The radical changes such as Eastman's dry plates (1879–80), Eastman and Walker's roll holders and paper film (1884–85), and Eastman's Kodak camera accompanied by Eastman and Reichenbach's nitrocellulose film (1888–89) represented radical discontinuity in products and methods of production in the photographic industry. Such innovations required faith in the viability of the new products and conviction that they would work and sell better than the old ones. Yet, important innovations of a less radical and spectacular nature were continuously pursued during the interim periods of 1881–83 and 1886–87, and in fact, during the periods of radical change themselves. It was this continuous innovation that brought the realization of the potential of the radical innovations.

The focus of George Eastman's innovative efforts seemed to shift. During the first decade of his activity, he personally provided leadership in radical technological innovation, leadership that was important in securing and developing a protected segment of the market. At the close of the first decade, after the basic elements of the roll film system had been created, patented, and successfully marketed, the potential of radical technological innovation for changing the industrial structure greatly diminished, and Eastman's attention shifted during the next decade to realizing the full potential of the newly developed technology by focusing upon the production function and initiating the necessary institutional changes to lay the foundations for a large corporation with domestic and worldwide influence. In successfully making the shift from engineer-entrepreneur to institutional-financial-entrepreneur, Eastman proved to be unusual. Successfully shedding the cloak of inventor-engineer, he gradually institutionalized his former function.

During the next decade in the photographic industry, fewer of the major technological innovations came from the Eastman company and more came from the new and struggling firms that were seeking access to Eastman's markets. Those innovations that provided modifications of the roll film system played a significant role in the technical character and popular success of the roll film system of photography.

[39] Carl W. Ackerman, *George Eastman* (London: Constable, 1930), p. 68.

6 / Modifications of the Roll Film System, 1890–95

The Blair Companies

At the time that George Eastman was creating and introducing the Kodak system of photography for the mass market, a group of firms located mainly in the Boston area was directing itself primarily to the amateur market and in some respects paralleling the development of the Eastman Company. These firms represented the only real competition in the American industry for the market Eastman had created. Although the companies were initially imitative of Eastman, they did introduce several important innovations that contributed significant improvements to the roll film system of photography.

Eastmans competition

The principal figure in the Boston group of firms was Thomas Henry Blair (1855–1919), an immigrant from Nova Scotia who settled in southwestern Massachusetts where he served as a photographer from 1873.[1] He developed in 1877 or 1878 a portable system of wet collodion photography which he called the Tourograph (fig. 6.1). It consisted of a moderately large wooden box which, when collapsed, could be carried like a suitcase. The camera was built into the box so that when the photographer was ready to take a picture, he placed the box on legs, stuck his head in it, and covered himself with the black, light-tight cloth. The inside of the box was equipped like a small photographer's tent, with all the requisite chemicals and supplies, so the photographer could prepare his plates, put them in

[1] "Northboro," *Shrewsbury Chronotype* (Mass.), 4 April 1919; "Death Records, Town of Northborough" (Town Clerk's Office, Northborough, Mass.), vol. 4 (1893–1942), p. 72; Rev. J. C. Kent, "Current Events in Northborough, Mass.," (a memoir) in the Northborough Historical Society Museum, Record of 4 April 1919; and Thomas H. Blair testimony, *U.S.* v. *EKCo.*, pp. 2–3.

place, expose the pictures, and develop the plates all at one time. Sensing a market among field photographers and amateurs for this portable collodion system, Blair filed an application for it with the U.S. Patent Office on 21 August 1878, entertaining the idea of entering upon its production himself.[2] He introduced the Tourograph in 1878, with the American Optical Company Division of Scovill Manufacturing Company producing and marketing it. Soon Blair filed incorporation papers for his enterprise, the Blair Tourograph Company, in the state of Connecticut.[3]

The introduction of the Tourograph came at an inauspicious time at the close of the wet collodion era; but Blair, with little investment in production facilities, quickly made the shift from wet collodion to dry gelatin photography by initiating the production of dry plates and by modifying the Tourograph for use with dry plates.[4] The shift in technology and the desire for independence stimulated Blair to seek additional capital and break his ties with Scovill. Late in 1881 the company moved to the Boston area and incorporated in Massachusetts as the Blair Tourograph and Dry Plate Company.[5]

[2] U.S. Patent #211,957 (4 February 1879); and Blair testimony, *U.S.* v. *EKCo.*, pp. 2–3.

[3] Blair testimony, *U.S.* v. *EKCo.*, pp. 2–3; and *Photog. T.* 9 (1879): 187, 202 (advertisement), 209–12, 234, and 262 (advertisement); and ibid. 10 (1880): 33.

[4] Blair testimony, *U.S.* v. *EKCo.*, pp. 2–3.

[5] Ibid.; "Joint Stock Corporations, Certificates of Organization," and "Certificates of Payment of Capital," Corporation Records, Commonwealth of Mass., vol. 76, pp. 543–44, and vol. 4, no. 89 (1873–): 487; *Boston City Directory*, 1879–1883; and D & B, Mass., Suffolk County, vol. 20, p. 218.

PACKED FOR TRAVEL.

Fig. 6.1
The Blair Tourograph. From Anthony Photog. Bul., *1880.*

Blair envisioned a broad product line covering apparatus and materials and including both production and sales. Initially, however, the production, the facilities for which were located in Cambridge, was confined to the Tourograph, dry plates, and plate holders, and the sales were conducted through Anthony, who acted as sale agents, and through the company's own small store in downtown Boston.[6] After about a year, a major shift in ownership of the small corporation occurred. The Goff family of Pawtucket, one of the most prominent textile and banking families of Rhode Island, assumed a large share of the stock. The combination of the entry of the Goff interest and the gradual accumulation of profits substantially raised the credit standing of the fledgling firm.[7]

During the mid and late 1880s the product line moved gradually away from photographic materials and increasingly toward apparatus, especially cameras. The dry plate sales faltered as competition became keen in that market in the mid-1880s, and the Tourograph was dropped as an anachronism in the day of small dry plate and hand detective cameras. Meanwhile, Blair himself contributed a number of new products out of his own experimentation.[8]

Like George Eastman, Thomas Blair possessed a firm respect for patented innovations as a principal means of entry into photographic apparatus manufacture. In less than a decade after beginning business, he filed and received eleven patents, including seven camera patents. Unlike Eastman's patents, however, these did not form a system of protection but rather a collection of improvements in camera features. Like Eastman, Blair had a worldwide perspective, patenting some of the improvements in England, France, Belgium, Germany, and Canada. At about the same time that Eastman began enforcing the paper machinery patents against Anthony, Blair likewise found it necessary to begin a program of enforcement, beginning with an infringement suit against Scovill Manufacturing Company in 1887.[9] Blair's new patent-controlled apparatus was directed largely toward the amateur photographer who wanted a simple hand camera but was willing to do his own developing and printing (fig. 6.2). Cameras such as the Lucidograph, on which Blair worked for two years, featured a swing front and folding bellows feature. Accessories included drying racks, plate holders, darkroom lanterns, plate boxes, developing boxes, and tripods.[10]

[6] "Joint Stock Companies: Charters," and "Certificates of Condition," Corporation Records, Commonwealth of Mass., vol. 63 (1882): 1470, and vol. 33 (1883): 283; *Cambridge City Directory*, 1881–82; *Anthony Photog. Bul.* 11 (1880): 212; D & B, Mass., Suffolk Co., vol. 20, p. 218; and "Blair Camera Mfg. Co. (Mass.)" (GEN).

[7] Darius L. Goff obituary, *Photog. T.* 21 (1891): 212; "Certificates of Condition," Corporation Records, Commonwealth of Mass., vol. 33 (1883): 283–84, vol. 37 (1884): 223–24, vol. 38 (1885): 274–75, vol. 42 (1886): 133, and vol. 45 (1887): 28; and D & B, Mass., Suffolk Co., vol. 20, p. 218.

[8] Blair Camera Company, *Catalogue, Photographic Apparatus*, 5th ed. (Boston, 1886), copy located at Society for the Preservation of New England Antiquities, Boston.

[9] "Blair Camera Mfg. Co. (Mass.)" (GEN).

[10] Blair Camera Co., "The Lucidograph," Trade sheet, MS Room, Baker Library, Harvard University, Cambridge, Mass.; and Blair testimony, *U.S. v. EKCo.*, p. 14.

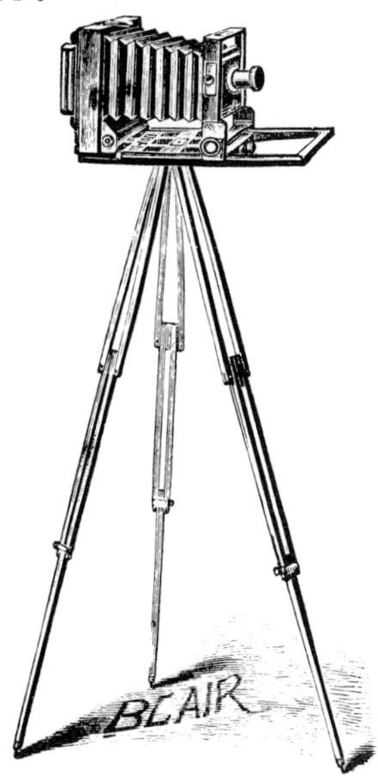

Fig. 6.2
Blair tripod camera. From American Annual of Photography, *1888.*

In the late 1880s the company began forward vertical integration by establishing supply houses and agencies first in Philadelphia and later in New York, Chicago, and Cincinnati. To facilitate its advertising and printing, the corporation name was shortened to Blair Camera Company. However, the ties with the Anthony company were retained. While sales and gross profits grew dramatically in the late 1880s, the demand for capital for production and sales expansion and for patents absorbed all the profits. Thus, dividends were confined to those paid in stock and bonds. The firm was, in general, financially weak.[11]

During the late 1880s Darius Goff, president of Blair Camera, and Thomas Blair, treasurer and general manager, moved the company toward greater diversification in product line and, specifically, toward an imitation of the Eastman company. As the Blair company became increasingly sales oriented, Blair, an extroverted and convivial businessman, became more intimately involved with some of the local Boston photographic manufacturing firms, especially with the Boston Camera Company and with the Allen & Rowell Company.

The Boston Camera Company was a small, single-proprietorship belonging to Samuel N. Turner. Turner started his business in the summer of 1884, initially serving as a mail order agency for an English camera firm. Soon he developed his own production capacity for cameras,[12] and by the late 1880s the Blair Camera Company was acting as both his purchasing and his sales agent in an arrangement similar to the one that existed between the Eastman and Brownell firms. About 1888 Turner designed and introduced a roll film camera called the Hawk-Eye. Initially he produced, or had produced for him by the Blair Camera Company, the roll holder mechanisms for the box camera. When the Eastman company learned that Blair was marketing this camera in what appeared to be infringement of the Walker and Houston patents, it threatened to institute court action against the Blair Camera Company. In June of 1889 George Eastman and Thomas Blair agreed that the Eastman company would supply the roll holders for the Hawk-Eye camera, even though it was a competitor of the Kodak camera.[13] At about the same time the Eastman company licensed the Scovill and Adams Company to produce roll holders for its roll film cameras. From Eastman's point of view, this arrangement with the Blair and Turner firms gave him control over who was making roll holders, provided the Eastman Company with a royalty on each roll holder made, and supplied the market with additional instruments that would consume the film, a commodity produced exclusively by the Eastman firm. However, he would have preferred an arrangement by which both Blair and Anthony would recognize the validity of the Eastman

[11] Blair testimony, *U.S.* v. *EKCo.*, pp. 4 and 19; "The Lucidograph"; "Certificates of Condition," Corporation Records, Commonwealth of Mass., vol. 42 (1886): 133; vol. 45 (1887): 28; vol. 47 (1888): 409; vol. 52 (1889): 40; and vol. (1890): A–I; and "Blair Camera Mfg. Co." (Mass.) (GEN).

[12] D & B, Mass., Suffolk County, vol. 21, p. 90; and *Boston City Directory*, 1883–91.

[13] *Eastman* v. *Blair*, vol. 1, pp. 97, 100, 222–23, and 231; and GE affidavit, filed 23 May 1891, in ibid.

company patents and agree to handle Eastman products exclusively in return for an extra 15 percent discount on the Eastman products.[14]

Eastman continued to supply the roll holders to Boston Camera Company, through Blair Camera, until October of 1890.[15] However, the popularity of the roll film system of photography placed enormous demands on both companies. Because the Eastman company had difficulty meeting the demand for film, Eastman carefully restricted the flow of roll holders to the market in order to avoid customer dissatisfaction with the inadequate supply of film. Blair failed to understand the restriction on the number of roll holders and the failure of Eastman to meet the needs of the Blair company for roll film and, therefore, began to lay plans for production of both roll holders and roll film. When Eastman heard rumors of Blair's plans, he discontinued shipping roll holders in October of 1890.[16]

The Blair company attained production capacity for the roll film system in 1889 with the acquisition of some interest in the Boston Camera Company. Then, in the summer of 1890, at the time when Blair was receiving a supply of roll holders and film only from Rochester, Blair sought to interest the Allen & Rowell Company of Boston—producers of photosensitized materials, including sheet celluloid plates—in producing celluloid roll film. Although the company refused initially, when the Blair Camera Company acquired controlling interest in the firm in August 1890[17] plans were laid for developing a production capacity for celluloid roll film. The Blair company purchased the celluloid support from the Celluloid Company and then coated it with the photosensitive emulsion in Boston. Commercial production began late in 1891, and the film sold fairly well in 1892 and 1893; however, the reason for its popularity had to do more with the Eastman company's difficulty in supplying the public than with the superiority of the Blair product.[18] Following the example of the Eastman Company, the Blair company introduced a developing and printing service for its film customers.

The second phase of the Blair Camera Company's imitation of the Eastman Company witnessed Blair's effort to extend its product line to nearly all photographic products but particularly to photographic paper. To accomplish this the Blair Camera Company acquired two firms: E. & H. T. Anthony & Company and Buffalo Argentic Paper Company. The Anthony company, plagued with difficulties during the 1880s, suffered at the end of the decade the death of its founder and the loss, due to fire, of part of its manufacturing capacity. The

[14] Ibid., pp. 229, 232, and 233; and GE to Moritz B. Philipp, 26 May 1889 (GEC).

[15] The Eastman company during the period from June 1889 to October 1890 supplied the Boston company with about fifteen to sixteen hundred roll holders.

[16] *Eastman* v. *Blair*, vol. 1, pp. 223–34.

[17] *American Annual of Photography Almanac* 4 (1890): 88 (advertisement); and "Allen & Rowell Co." (GEN).

[18] Sales of celluloid support to Blair Camera by Celluloid Company (in feet): 1891—7,335; 1892—79,030; 1893—68,104; 1894—31,962; 1895—49,024; 1896—45,291 (Blair testimony, *U.S.* v. *EKCo.*, p. 4; and Marshall C. Lefferts testimony, *Goodwin* v. *EKCo.*, pp. 2429–35).

financial problems concerning the estate and company stock holdings of Edward Anthony led, in the fall of 1889, to reports of the company's financial embarrassment.[19] Early in 1891, the Anthony company transferred to the Blair firm, in exchange for $76,000 worth of Blair stock, the Victoria Manufacturing Company, Anthony's chloride paper factory on Duane Street in New York City; the Greenpoint Optical Company, the camera and camera accessory factory at Greenpoint, Long Island; and several patents, valued at $40,000.[20]

Within a year the Blair company increased its nominal capitalization to $500,000 and also added a controlling interest in the Buffalo Argentic Paper Company, a producer of bromide paper.[21] Thus, in the years from 1889 to 1892, the Blair Camera Company had greatly expanded its product line and entered into the market for the roll film photography, largely through acquisition and merger, that is, through horizontal integration. In that same period the nominal capitalization had increased more than tenfold; yet, in spite of the controlling interest having fallen into the hands of the Anthony company, Blair and Goff were still very much in command, their combined holdings more than counterbalancing the Anthony interest. In terms of product line, the Blair company matched the Eastman Company, and Blair had the advantage of owning a small chain of company-controlled supply houses in major eastern and middle western cities. The Eastman Company did possess a substantial foreign business, an asset Blair could not boast of, but that deficiency was recognized and addressed within a few years.

The pattern of development of the Blair Camera Company from its beginnings in the late 1870s until the early 1890s may be understood in terms of its goals and strategies. The generally accepted goals of Goff and Blair during this period were growth and expansion rather than cash dividends. This is in contrast with the goals of Strong and Eastman after 1885. The operating capital of the Blair firm came largely from the retained profits of the company. Only three cash dividends were ever paid. Patented technological innovation, adopted initially as a strategy of entry into the industry when Blair had no capital, allowed him to move from dependence on Scovill to the establishment of his own independent company. Thereafter, technological innovation, particularly with regard to the introduction and promotion of new cameras and plate holders, played an important competitive role in those markets.

The pattern of development may also be understood largely in terms of the personal orientation of Thomas Blair and the economic environment in which he and the Blair company operated. Blair's technical ability and inventive talent, so significant for the company's existence, was, like Eastman's, largely mechanical. However,

19 *Goodwin v. EKCo.*, pp. 365 and 442; *Photog. N.* 32 (1888): 807; and GE to M. B. Philipp, 14 October 1889 (GECP).

20 Blair testimony, *U.S.* v. *EKCo.*, p. 19; and "Blair Camera Mfg. Co. (Mass.)" (GEN).

21 Blair testimony, *U.S.* v. *EKCo.*, p. 16; *Buffalo City Directory*, 1893 and 1894; and "Certificates of Condition," Corporation Records, Commonwealth of Mass., 1893.

unlike Eastman, Blair was no emulsion maker or amateur chemist. Although Blair initially produced dry plates as well as his Touro-graph, he moved quickly and almost exclusively into the apparatus field when the dry plate market became saturated in the mid-1880s. When the need for an ability to produce photographic materials arose late in the decade, the Blair Camera Company had to meet it through company acquisition.

Another aspect of Blair's initial and continuing orientation was his commitment to the amateur market. Although he did little to create an amateur market in the early 1880s, he responded to its needs with simple cameras and outfits. As George Eastman shifted his focus from the professional and serious amateur to the novice in the late 1880s, Blair perceived the importance of that shift and its relevance to the Blair market and was quick to follow Eastman into that newly created market. The pattern of the Blair company development in the late 1880s and early 1890s was a consequence of its effort to become independent of its competitor. George Eastman was not pleased with this growing Yankee threat on the eastern horizon, and soon he and his associates began to address themselves to it; but the threats to the Blair company were as great from within as from without.

The Technical and Economic Vicissitudes of the Blair and Eastman Companies: 1890–95

During the first half of the last decade of the century, the Eastman and Blair companies faced common problems and internal difficulties that required astute leadership. Characteristic of the photographic industry in the late nineteenth century—and of the chemical and electrical industries—was the importance of patents in structuring the industry; but patents were meaningless unless they were enforced. The control of fundamental patents was, therefore, hotly contested in the early 1890s by the firms involved in roll film photography, particularly Eastman and Blair. Moreover, the Blair companies succeeded in circumventing the well-designed Eastman patent barriers by introducing highly original camera designs and features, which ultimately helped to popularize roll film photography for amateurs.

Blair Companies

As George Eastman followed his conscious plan to use a system of patented inventions to deter competitors, he created a substantial technological barrier for the Blair Camera Company, which was trying to imitate the Eastman performance. Likewise, Blair's management responded to the constraints that gradually enveloped the firm's efforts to supply the roll film market with technological innovation. Upon dropping the association with the Eastman Company in 1890, the Blair company successfully circumvented the possible infringement on the celluloid support by purchasing it, at a substantial loss in profit, directly from the largest manufacturer of celluloid

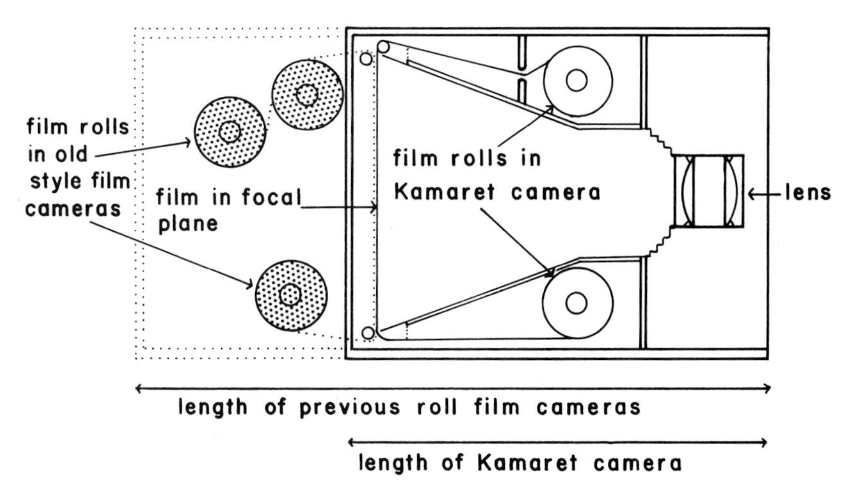

Fig. 6.3
The Kamaret camera.

in the country, the Celluloid Company. Blair thereby externalized the risk of infringement, the investment in production, and the legal expense of entanglements. With regard to the roll film camera, the Blair Camera Company drew upon some recent roll film camera inventions, no doubt stimulated by the commercial appearance of the Walker-Eastman roll holders and the Kodak camera, and introduced a very important roll film camera innovation: placement of the film rolls in front of the focal plane of the camera rather than behind it. The Kodak camera was essentially dominated by the roll *holder* conception and represented the encasement of an ordinary camera with the roll holders attached at the back. Other innovators, less dominated by the roll-holder-as-an-attachment idea, developed an arrangement whereby the film rolls were placed near the front of the camera (front roll camera). Patents granted to Whitney in 1891 and Houston in 1894 were the important controls on this invention.[22]

The Blair Camera Company obtained a license under the Whitney patent and began production of the Kamaret camera in 1891 (fig. 6.3).[23] The advantages of the front roll camera were (1) a reduction in required camera size of about one-third; (2) elimination of straining devices for the film; and (3) less distortion in the resulting prints. The third advantage occurred because the natural curvature of the film with the back rolls was in the opposite direction to the curvature of field of the lens, while with the front rolls the natural curvature of the film was parallel to, and therefore lessened the effect of, the curvature of field of the lens. The Kamaret camera received an excellent response. Moreover, the Blair Camera Company believed this innovation circumvented the Eastman Company's patent barrier; however, noting the effect of the Kamaret camera on

[22] Whitney: U.S. Patent #446,368 (10 February 1891) and #446,372 (10 February 1891); Houston: #526,445 (25 September 1894) and #526,446 (25 September 1894); these patents were linked to an abandoned application of 12 March 1886. Other related patents include: Gray and Stammers: #362,271 (3 May 1887); and Good: #421,923 (25 February 1890).

[23] Blair testimony, *U.S.* v. *EKCo.*, p. 23.

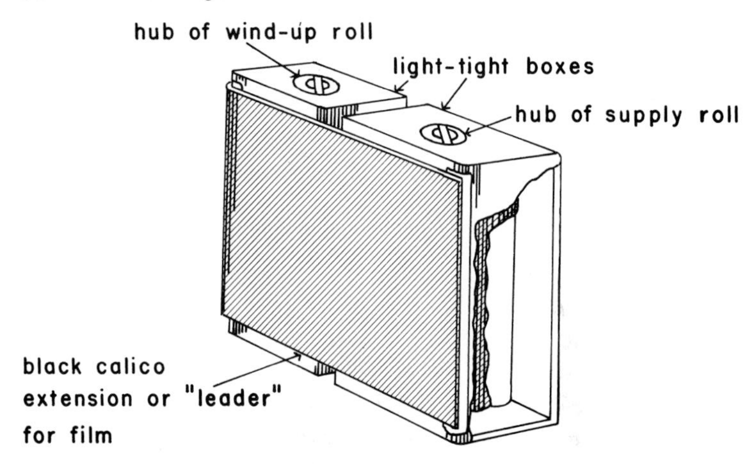

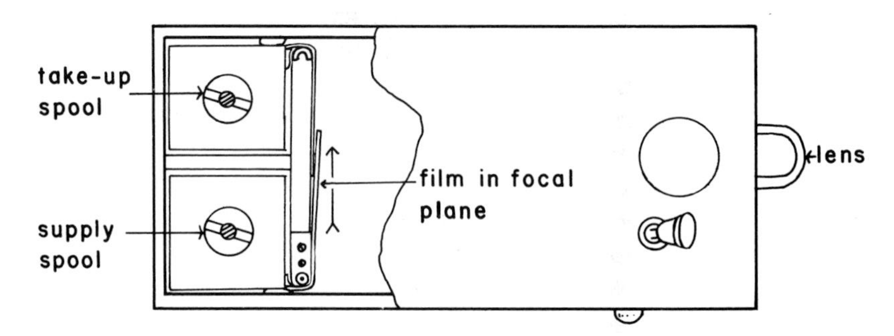

Fig. 6.4
Brownell's cartridge system.

Kodak camera sales, George Eastman initiated a suit against the Blair company. By successfully obtaining an injunction against production of the camera during the legal proceedings, the Eastman Company halted manufacture and sale of this camera from 1891 to 1894. The final ruling in the case upheld the Eastman allegations of infringement.[24]

Other interests in the Blair company did not, however, let this Eastman response deter them, and further efforts were made to invade the well-fortified Eastman patent barrier. Less than three months after the introduction of Brownell's daylight loading cartridge, which was designed to operate in the Kodak back roll cameras, Samuel N. Turner of Blair Camera filed a patent for a daylight loading cartridge system for front roll cameras.[25] In contrast to the Brownell system, in which two boxes contained the film with long, nonsensitized leaders and trailers on the film (fig. 6.4), Turner developed a system in which black paper was attached to the back of the celluloid film, flanged spools moved the film, markings on the

[24] GE to Philipp, 2 April and 21 May 1891 (GECP); Blair testimony, *U.S.* v. *EKCo.*, p. 6; and *Blair* v. *Eastman*, 64 *Fed. Rep.* 491 (1894).

[25] Brownell: U.S. Patent #477,243 (25 January 1892), application filed 28 December 1891; and Turner: U.S. Patent #539,713 (21 May 1895), application filed 21 April 1892.

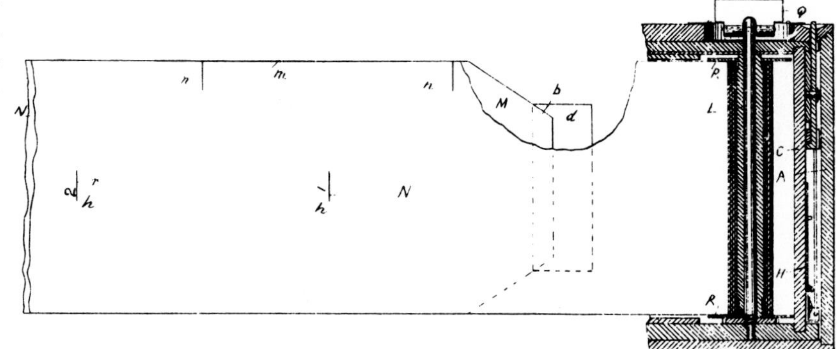

Fig. 6.5
Turner's daylight loading cartridge. From U.S. Patent # 539,713.

back of the paper indicated the exposure sequence and the places to cut the film between exposures, and the red peephole in the camera permitted the reading of the exposure sequence markings (fig. 6.5). Armed with the application for a patent on this system, in less than a month Turner was prepared to resume production.[26]

From 1891 the Blair Camera Company was beset by numerous problems that led to internal conflict among the management and stockholders and eventually to the loss of certain important interests in the firm. During 1891 the company encountered a number of technical problems, not the least of which was the Eastman injunction against the Kamaret camera. Because numerous difficulties plagued the introduction of production of the celluloid roll film that year, even though the firm did not have the responsibility for production of the celluloid support, total production was small; therefore, there was no opportunity to recover any of the losses. Although the small, financially weak company tried to produce a broad line of photographic goods—cameras, lenses, dry plates, film, chloride paper—and run a small chain of retail branches, none of the products obtained a sufficient market to produce substantial profit. The company suffered substantial deficits in 1891.[27]

The entry of the Anthony interests into the firm in 1891 added another dimension to the Blair problems. The Anthony contribution in terms of production facilities was of only limited value; however, full recognition of this did not come until after the merger, and the Blair interests felt that they had been deceived. Consequently, there was internal strife among the owners and managers, largely between the Anthony and Blair interests. Had the early years of the merger been years of outstanding profits and technical successes, the difficulties might have been mitigated, but the financial and technical problems simply exacerbated an already sensitive balance between the conservative and liberal investment and business policies of the Anthonys and of Blair.[28]

An immediate effect of the internal division at the Blair Camera Company was that each of the competing interests, unbeknownst to

[26] 19 May 1892, "Blair Camera Mfg. Co. (Mass.)" (GEN).
[27] GE to William H. Walker, 9 February 1892 (GECP).
[28] Ibid.; and GE to Walker, 11 May 1892 (GECP).

Table 6.1 Blair Camera Company Financial Data, 1891–97

Date	*Liabilities*	*Profit & Loss*	*Inventories*[a]	*Purchases of Celluloid Support*[b]
Jan. 1891	$ 42,186	+57,315	$ 92,239	7,335 ft.
Feb. 1892	101,994	+6,310	163,005	79,030 ft.
Jan. 1894	174,243	−64,104	177,380	68,104 ft.
Jan. 1895	168,479	−109,585	105,949	31,962 ft.
Jan. 1896	152,882	−131,299	61,099	49,024 ft.
Jan. 1897	148,046	−134,372	72,582	45,291 ft.

SOURCE: Financial statements submitted with "Certificates of Condition," Corporation Records, Boston, Commonwealth of Mass.

[a] Includes value of materials in process.

[b] According to testimony of Marshall Lefferts, president of Celluloid Co., in *Goodwin* v. *EKCo.*, p. 2429; these figures provide a rough estimate of the output of sensitized film.

the other, made overtures to the Eastman Company to sell controlling interest in the company. From late 1891 until 1893 negotiations continued more or less continuously on an informal basis, largely with Thomas Blair and Darius L. Goff. The expense of litigation over the roll camera suit represented a threat to both sides that could be eliminated by a merger, and Eastman sought Blair's recognition of the Eastman patents on papermaking machinery so that the paper market could be rationalized around them. The protracted negotiations, with the transmission of much data about the Blair Camera Company's financial status, was eventually regarded by Blair as an Eastman ruse to gain information about his competitor. No doubt the information was useful to Eastman, but he and the Board of Directors were quite serious about the negotiations. Because of disagreement over the form of payment and also some disagreements over evaluation of Blair's tangible assets and liabilities, the negotiations ceased in the spring of 1893.[29]

The Blair company's efforts to sell the firm to Eastman betrayed the technical, financial, and personnel problems that converged upon the firm in the early 1890s at about the same time as the depression. To the Blair Camera Company, a firm already in distress, the depression was simply the ultimate blow. Despite being the sole supplier of roll film during parts of 1892 and 1893 and being able, therefore, to raise film prices,[30] the company lost at least $70,000 between February 1892 and January 1894 (see table 6.1). The company continued to lose heavily during 1894 and 1895. Clearly, the depression created more financial problems for an already troubled corporation.

Internal conflict among the principal interest and leadership culminated in Thomas Blair's removal from leadership, the assumption of financial management by the Goff and Anthony interests,

[29] Ibid., 26 April and 11 May 1892 (GECP); GE to Blair, 25 February, 8 April, 23 September and 15 November 1892; Blair to GE, 13 September, 26 September [wire], and 18 November 1892 and 10 and 26 January and 10 March 1893; and Goff to GE, 16, 17, and 20 March, and 18 April 1893 (GEC).

[30] Blair Camera Company to The Trade, 19 November 1892 (GEC).

Fig. 6.6
Charles F. Ames. Courtesy of Eastman Kodak Company.

and the placement of supervision of production in the hands of a young employee, Charles Ames (fig. 6.6). Moreover, as the effects of the depression and the loss of two patent suits against infringers were felt, the film production facilities were moved to Pawtucket, near the home of the Goffs, and one of the major apparatus factories in Boston was closed.[31] Blair maintained a tenuous relationship with the company but went to London where he founded, in imitation of the Eastman company, an English branch, European Blair Camera Company, Limited (fig. 6.7).[32] Meanwhile, Turner also left the American company, acquired exclusive control of the Houston patents on the front roll principle, and resumed, under the name of the Boston Camera Manufacturing Company, production of the Bulls-Eye camera (see fig. 6.12). Into this camera he incorporated the front roll design and his daylight loading system. The Blair Camera Company made film to accommodate his new daylight loading feature.[33] Initially, Turner's camera output went to the European Blair Camera Company. Blair's role as a silent partner in Turner's company insured a cooperative relationship between the American producer and the European distributor.[34]

As a result of these changes the American firm lost its two principal technical personnel and innovators: Blair and Turner. Although Thomas Blair's visions and plans for the company may have been unrealistic in view of the company's resources, when Blair ceased active participation in the firm, the dynamic factor of technical innovation also ceased. That factor came to the fore in the companies with which Blair continued to associate himself, Boston Camera Manufacturing and American Camera Manufacturing, while Blair Camera sank into a five-year period of technological and financial stagnation.

Fig. 6.7
Blair International trademark.

The Eastman Company

Blair Camera Company's competitor in the production and sale of roll film and cameras, the Eastman company, also experienced a multitude of technical problems during the first half of the 1890s. After about three years of high sales of film and cameras, the company found its sales leveling off and then experienced a combination of technical problems and personnel defections and retirements just before the depression of 1893 struck.

The Eastman Dry Plate & Film Company, the first firm to produce celluloid roll film commercially, initiated production in August 1889. One of Eastman's employees who was directly involved in the manufacture of the film, F. C. Axtell, was pirated away to the Celluloid Company late in 1889—a not uncommon occurrence in industry in

[31] "Blair Camera Mfg. Co. (Mass.)" (GEN).

[32] "Blair Camera Mfg. Co." (GEN); and GE testimony, *Goodwin* v. *EKCo.*, p. 357.

[33] Blair testimony, *U.S.* v. *EKCo.*, pp. 6 and 17.

[34] GE testimony, *Goodwin* v. *EKCo.*, p. 357; Corporation Records, Commonwealth of Mass.

the late nineteenth century. At about the same time, the Celluloid Company, which had not previously engaged in producing thin celluloid film by spreading, flowing, and evaporating, was approached by the Blair Camera Company to produce for it thin, rollable film. As the Celluloid Company began to organize the production facilities to meet the Blair order, Axtell filed an affidavit implicating the Eastman Company in infringement of a Celluloid Company patent in the manufacture of celluloid film. The Celluloid Company film was produced differently from that of the Eastman Company. It was made from pyroxyline, acetone, and acetanilid, poured over a large, rotating copper wheel and stripped from it on a continuous basis. This method of production was a cheaper and more efficient method than Eastman's, but the resulting film was thicker and, because of its composition, translucent rather than transparent.[35]

The Celluloid Company suit was important to both Eastman and the Celluloid Company. The latter retained as its technical adviser professor of chemistry Charles F. Chandler of Columbia University,[36] and the Eastman Company hired Dr. Leonard Paget, a well-known celluloid chemist. The Celluloid Company began by requesting the court to grant an injunction against the Eastman Company, prohibiting it from continuing to manufacture the photosensitive film during the subsequent trial proceedings. The court refused to order the injunction,[37] and the trial proceeded through 1890 and 1891. The Celluloid Company argued that the Eastman Company was infringing upon certain claims that covered celluloid solvents listed in three of its patents. Early in 1892 the dispute was settled out of court with the Eastman Company agreeing to purchase its pyroxyline from Celluloid-Zapon Company, a subsidiary of the Celluloid Company, rather than from Eastman's supplier, Charles Copper & Company.[38]

Early in 1892 the Patent Office notified the Eastman Company that an interference had again been declared between the Goodwin application and the Reichenbach patent for manufacturing celluloid roll film for photography. Affidavits and testimony were taken in the interference matter, and the primary examiner refused to dissolve the interference on the grounds argued by the Eastman Company, that the burden of proof fell on Goodwin, not Reichenbach. In the spring of 1893, upon appeal to the commissioner of patents, the primary examiner's ruling was overturned, and the burden of proof then fell on Goodwin. Failing to make the necessary appearances, Goodwin again found his claims rejected by the Patent Office.[39] Again, it is likely that he was unable to afford the necessary legal

[35] Lefferts testimony, *Goodwin* v. *EKCo.*, vol. 3: 2429, 2435, 2437, 2453, and 2454.

[36] Chandler's Retainers, "Chandler Clippings Box," in Envelope: "Celluloid Co., 1881–91," Chandler Collection, Butler Library, Columbia University, New York.

[37] *Celluloid Manufacturing Company* v. *Eastman Dry Plate & Film Co.*, 42 *Fed. Rep.* 159 (1890).

[38] Lefferts to EKCo., 14 March, 1 and 11 November 1893 (GEC).

[39] "Goodwin File Wrapper," *Goodwin* v. *EKCo.*, pp. 3824–30; and Goodwin-Reichenbach Interference, #15,529, in ibid., pp. 3313–28.

Fig. 6.8
Kodak Park, c. 1891.
Courtesy of Eastman Kodak Company.

counsel. By the fall of 1893, it appeared to George Eastman that the Goodwin matter was closed.

After the introduction of the Kodak camera in 1888 and the celluloid film in 1889, the Eastman organization sought to acquire the resources and capacity to meet the overwhelming demand for cameras and photographic film. The Court Street film factory, with its twelve fifty-foot-long coating tables, was simply inadequate to meet the demand from both domestic and foreign markets. Soon George Eastman realized that the company's manufacturing location near downtown Rochester was not suitable for expansion. Quietly, negotiations were conducted by third parties, and a very substantial tract of farm land three miles north of the center of Rochester, near the Genesee River and Lake Ontario, was acquired. The decision to move the materials-manufacturing site to the country was influenced by such factors as the independent water supply, air free from the dust and dirt of the other Rochester factories, and the remoteness from residential areas, where complaints regarding the evaporated solvents for the celluloid film originated. Cleanliness, the hallmark of the dry plate and photographic paper business, now became equally important for the vast new film business.

On 1 October 1890, the ground was broken for the first three buildings at Kodak Park: a power plant, a film factory, and significantly, a laboratory (fig. 6.8). The laboratory, by far the smallest building, was not a research laboratory but a testing laboratory for the raw materials and finished products—more or less a quality-control center. The new film factory, completed in June 1891, initially contained twelve coating tables, each two hundred feet in length. The plant was very modern, with all the lighting and motors operated by electricity and the film and emulsion rooms air-conditioned by a fifty-ton ice machine. Less than six months after completion of the film factory, plans were already underway to enlarge it so as to double its capacity the next year. Moreover, following a small fire in the finishing room at Brownell's factory in the spring of 1892, a new fireproof six-story factory for camera production was completed early in 1893 (fig. 6.9). The cost, including the ma-

Fig. 6.9 Eastman Company's new camera factory. Photograph taken c. 1897. Courtesy of Eastman Kodak Company.

chinery and tools, was about $60,000.[40] Therefore, by early 1893 the Eastman company had rapidly expanded its production capacity in the United States for both cameras and celluloid film, with a capital investment in excess of $350,000.[41]

George Eastman, growing ever more aware of the need for continued product innovation in order to sustain demand for new cameras and as a way of "keeping ahead of the competition," introduced in 1890 four new models of the Kodak and three models of a new style camera, the folding Kodak.[42] His own personal interest shifted from designing new cameras to supervising the new manufacturing facilities and implementing improvements in production. He developed a double-coated celluloid film with plain gelatin on one side and gelatin emulsion on the other. The double-coating of the film prevented buckling of the film caused by changes in humidity or the different rates of absorption of moisture by gelatin and celluloid. Because of manufacturing difficulties, the film itself was not introduced for more than a decade. Working together, Eastman and Reichenbach successfuly eliminated the problem of streaks on the photosensitive emulsion (these were caused by the discharge of static electricity that occurred when the film was stripped from the glass tables), when, late in 1891, they added small quantities of

[40] GE to Walker, 4 April 1892 (GECP); GE to Eastman Photo Materials Co., 19 December 1892 (GEC); and *Rochester Union and Advertiser*, 27 June 1892, p. 5, and 14 February 1893, p. 5.

[41] The agreements with Brownell, whereby the Eastman company assumed ownership of all of Brownell's tools and machinery in exchange for payment in company stock, were also renegotiated at this time. Brownell then paid rent on the investment (6 percent) and was to be responsible for producing Kodak goods at agreed prices. It is likely that at this point it would have been more convenient for the Eastman company simply to have taken over the entire operation, especially in light of George Eastman's criticism of Brownell's capacity to organize large-scale production, but Eastman and his associates were rewarding a faithful ally from the early days of the company. See GE to Eastman Photo Materials Co., 19 December 1892 (GEC).

[42] *The Kodak Museum* (Harrow, Eng.: Kodak Ltd., 1947), p. 25.

electrolytic salts to the gelatin and thereby made it a conductor of electricity.[43]

While the company was making an outstanding profit on the celluloid film and encountered relatively little competition from its only rival, the Blair Camera Company, Eastman still concerned himself with cost-cutting innovations in the production area. For example, different lengths of film were placed on the market, including rolls of film with only six, eight, or twelve exposures, so that there would be less scrap on the ends of two-hundred-foot sheets of celluloid. Eastman also decided to market celluloid sheet film, thus making a profit from coated celluloid film that would otherwise have been wasted.[44]

Although Eastman maintained an interest in the manufacturing part of the business and recognized the importance of product innovations, the business had grown sufficiently, especially with the extension of manufacture and distribution to Europe and the rest of the world, that Eastman's direct attention shifted away from the technical details of production. Recognizing the need for well-trained engineers and chemists, he tried to recruit college-trained technicians and give them the responsibility for the technical duties he had previously performed. In the middle 1880s he hired Henry Reichenbach and in 1891 named him not only chief emulsion maker but also manager of the entire new Kodak Park plant. He also hired Dr. Passavant, an analytical chemist, to assist Reichenbach. Moreover, Eastman engaged a young mechanical engineer, Darragh de Lancey, upon de Lancey's graduation from the Massachusetts Institute of Technology and gave him the responsibility for planning and supervising the construction of Kodak Park. This signaled the beginning of a very important link between George Eastman and M.I.T. and the beginning of a shift in the photographic industry from empirical trial-and-error methods to theoretically based methods of production.

Not all the technical work was turned over to young college graduates, however. Camera design and product improvements were encouraged from the seasoned workers in the company's employ. Frank Brownell himself became increasingly active in designing new cameras and introducing small improvements. Late in 1891 the company introduced a daylight loading cartridge and cartridge camera, developed by Brownell. The film was placed in a light-tight spool carton with a velvet slot through which a long leader of black paper or cloth protruded; this could be threaded through the focal plane of the camera. The end of the film also had a long paper or cloth tail that allowed the exposed film to be removed from the focal plane. In this way, cameras could be loaded and unloaded in subdued light rather than in a darkroom. From this time forward, the amateur photographer was freed from having to send the camera to Kodak or take it to a nearby photographer. Only the small cartridges

[43] GE to Walker, 16 and 21 October 1890, 30 December 1891, and 18 April 1892 (GECP).

[44] GE to Strong, 31 March 1892 (GECP); and GE to George Dickman, 31 May 1894 (GEC).

needed to be handled in the developing and printing operations; yet, these ABC daylight cartridges, as they were known, were not fully satisfactory, and numerous improvements were to be made in them during the mid-1890s, the most significant of which was made by Samuel N. Turner.[45]

Furthermore, Eastman did not depend upon internal sources alone, anymore than he had in the previous decade when he, Walker, and Reichenbach had been the principal inventors in the company. Eastman followed the trade journals and the patent literature closely. Small sums were invested in purchasing outright or gaining licensing agreements on newly patented inventions, and Eastman pursued this path as assiduously as he had pursued the institutionalization of innovation, i.e., the institutionalization of the technical part of his earlier entrepreneurial function.

Eastman's other functions in the business were also becoming institutionalized during the early 1890s. In the late 1880s when Eastman dropped Selden as his patent attorney, he added to the company a local attorney, Walter S. Hubbell, and turned to him for counsel more and more each year.[46] Eastman also hired from the University of Rochester Lewis Bunnell Jones, who became the director of the Advertising Department in 1892, no minor post in a company that was selling a new consumer product.[47] Also, in 1890 he hired his first and only personal secretary, Alice Whitney. Eastman highly respected Miss Whitney's judgment and left important decisions to her. She served as assistant secretary of the Eastman company from 1900 to 1934 and in many other posts in the numerous companies that the Eastman company acquired over the years. Increasingly, much of the load of business and many personal interruptions were removed from Eastman's shoulders as Miss Whitney loyally protected him.[48] Hubbell, Jones, and Whitney were only the first of many persons who became George Eastman's eyes, ears, arms, and legs as he continued to direct the company for the next twenty-five years much as he had for the first ten.

The direct responsibility for the non-American operations of the company fell upon William Walker. Increasingly, as the company's business expanded to the world and as the London company moved into production of film and paper (fig. 6.10), George Eastman became critical of Walker's management. At times their communication was quite bitter in tone, and at one point Eastman even threatened Strong with quitting if further concessions were made to Walker;[49] yet, Eastman recognized that he needed Walker in London. In late 1891, when Walker wished to retire because of poor health, Eastman urged him to take a vacation, carry on, and start training a replacement. In the spring of 1892 George Dickman

[45] GE to Davison, 5 December 1902 (GEC); *The Kodak Museum*, p. 26; and "Supplemental Brief and Argument for Defendants," *U.S.* v. *EKCo.*

[46] GE to Walker, 14 July 1890 (GECP); and Hubbell obituary, *Rochester Democrat and Chronicle*, 2 January 1932.

[47] *University of Rochester Catalogue* (Class of 1891).

[48] Alice Whitney Hutchison obituary, *Rochester Democrat and Chronicle*, 14 April 1937.

[49] GE to Strong, 17 February 1893 (GECP).

Fig. 6.10 Eastman production facilities, Harrow, England, c. 1897. Courtesy of Eastman Kodak Company.

joined the company to learn the business from Walker; he replaced Walker in 1893, successfully assuming the new responsibilities and facing the problems of the worldwide depression with equanimity. Eastman's business advice to his new manager revealed much about George Eastman's business policy and his conduct of the company during his tenure:

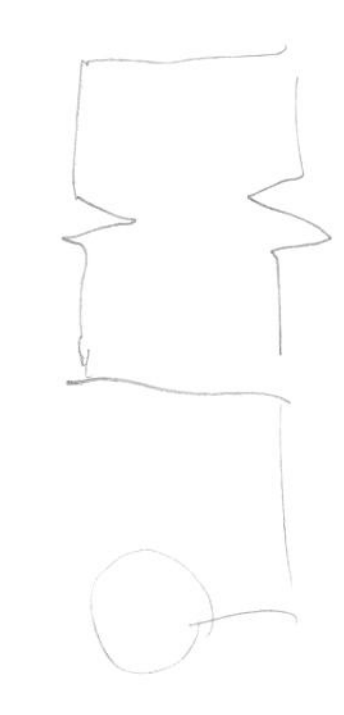

I am a believer of one man management and that a board of directors is valuable only as an advisory instrument to a good manager. I would strongly advise you to adopt the same course that I have, viz: to call on your board for advice only as to general policy and never as to details. If you cannot run the business alone as far as these matters are concerned you will certainly be no better off by letting four or five others dabble at it.[50]

The unmentioned corollary to this policy was that the manager must know more about the business than anyone else. Eastman's vast knowledge of the technical and financial aspects of the business, combined with the respect he had gained because of his early and continuing achievements, allowed him to dominate the policy decisions of the board and run the company nearly single-handed for a generation and a half.

One of the areas of disagreement between Walker and Eastman was that of financing. Walker encouraged the policy of inflating the value of the stock and then pulling out; but Eastman, careful planner that he was, was more interested in building a sound foundation and reaping the profits from the sturdy frame built on top of that foundation. The patented innovations were a large part of that foundation, but the image as a manufacturer of quality products was also important. A third and vitally important element, especially for an enterprise with an ambitious growth goal, was a sound financial reputation. While George Eastman was busy in the film plant at Kodak Park trying to devise a way to cut the waste in the celluloid film, he was also paying careful attention to the company's balance sheet, sales figures, and gross profits. He knew and reported to his confidants in the company daily sales and profit figures and how they compared with the previous year's figures for that month.[51] He paid particular care to maintaining the integrity of Eastman stock, because he knew that its reputation and salability were the key to obtaining expansion capital. Indication of the careful nurturing of the stock is that it nearly always sold at above par.

A large part of the program to maintain the integrity of the stock

[50] GE to Dickman, 18 March 1893 (GEC).

[51] The seasonal nature of photographic sales, with peaks in the spring, summer, and Christmas periods, made annual comparisons by months a more meaningful measure than a strict, linear, month-by-month comparison.

was the regular payment of generous monthly dividends, providing the stockholders with frequent indicators of the financial condition of the firm. This policy did not deprive the company of needed capital. From the time of the introduction of the Kodak camera, the profits were usually sufficiently great to allow both generous dividends and retention of a significant part of the profits for reinvestment in the business. The combination of producing a highly profitable set of products and paying good dividends regularly did much to make the Eastman stock highly respected.

The rapid expansion program between 1889 and 1892 required a substantial amount of new capital and stimulated a reorganization and recapitalization of the company in 1892. Essentially, there was a four-for-one stock split, the capitalization was increased five-fold, and the corporate name was changed to the Eastman Kodak Company. This provided nearly a million dollars worth of new stock for issue, but only a small portion of it was sold.[52]

At the end of the successful year, 1891, when plans were being made to reorganize the company and build the new camera works, George Eastman was bombarded with problems: the Celluloid Company suit, the Blair Camera Company negotiations, the rising competition in the photographic paper market, Walker's desire to retire, and growing production problems at Kodak Park. In addition, a major defection heralded a three-year period of considerable technical and financial difficulties for the company. The major problem for the company shifted from trying to create sufficient capacity to meet consumer demand, to trying to (1) produce photographic materials of satisfactory quality and (2) create sufficient demand to operate the plants profitably.

On the last day of 1891, George Eastman learned that three of his most trusted employees, Henry Reichenbach, S. Carl Passavant, and Gustave D. Milburn, were organizing their own company to produce and distribute a line of photographic materials and apparatus similar to that of the Eastman company. Reichenbach, developer of the celluloid film support, emulsion maker, research chemist, manager of the entire Kodak Park Works, was Eastman's key employee. Reichenbach's assistant, Passavant, was a chemist and a superintendent at Kodak Park. Milburn was the supervisor of the company's traveling salesmen and demonstrators. Eastman, incensed at the disloyalty, summarily discharged the three the following day.[53]

The planned departure was not just a matter of disloyalty, however. Eastman was a hard taskmaster. A perfectionist himself, he possessed an enormous capacity for long hours of hard work. He expected personal sacrifice and near perfection not only from himself but from his employees as well. He rewarded well those who in the long run met his exacting standards, but in the short run, he irritated many of his subordinates with his pointed criticisms and the parsimony with which he distributed compliments. Reichenbach

[52] "Formation of Eastman Kodak Company," *Rochester Union and Advertiser*, 24 May 1892, p. 8; GE to Walker, 5 March 1892 (GECP); and GE to Mrs. Cornelia J. Hagan, 27 January 1892 (GEC).

[53] GE to Philipp, 2 January 1892 (GECP); and *Eastman Company* v. *Reichenbach*, 20 *N.Y.S.* 110 (1892).

felt that he was not fully appreciated. During the summer and fall of 1891, Eastman was bluntly critical of Reichenbach's apparently wasteful employment of the company's resources. This atmosphere may have contributed to the planned departure of Reichenbach, Passavant, and Milburn.[54] Of course, the prospects of obtaining large profits by imitating the Eastman Company's success was also a major consideration.

At the same time that Eastman discovered his employees' plans, he also found that these men expected support in their new endeavor from certain local capitalists and also from the Celluloid Company, with whom the Eastman Company was involved in litigation. Most frightening to Eastman and the directors of the company was the open admission of intent on the part of the defectors to make use of Eastman secrets in producing bromide paper and celluloid support. Apparently, the employees' plans were discovered prematurely, and it was necessary for them hurriedly to seek support for their endeavor.[55] The support from the Celluloid Company, which concluded a favorable settlement of its suit against the Eastman Company during January 1892, did not materialize. The Eastman Company, fearing the loss of its production secrets, at once began legal action against the three former employees, eventually securing a permanent injunction prohibiting them from utilizing the trade secrets of the Eastman Company in their new venture, the Photo Materials Company.[56]

Soon after Reichenbach had departed, Eastman, needing a good emulsion maker to get the plant into operation again, persuaded his former photography teacher, George Monroe, to join the company. Monroe, who had been employed in the dry plate industry in St. Louis, brought with him a good deal of knowledge of the Cramer emulsion formulae and also some knowledge of the techniques employed at the Seed Dry Plate Company.[57] At first everyone was delighted with the emulsion that Monroe produced, but rivalries among employees created difficulties between Monroe and de Lancey, the new manager of Kodak Park. When suddenly, late in June, reports were received about film that was bad and was deteriorating on the dealers' shelves, Eastman promptly fired Monroe and replaced him with a man who had been working with Monroe and de Lancey. The promptness with which Monroe was discharged may have been largely a consequence of the high level of suspicion Eastman felt as a result of the Reichenbach situation.[58]

The deterioration of the film sensitivity required the recall and replacement of a considerable amount of film. Ultimately, production of film was discontinued during part of 1893, while emulsion and support experiments continued. An almost simultaneous change in a number of variables led to complications in determining the

[54] GE to Walker, 16 February 1892 (GECP).
[55] GE to Philipp, 2 January 1892; and GE to J. W. Millington, 4 January 1892 (GECP).
[56] GE to Henry Verden, 21 August 1899 (GECP); and *Eastman Company* v. *Reichenbach*, 20 *N.Y.S.* 110 (1892).
[57] GE to Walker, 9 and 16 February 1892 (GECP).
[58] GE to Strong, 27 June 1892 (GECP).

cause of the problems. Emulsion makers had been changed, and emulsion making was very much an art, not a science. The slightest variation in procedure could be significant to the final results. Also, the basic pyroxylin "dope" (the dissolved nitrocellulose) from which the celluloid base was made came from a new source, the Celluloid-Zapon Company, in accordance with the settlement of the Celluloid suit against the Eastman company. The change in source of pyroxylin alone could have introduced significant variations in the quality of the photosensitive coating. In addition, in an effort to become independent of the Celluloid patents, significant changes in the celluloid support formula were introduced in 1892, including a reduction in the quantities of camphor, fusel oil, and amyl acetate.[59] Finally, the remedy for the electrical discharge problem, the utilization of nitrate salts, added another variable.[60] At the time, Eastman blamed the problem on the emulsion, but he later regarded the changes in source of dope and the introduction of nitrates as the principal causes. During the Goodwin trial, the complainants argued that the high portion of camphor in the Reichenbach celluloid film was the cause of the problem, but the success of the Reichenbach formula from the fall of 1889 to the end of 1891 indicates otherwise.

While the company sought feverishly to improve the film and renew production, it suffered great losses in film sales. Furthermore, the sales in Kodak cameras also sharply declined because of (1) the unavailability of film except for the Blair film;[61] (2) the general economic depression that began in 1893; and (3) the failure of the Eastman company to introduce any new camera models in either 1892 or 1893.

Eastman, continuing to blame the film problem on the emulsion, sought a new emulsion maker of known reputation. In general, the reputation of the Eastman emulsion had not been outstanding after the mid-1880s, and at this time the Seed, Cramer, and Stanley plates had a far superior reputation to those of Eastman and, consequently, substantially greater sales. Therefore, Eastman sought a good emulsion not only for the film but also for the dry plates. Late in 1893, he induced William G. Stuber, a nationally renowned photographer, to come to Rochester to take charge of the dry plate emulsion department (fig. 6.11). In preparation for entering the manufacture of dry plates in this country, Stuber had gone to Zurich, Switzerland, where he had studied emulsion making with Dr. John Henry Smith, a manufacturer of dry plates and paper who had an outstanding reputation in Europe. Stuber also gained rights from Smith on a new coating machine that reduced the cost of plate production. Therefore, the Eastman company gained both the American rights to the Smith machine and knowledge of Smith's

[59] It was this reduction in camphor that later led to difficulties in connection with the Goodwin patent. It seems clear that the Eastman company cut the amount of camphor in the film in 1892 not to improve sensitivity of the emulsion but to reduce the shrinkage after coating, in an effort to obtain flatness in the focal plane in the roll film cameras and to avoid the celluloid patents. See GE testimony and de Lancey testimony in *Goodwin* v. *EKCo.*, pp. 355 and 2337, respectively.

[60] GE testimony, in ibid., p. 356.

[61] GE testimony, *U.S.* v. *EKCo.*, p. 240.

Fig. 6.11 William G. Stuber. Courtesy of Eastman Kodak Company.

emulsions, a direct input of European technology to the American industry.[62]

George Eastman quickly recognized the exceptional emulsion-making ability of this young man and renegotiated his contract, placed him in charge of producing plate and film emulsions, and began laying plans for building a new dry plate plant.[63] Stuber's appointment solved the basic emulsion problems of the Kodak company and stimulated Eastman to return his attention to photographic materials for the professional photographer. While the precise details of Stuber's emulsion and of the modifications made in the constitution of the celluloid support are not known, it is clear that with the addition in 1894 of Stuber and of Miss Harriet Gallup, a young chemist just graduated from the Massachusetts Institute of Technology, the quality and reliability of the photographic film and dry plates improved markedly.[64] In a market with only one film competitor, the company had little difficulty in regaining the confidence of the photographic materials market in the middle 1890s.

The vicissitudes of the Eastman company during the early 1890s were not independent of the general economic climate of the United States and the world. The Eastman companies had originated and grown during relatively prosperous times; however, the prosperity peaked in the very early 1890s, and in the spring of 1893 a severe depression began which lasted until about 1897, although recovery began in 1895. The photographic industry, producer of nonnecessities, was hard hit as rising unemployment cut personal income.

With the Eastman Kodak Company it is difficult to separate the effects of the depression from the inherent difficulties of producing satisfactory sensitive film. Yet, the financial problems of the company during 1893 and 1894 cannot be attributed solely to the internal technical problems and debts for plant expansion. Eastman was more than willing to attribute a large part of the company's financial problem to the depression: ". . . Our business, as we suppose that of other manufacturers of goods that are not necessities but in the nature of luxury, has suffered severely, especially the roll holder and camera part of it, by the financial depression throughout the country. We have been compelled to retrench in every direction. . . ."[65] The Eastman Kodak Company carried a debt of well over $165,000 into 1893 and reduced it to about $130,000 during that year. George Eastman carried the bulk of the debt on his own account at 6 percent interest so that the company would be free of the banks.[66] The company also received some temporary financial support from

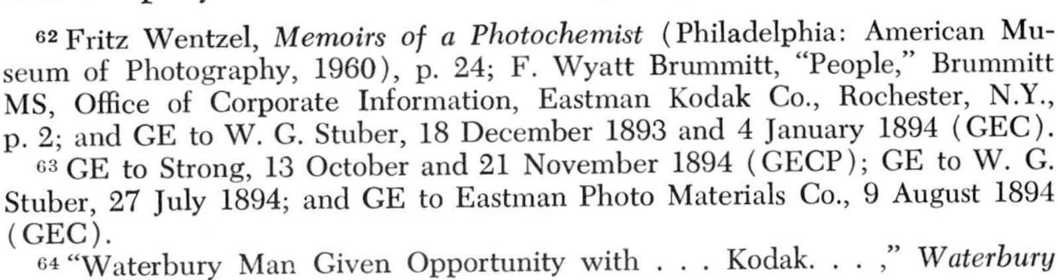

62 Fritz Wentzel, *Memoirs of a Photochemist* (Philadelphia: American Museum of Photography, 1960), p. 24; F. Wyatt Brummitt, "People," Brummitt MS, Office of Corporate Information, Eastman Kodak Co., Rochester, N.Y., p. 2; and GE to W. G. Stuber, 18 December 1893 and 4 January 1894 (GEC).

63 GE to Strong, 13 October and 21 November 1894 (GECP); GE to W. G. Stuber, 27 July 1894; and GE to Eastman Photo Materials Co., 9 August 1894 (GEC).

64 "Waterbury Man Given Opportunity with . . . Kodak. . . ," *Waterbury* [Conn.] *Republican*, 12 April 1936.

65 GE to Chauncey Smith, 18 June 1894 (GEC).

66 "Eastman Kodak Annual Report," 2 December 1893; GE to Strong, 22 February 1893; GE to Edmond Seymour, 2 January 1894 (GECP); and GE to Myron G. Peck, 8 September 1893 (GEC).

its English subsidiary. While the company suspended payment of dividends in 1893 and applied that year's modest profit to reducing the debt, the budget for 1894 showed modest increases and dividends were resumed, thereby indicating a degree of recovery (see table 6.2).[67]

Perhaps most significantly, renewed efforts were exerted in product innovation. The lull in innovation that occurred during 1892–94, while efforts were turned to the sensitized film, proved to George Eastman the importance and commercial necessity of continuous product innovation. In speaking of the effects of the depression he observed: ". . . If it were not for certain new articles we are now manufacturing, we would feel more than discouraged at the present outlook."[68] As business improved for the Eastman company during the latter part of 1894 and 1895, retained profits were sufficient to permit the company to expand its plant during the nadir of the depression (see table 6.2). The lack of a generally profitable line of products and of adequate capital precluded the Blair company from making a similar move.

Clearly, both the Blair and Eastman companies suffered from the depression; however, the overall technological and financial strength of Kodak allowed it to rebound from both its technical problems and the depression even before the depression had reached its nadir. The strength of the Eastman photographic paper business did much to carry the company through its period of technical difficulties, much as it had done during the trying days of the introduction of the roll holders and American film. In contrast, the Blair company, which also had a diversified product line, had no product of substantial market strength upon which to rely. Consequently, the firm just barely survived the depression.

Even though there were only two manufacturers of products for the roll film system of photography, the first half of the 1890s witnessed numerous competitive battles on several fronts. The Eastman company faced a number of threats and restraints, both external and internal, and responded largely by further technological innovation. The suit and attempted injunction brought by the Celluloid Company in 1890 posed an immediate threat to the continued production of the celluloid film. Vigorous legal defense defeated the injunction request, and the suit itself was settled out of court through a purchasing accommodation; yet, the realization that the Eastman Company was dependent upon the terms of that agreement set in motion a series of experiments and changes in the Eastman celluloid support formula that were aimed at avoiding the claims of the rival's patent.[69] Meanwhile, the Goodwin patent application continued to provide another dimension of threat and restraint in the preparation of the celluloid film. In reducing in 1892 the quantites of camphor employed in making celluloid film, the Eastman support formula did

[67] "Eastman Kodak Annual Report," 2 December 1893; Expense memo for 1893 and 1894, 22 January 1895 (GEC: incoming); GE to M. B. Philipp, 2 August 1894; and GE to Walker, 19 September 1894 (GECP).

[68] GE to Chauncey Smith, 18 June 1894 (GEC).

[69] GE testimony and de Lancey testimony, in *Goodwin* v. *EKCo.*, pp. 355 and 2337, respectively.

Table 6.2 Eastman Companies Financial Data, 1886–95

Category	*1886*	*1887*	*1888*	*1889*	*1890*	*1891*	*1892*	*1893*	*1894*	*1895*
Dividends	$21,000	$36,000	$18,000	$ 72,000	$200,000[a]	—		$ 40,310	$ 80,620	$ 40,324
Net profit								87,718	90,273	184,292
Gross profit				120,000[b]	240,000[b]			97,324	110,350	185,338
Total sales				~510,000[c]				541,205	630,918	899,496
Sales to English company	$32,815[b]	$42,272[b]	$51,150[b]	57,457[b]				57,875	62,458	98,534
Net U.S. sales				~450,000			$497,824	483,330	568,460	800,961

SOURCES: Figures for dividends for 1886–89 from "Eastman Companies" (GEN); for 1893–95, from Treasurer's Report, Eastman Kodak Co., Annual Meeting, 25 January 1897. Figures for sales to English company, 1886–89, from "Eastman Companies" (GEN); for net U.S. Sales 1892–95, from *U.S.* v. *EKCo.*, p. 2565.

[a] See GE to Henry A. Strong, 16 October and 29 September 1889; GE to William H. Walker, 12 October 1889 and 1 January 1891; GE to Moritz B. Philipp, 22 September 1889; GE to Harris Hayden, 4 February 1891 (GECP).

[b] Year ending 30 September.

[c] Total sales are calculated by adding (1) net U.S. sales obtained from census reports for 1889–90 and gross profit levels and (2) sales to English company (~$450,000 + ~$60,000 = $510,000).

Fig. 6.12
Bulls-Eye camera. From Scientific American, *31 March 1894.*

Fig. 6.13
No. 2 Bullet camera. Courtesy of Don Ryon, Eastman Kodak Company.

begin to differ substantially from the formula specified in the Reichenbach patent. While facing the Celluloid and Goodwin problems, the Eastman Company pressed patent infringement claims against the Buffalo Argentic Paper Company (Hoover and Getz) and also successfully prosecuted infringement proceedings against the Blair Camera Company.[70]

Eastman next turned his attention to a new threat and, in July 1894, sought an injunction against the Boston Camera Manufacturing Company's production of the Bulls-Eye camera, Turner's new design based on the front roll and daylight loading features (fig. 6.12). Unlike the decision in the case with the Blair company, the judge denied the injunction. At this point Eastman changed his strategy and began development of a similar competing camera, the Bullet. The name of the camera was consciously indicative of the Eastman company's intentions: the Bullet was intended to hit the Bulls-Eye (fig. 6.13). Although the Bullet had been patented by Brownell, it was very similar in appearance, design, and operation to the Bulls-Eye. The Bullet was introduced in March 1895 and was extensively promoted during the remainder of the year.[71]

George Eastman learned of the issuance of the Turner daylight cartridge patent, the application for which he had noted much earlier.[72] Early in June the attorney for the Boston Camera Manufacturing Company informed the Eastman Kodak Company that its camera, the Bullet, was infringing on the claims of two patents controlled by the Turner company: (1) Turner's newly issued daylight loading system; and (2) Houston's front roll camera design.[73] Consulting with his patent attorneys, Eastman learned that they, too, thought the Bullet infringed. Late in June, in Boston, Eastman successfully negotiated with Turner an agreement that gave the Eastman Kodak Company a sole and exclusive license on the Turner patent (excepting the Boston Camera Manufacturing Company). Reflecting the heightened sensitivity to the importance of litigation, the agreement provided that the Eastman company could take the responsibility for prosecuting any violators of the patent. Also, a part of this licensing agreement was a separate contract that named a minimum price level for the film produced under the patent. This was to the mutual advantage of both parties.[74] In early July the Pocket Kodak, utilizing the front roll design, the Turner cartridge system, and miniaturization of parts, was introduced to the public and received an enthusiastic reception.[75] Early in August Eastman successfully concluded a contract for the purchase of the Boston firm and all its assets. Furthermore, Turner agreed not to engage in the

[70] The Blair Camera case was concerned with the Kamaret camera's infringing the Houston and Walker-Eastman patents.

[71] *Eastman* v. *Blair*, pp. 5–7 and 16–18; Brownell: U.S. Patent #573,208; and "Supplemental Brief and Argument for Defendants," *U.S.* v. *EKCo.*, pp. 58–60.

[72] GE to Strong, 27 June 1895 (GECP).

[73] John L. S. Roberts to EKCo., 4 June 1895, in *U.S.* v. *EKCo.*, p. 3201.

[74] GE to Strong, 27 June 1895 (GECP) and agreements between S. N. Turner and EKCo., 25 June 1895, *U.S.* v. *EKCo.*, pp. 2164–67 and 2176–78.

[75] "Supplemental Brief and Argument for Defendants," *U.S.* v. *EKCo.*, pp. 58–60; and GE to Strong, 27 June 1895 (GECP).

manufacture of cameras or apparatus for a period of five years, thereby insuring that the nonpatentable camera technology—the "trade-secret" technology purchased with the company would not then be used again in competition with the Eastman Kodak Company.[76]

Significantly, Samuel N. Turner had succeeded, even where the Blair Camera Company had not, in developing a new system of roll film photography that did not infringe upon the patents indigenous to the Eastman system. The Eastman Kodak patent barrier had been penetrated, but George Eastman perceptively acquired the patent from a threatening company while that company was still very small. When Houston's front roll camera patent, which Eastman had hoped to obtain in the purchase, was also unavailable because of contractual complications, Kodak continued to produce front roll cameras, ignoring the patent until it received warnings in 1897.

The contrast between the strategies followed by the Blair and Eastman companies during the first half of the 1890s was significant. By the late 1880s and early years of the 1890s, both companies were moving into roll film amateur photography; but by the mid-1890s, even though both companies had experienced serious difficulties during the immediately preceding years, Eastman Kodak, the pioneer innovating firm, was moving into the dominant position and the Blair Camera Company was stagnating and disintegrating. The key element in analyzing the two firms and their evolution is entrepreneurship. Schumpeter defines the entrepreneurial function as "the carrying out of new combinations" and points out that the function "consists precisely in breaking up old, and creating new, traditions."[77] Of the leaders of the photographic industry in the period 1880–95, George Eastman and Thomas Blair most nearly approach the ideal of the entrepreneur. Both adopted a similar goal, i.e., to enter and maintain dominance in the industry through innovations protected by patents. However, differences in the sophistication and execution of their strategies and in their abilities to obtain sustained, loyal support from sources of capital account for the differences in success of their two firms.

Nevertheless, the Eastman and Blair companies represented the institutional setting for the introduction of the new technological framework of the industry. While film photography emerged during the decade following the middle 1880s, the economic and technological implications for the industry as a whole were realized only from the middle 1890s onward. In the meantime, the traditional marketing and production leaders of the industry, Anthony and Scovill, faced a rapidly changing business environment. They confronted not only the depression of the middle 1890s but, more fundamentally, two successive waves of technical change that brought in their wake new problems, rivals, and allies.

[76] "Agreement A," between S. N. Turner and Albert O. Fenn, 30 October 1895, in *U.S.* v. *EKCo.*, pp. 2169–74.

[77] Joseph A. Schumpeter, *The Theory of Economic Development*, trans. R. Opie (New York: Oxford University Press, 1962), p. 74.

7 / Responses of the Traditional Leadership to the Technical Revolution, 1880-95

With the advent of the gelatin revolution about 1880 and the roll film revolution soon thereafter, the dominant firms in the American photographic industry, Anthony and Scovill, were confronted with two radical changes in the technological mind-set. While they were better able to cope with the former revolution than with the latter, their slowness in perceiving the implications of the gelatin technology precluded their successful adaptation to the use of roll film.

The gelatin dry plate brought not only a revolution in the products and in production in photographic materials but also defined the direction for much of the technological development in the apparatus sector of the industry. While the dry plate revolutionized the photographic materials sector, the substantially increased sensitivity of the new plates introduced a new set of technical problems for the production of apparatus. Because the photographer had to exercise considerably greater caution in using the new photosensitive materials, darkroom lanterns were introduced and camera boxes were made increasingly light-tight. One of the most serious problems was the shutter. With the relatively slow wet collodion process, the shutter was of little concern to the photographer. The lens cap was the most common shutter prior to the gelatin dry plate, although simple shutters of the hinged flap and drop shutter types were introduced during the 1860s. With the sharp increase of sensitivity with the gelatin emulsions, shutters became one of the principal areas of technological change. During the 1880s various types of between-the-lens and focal-plane shutters were developed. The most common were shutters that required resetting immediately after release (e.g., drop shutters). In order to avoid a second exposure being made, the lens was capped during the resetting. During the late 1880s and early 1890s, advances in shutter technology brought forth the blade and rotary shutters, self-setting shutters that did not require the capping and resetting operation. Furthermore, the increased sensitivity of the photosensitive material lessened the importance of lenses of great light-gathering power in the new cameras.

Developments in apparatus resulting from the new technology simplified the photographer's task. He no longer needed to produce the photosensitive material. Also, with the introduction of bromide paper and the ability to produce enlarged prints, large plates and cameras were no longer necessary. The camera box therefore underwent miniaturization, which relieved the field photographer of bulk and weight.

During the 1880s the apparatus for the professional photographer continued to be supplied from the traditional sources: Scovill and Anthony for camera boxes, and European suppliers for high quality optics. The traditional American leaders also entered upon the production of complete camera and developing outfits for the newly emerging amateur market, but this new sector of the market also witnessed the entry of a number of new camera manufacturers. Although Anthony and Scovill viewed these new manufacturers with suspicion, in keeping with their traditional strategy they did serve as marketing agents for them, thereby guaranteeing themselves the jobbing profits, at least. Simultaneously, during the 1880s

Fig. 7.1
Scovill & Adams Company offices, New York City. From American Annual of Photography, *1897.*

and early 1890s the focus of capital, profits, and industrial control began to shift from the marketing function to the production function and, within the production function, from apparatus to photosensitive materials. By the time these trends were recognized by Scovill and Anthony, trade secrets (emulsions), patents, and trademark identification and reputation provided almost insurmountable barriers that even the former industrial leaders were unable to penetrate successfully. Although both of the leaders in the industry underwent major organizational and leadership changes during the 1880s and early 1890s, the older marketing conceptions of the industry and of their company's role in it dominated the new leadership.

The first main organizational change came to Scovill when in the late 1880s, the company decided to separate the photographic area from the rest of the company and establish it in New York City as the Scovill & Adams Company (fig. 7.1).[1] The president, treasurer, and manager of the new organization was Washington Irving Adams, who had been a director of Scovill and manager of the New York City business since 1878.[2] During the 1880s Washington Irving Adams had begun to groom his son, Washington Irving Lincoln Adams, to assume the leadership of the company.[3] Other younger men who began to emerge in leadership positions included Alexander C. Lamoutte and Carl Bornmann. Lamoutte, born in Puerto Rico, joined Scovill in 1885, and as a result of his success in promoting a Spanish-speaking department that developed a large business with Central and South America, at the time of the organization of Scovill & Adams he was placed in charge of the foreign and the order departments of the company.[4]

Under the leadership of W. I. Adams, the Scovill & Adams Company pursued a series of policies that reflected a commitment to the traditional strategies of the company with only a limited recognition of the new trends in the American photographic industry. The company continued to rely heavily upon its profits from the jobbing function and from general retailing of photographic apparatus and supplies. In the manufacturing sector, it continued to concentrate upon photographic apparatus, especially cameras, plate holders, and inexpensive lenses. With the advent of gelatin dry plate photography, Adams developed a real interest in promoting and exploiting the amateur market, but his approach to it was largely traditional: the introduction of small, light-weight, plate cameras and developing and printing outfits. Scovill inaugurated production and sale of complete amateur outfits as early as 1881 under their American Optical Company brand. These ranged in price from ten to forty dollars. The cameras were somewhat smaller than traditional styles, with inexpensive lenses and a cap for a shutter. They required the

[1] P. W. Bishop, "Scovill and Photography: The Tail That Almost Wagged the Dog" (unpublished manuscript), pp. 452–53.
[2] Ibid.; and D & B, Conn., vol. 35, p. g.
[3] *Photog. T.* 23 (1893): 725–28; and 28 (1896): 65–66.
[4] Alexander C. Lamoutte testimony, *Goodwin* v. *EKCo.*, pp. 1299–1320.

use of a tripod and therefore could hardly be called hand cameras.[5] Trying to stimulate interest in photography, Adams published books and pamphlets for serious amateurs, providing instruction in the art and techniques of photography; but there is no indication that Adams recognized the potential of the mass market for photography until George Eastman began to exploit it.

The Scovill company was much more responsive than the Anthony company to the system of roll film photography. By the mid-1880s Scovill had two popular amateur outfits, one of which contained the Scovill detective camera, a fairly expensive ($50–$75) leather-covered camera with bellows contained within the box. The other contained the Waterbury detective camera, which was a much cheaper ($25) wooden camera box. These cameras utilized shutters that were much faster than the capping of a lens, but they had to be reset after each exposure. Both cameras were constructed to carry a Walker-Eastman roll film holder.[6]

On the other hand, the first indications of the Eastman firm's entering upon camera manufacture in 1887 caused Adams and the management at Scovill concern.[7] Similar fears were expressed when the Kodak camera was first marketed in the summer of 1888. George Eastman tried to allay the concern at Scovill, arranging for the London branch of the Eastman company to act as the exclusive agents for Scovill in Great Britain during the late 1880s and early 1890s and assuring Adams and Lewis that the Eastman company would not enter the market in cameras for any but roll film cameras. He indicated that Eastman had no intention to start cutting prices on cameras and studiously avoided opening an Eastman agency in New York, which would have competed with the Scovill store.[8] Despite these conciliatory moves from the burgeoning Rochester firm, the Scovill management recognized the threat of the roll film system to their efforts to woo the serious amateur photographer and, as early as the fall of 1887, made overtures regarding a "coalition of interest" with the Eastman company.[9] In January of 1888, the Scovill Manufacturing Company formally proposed to the Eastman Board of Directors that the camera plants of the two companies be consolidated in Waterbury. The directors rejected the proposal on the grounds that George Eastman needed to be at the camera works because of the experiments being conducted there under his personal supervision.[10] While Adams's efforts were not ultimately successful, he at least recognized at an early point the potential of the

[5] *Photog. T.* 11 (April 1881): 158 and special insert; Robert Taft, *Photography and the American Scene* (New York: Macmillan Co., 1938), p. 376; and Lamoutte testimony, *U.S.* v. *EKCo.*, pp. 539–40.

[6] Lamoutte testimony and Carl Bornmann testimony, in *U.S.* v. *EKCo.*, pp. 541–42, 544–45 and 511–12, 515, respectively; and "The Scovill Detective Camera" and "The Waterbury Detective Camera," *Photographic Amateur* (1885) in Government Exhibits # 301 and #302, *U.S.* v. *EKCo.*, pp. 3230–32.

[7] GE to W. J. Stillman, 22 and 26 October 1887 (GEC).

[8] GE to Hunt, 6 September 1888; GE to Walmsley, 14 September 1888; GE to R. S. Lewis, 7 December 1887; GE to Entreken, 19 March 1888; GE to Stillman, 22 October 1887; and GE to Scovill Mfg. Co., 10 January 1888 (GEC).

[9] GE to Stillman, 22 October 1887 (GEC).

[10] GE to Scovill Mfg. Co., 10 January 1888 (GEC).

Fig. 7.2
Richard A. Anthony. From Taft, Photography and the American Scene, *A Social History, 1839–1889* (New York: Macmillan Co., 1938; reprint ed., New York: Dover Publications, 1964). *Courtesy of Dover Publications, Inc.*

Fig. 7.3
Frederick A. Anthony. From Taft, Photography and the American Scene. *Courtesy of Dover Publications, Inc.*

Fig. 7.4
Anthony view camera, c. 1884.

Walker-Eastman roll film system of photography and made numerous efforts to become part of the new trend in photography.

Just as the Scovill company experienced major organizational changes, the leader of the American photographic industry, the E. & H. T. Anthony & Co., also underwent major ownership and organizational changes during the 1880s. Early in the decade, the Anthony brothers had been the principal stockholders and held the principal offices in the firm; a few years later, Colonel Wilcox was serving as treasurer, secretary, and manager of the firm and clearly conducting the business both in terms of day-to-day affairs and policy matters.[11] During the 1880s the Anthony brothers died, leaving the leadership of the firm fully in the hands of Wilcox, who was in his late fifties, and two family members, Richard and Frederick Anthony (figs. 7.2 and 7.3). The youth and inexperience of the two Anthonys left them in a weak position to challenge the mature and experienced leadership of Colonel Wilcox.[12] Therefore, the company activities continued to be dominated by the strategy that had been successful during the 1860s and 1870s: reliance largely on the jobber's profits and externalization of the risk of production by having the manufacturing done by small companies tied to Anthony by trade agreements. In the area of manufacturing, the firm continued to focus on areas of traditional strength: chemicals, albums, paper, and camera apparatus.

Despite the continuance of the traditional strategy, manufacturing in some new directions was attempted. Although the Anthony company was less enthusiastic about the amateur market for apparatus and was, therefore, somewhat less responsive to it than was Scovill, shortly after the Scovill company introduced amateur outfits, Anthony followed with cameras produced by its suppliers.[13] Anthony also sought to produce gelatin dry plates in early 1880 and to produce photosensitive paper in the late 1880s.

The introduction of roll film photography stimulated some new directions of endeavor. Though the Anthony company refused to cooperate with the Eastman company in the promotion of roll film photography, Erastus B. Barker and William H. Lewis, two of the Anthonys' best camera and apparatus designers, patented several modifications of the roll holder during the mid-1880s.[14] Because the improvements were not of great importance and no effort was made to create a system of improvements, the patents proved to be of little value. Also, during the late 1880s the E. & H. T. Anthony & Company imitated John Carbutt in introducing cut celluloid film plates. The Climax plates did not generate a revolution in photography, but even if they had, in the new environment the Anthonys would not have been able to make a large profit from them without

[11] D & B, N.Y.C. Series 8, p. 400°.

[12] F. A. Anthony testimony, *Goodwin* v. *EKCo.*, pp. 365–419 and 1314; and GE to Moritz B. Philipp, 14 October 1889 (GEC).

[13] William H. Lewis and Henry Dietrich (Success Camera Company), D & B, N.Y.C. Series 8, p. 400xx.

[14] U.S. Patents: Barker: #367,176 (26 July 1887), #407,050 (16 July 1889), and #408,451 (6 August 1889); and Lewis: #391,167 (16 October 1888).

encountering considerable competition.[15] While the Anthony company was more innovative than the Scovill company and obtained substantially more patents, its weakness in planning innovation and its inability to execute effectively the manufacturing function with either traditional products or the new products made much of its innovative effort worthless.

The Anthonys were much more responsive to science and its role in business than was the Scovill & Adams Company. Dr. Charles Chandler, the renowned analytical and industrial chemist at Columbia University, was retained by the Anthonys as a consultant for nearly three decades and as editor of *Anthony's Photographic Bulletin* for more than a decade.[16] Late in 1889 the Anthony company hired the young Belgian chemist, Dr. Leo Baekeland, who had just immigrated to the United States in search of industrial employment. While most of his work for the company during the more than a year he was with it (1889–90) was routine chemical analysis, he sought to improve the celluloid cut sheet film base and worked with the film emulsions.[17] While these associations with major chemists and attempts at technological innovation placed the Anthony company in no better position than did the numerous patents it obtained during the 1880s, they nevertheless demonstrated a level of sensitivity to general industrial and technical trends. However, the real potentials of scientific research, invention, and patents were never realized, perhaps because of the gulf in the firm between the technical men and the men in positions of leadership. Edward Anthony, with his background in science and engineering, had built a strong business based on the union of technical knowledge and business acumen. The new generation of leaders, Colonel Wilcox, Richard Anthony, and Frederick Anthony, represented considerable business, banking, and economic knowledge but failed to represent the technical understanding and practical energy that the new leadership in the industry provided.

In all the new endeavors, the Anthony company encountered considerable difficulty and was basically unsuccessful. During the years of financial stringency following the death of Edward Anthony (1888), the company turned over much of its manufacturing capacity and some of its patents to the Blair Camera Company, thereby relinquishing any production leadership to the newer firms in the industry.[18] At that time, also, it entered into exclusive trade agreements with the American Aristotype Company and with the Hammer Dry Plate Company.[19] Hence, the strategy followed was to return to the jobbing and marketing function and externalize the

[15] F. A. Anthony testimony, *Goodwin* v. *EKCo.*, pp. 424–28.

[16] Actually, Chandler was editor in name only, and most of the editorial work was done by the assistant editor, Dr. Arthur H. Elliott (1851–1918), an associate and former student of Chandler. Arthur H. Elliott to C. F. Chandler, 1 August 1885, Chandler Collection, Butler Library, Columbia University, New York.

[17] F. A. Anthony testimony, *Goodwin* v. *EKCo.*, pp. 387–88 and 426.

[18] R. A. Anthony testimony, *U.S.* v. *EKCo.*, pp. 560–63.

[19] R. A. Anthony to C. F. Chandler, 7 September 1892, Chandler Collection; and "American Aristotype Co." (GEN).

manufacturing. Token ownership in these manufacturing firms insured some profit return from that quarter but provided no opportunity to exercise leadership in the industry. Consequently, the Anthony company became more committed to marketing when the industry trend was to production; it thereby lost, by the middle of the 1890s, the leadership position it had held in the late 1870s.

During the 1880s several new manufacturers focused on the serious-amateur market and allowed the traditional producers to continue dominating the professional market. One of the earliest and most important companies in this sector of business was the Rochester Optical Company. Firms operated by William H. Walker, Franklin M. Cossitt, and William F. Carlton ultimately joined to form this company. While Walker and Cossitt left the Rochester Optical Company in the middle 1880s to join the Eastman company, Carlton continued the business alone, gaining a reputation for good quality cameras and accessories and making a specialty of the manufacture of view cameras for amateurs and professionals. Like the Blair firm, Rochester Optical Company relied on the three leading jobbers in the industry to distribute the bulk of its production.[20]

The first Kodak camera presented a new challenge to all the leading camera manufacturers, for it not only addressed the mass amateur market but also struck at the heart of the new serious amateur market. The plate camera industry during the next seven years responded in five ways. The first was the outright rejection of the roll film system and the continued promotion of traditional plate cameras. The E. & H. T. Anthony & Company typified this response. The reasons for this reaction were no doubt the personal and business rivalry between the Anthonys and Eastman stemming from the severance of their marketing arrangement in 1885 and rivalry in photosensitive paper production from 1887.

In contrast to the response of rejection, a second initial reaction was cooperation and coordination. Scovill Manufacturing and, later, Scovill & Adams typified this response. They responded favorably when the roll holders were first marketed, by promoting them, equipping their cameras for them, and obtaining a license to produce them for their own cameras. They likewise responded favorably to the Kodak camera until its inroads into their own camera business became quite manifest. Blair Camera Company also followed this pattern of cooperation at first, even putting on the market, under license, a camera similar to the Kodak camera. The pressures of demand upon the Eastman Company for Kodak cameras and film were responsible for the deterioration in this coordination and cooperation with the Blair Camera Company.

A third response, which characterized all responses made after the Kodak camera had manifested its truly challenging character, was imitation. The responses of Blair and Turner, as described above, characterize the various attempts on the part of plate camera producers to shift to the manufacture of roll film cameras through outright imitation or through the introduction of innovative improve-

[20] EDPF to A. Henderson, 21 April 1887 (GEC).

Fig. 7.5
Magazine camera, c. 1900. From American Annual of Photography, *1900.*

ments in roll film cameras in the hope of circumventing the strong Eastman patent barrier on the roll film system.

A fourth response to the roll film system by the plate manufacturers was innovation in plate cameras in order to meet the challenging characteristics of the roll film system. The introduction and active promotion of self-setting shutters, magazine cameras, celluloid plate film cameras, film packs, and very inexpensive, small plate cameras represent efforts to meet the Kodak camera challenge through simplifying the operation of the camera, lightening its weight, decreasing its size, and providing for a method of consecutive exposure without having to remove the plates (fig. 7.5). Hetherington & Hibben, of Indianapolis,[21] and Scovill promoted the magazine camera (which had been introduced in Europe several years before), while Anthony and the Photo Materials Company introduced film plate cameras.[22]

The fifth response to the success of the Kodak camera was an influx of new camera manufacturers to the industry. Observing the success of the Eastman Company but perhaps unaware of the exact elements of that success, many new companies began manufacturing apparatus in the early 1890s, choosing the plate camera sector because of the patent barrier against entry into the truly successful sector, that of the roll film camera. Numerous small and poorly financed companies, such as the Student Camera Company of New York City, tried to meet the Eastman Kodak challenge by offering very inexpensive cameras.[23] They met with at least three major problems. First, their products still required the photographer to develop and produce his own prints. Second, by cutting the price drastically they made their margin of profit very low and therefore provided themselves with little capital for promotion and production economies. Third, with a narrow market segment, little capital, and little profit, when the depression struck in 1893 most of these firms disappeared. They may have benefited for a short time from the difficulties experienced at the Eastman company in the early 1890s, but in the long run their conceptions and strategies proved unsuccessful.

A few firms that entered the industry during the early 1890s with a different strategy fared somewhat better and were able to survive the depression. The key to survival seemed to be diversification and innovation—to say nothing of sound management and adequate capital. The Gundlach Optical Company (Rochester) and the Manhatten Optical Company concentrated on lenses for cameras but also supplied some cameras. The latter company benefited in 1894, when Dr. Hugo Schroeder became the manager of its optical department. Schroeder had been the director of the Ross and Sons'

[21] Hetherington & Hibben, Indianapolis, Indiana, "Magazine Camera," *Photog. T.* 21 (29 June 1891), reproduced as Government Exhibit #305, *U.S.* v. *EKCo.*, pp. 3232–33.

[22] Helmut Gernsheim and Alison Gernsheim, *History of Photography: From the Earliest Use of the Camera Obscura in the Eleventh Century up to 1914* (London: Oxford University Press, 1955), pp. 297–99.

[23] "Picture Making with the Student Camera" (New York: Student Camera Co., 1891), Archives, Baker Library, Harvard University, Cambridge, Mass.

Fig. 7.6
A.G.F.A. trademark.

optical works in London for the previous thirteen years.[24] Folmer & Schwing, a New York City company that produced gas illuminating fixtures, entered the photographic industry in 1891 but had its own design of cameras produced by Scovill & Adams Company and Flammang Camera Company in New York City and only later entered into production itself.[25] The Rochester Camera and Supply Company, started in 1891 as a partnership of Harvey B. Carlton, a relative of William H. Carlton of the Rochester Optical Company, and Charles F. Hovey, the son of Douglas Hovey of the American Albumenizing Company, initially produced hand plate cameras but by 1895 began to produce the Poco brand cameras, which included regular plate cameras, magazine cameras, and a limited number of roll film cameras.[26] The Photo Materials Company, started by Reichenbach, Passavant, and Milburn, entered upon the production of a celluloid film plate camera, the Trochenette. While the production facilities were quite elaborate and required considerable capital, the product was not a success and within a year the camera production was halted, and the company concentrated on the production of photographic paper.[27]

The dynamic factors in the apparatus sector of the industry changed from the early 1880s to the early 1890s. The serious amateur sector of the market stimulated technological change and the entry of new firms into the industry during the early and mid-1880s, while during the early 1890s the Kodak camera not only addressed the new mass market but also challenged the plate camera manufacturers' position in the serious amateur market.

The major change in technology also influenced the photochemical sector of the industry. The demand for chemicals associated with the production of collodion plates and sensitized albumen paper fell sharply. The market for silver nitrate and halogen salts, which had once included the widely distributed production agents, the professional photographers, was now limited to producers of dry plates and sensitized papers. At the same time, new organic developers created by aniline dye researchers—Agfa and Schering in Germany, Lumière in France—began to make inroads into the American photochemical industry. Traditional manufacturers, such as Mallinckrodt, Rosengarten, and Powers & Weightman, responded with rapidly expanding product lines and diversification into nonphotographic lines.

The technological mind-set changed, and, with it, so did many of the technological and business conditions. The established leadership of the industry was slow to perceive the change or its implications. When the traditional firms did seek to change and adjust, their leadership failed to combine both the business and technical talents

[24] *Photog. T.* 25 (1894): 230.

[25] "Folmer and Schwing Mfg. Co." (GEN); and *Grafolks* 13 (October 1957).

[26] "Photographic Apparatus" (Rochester: Rochester Camera Co., May 1898), copy at EKCo. Patent Museum; *Rochester City Directory*, 1892.

[27] GE to George Dickman, 16 August 1893 (GEC); Abram J. Katz testimony and Max Brickner testimony, in *U.S.* v. *EKCo.*, pp. 433 and 436, respectively; and *Rochester Union and Advertiser*, 31 August 1892, p. 5.

and insights that were so vital to successful innovation. The dynamic and innovative leaders—Eastman and Blair—combined a number of assets that overwhelmed the traditional leaders. They were responsive to new opportunities and new conceptions of the industry and the market, they sought new ways of protecting or creating protected markets, they appreciated the potential of new technical knowledge and the importance of technically-trained engineers and chemists in implementing the production of new products and developing new modes of production, and they appreciated the significance for business in the new environment at the end of the century of the system of invention-patent-innovation-litigation. It was leadership sensitive to and appreciative of these new considerations that wrested control of the industry from the Anthonys and the Scovills and set the industry on a new path at home and abroad.

The manifest destiny of the Eastman Kodak Company is to be the largest manufacturer of photographic materials in the world or else to go to pot.

George Eastman
1894

IV

Period of the Amateur Roll Film System 1895 to 1909

8 / The Corporate, Industrial, and Technological Setting

The Corporate Revolution in America

During the last two decades of the nineteenth and the first few years of the twentieth centuries—a period during which prices fell, the standard of living increased, and disposable income rose[1]—there was a general movement away from decentralized, single-function, family-operated manufacturing firms to large-scale, centralized, vertically integrated,[2] professionally managed corporations. Although this movement, often referred to as the Corporate Revolution, or the Rise of Big Business, was not clearly and broadly discerned until the late 1880s, its roots lay in an earlier period. Traditional manufacturers—producers of textiles, sewing machines, clocks, watches, reapers, shoes, guns, and pistols—responded slowly to the new forces. Yet, enterprises in the newer industries—railroads, iron and steel, petroleum—served as models for the revolution. The railroads,

[1] Bureau of the Census, U.S., Department of Commerce, *Historical Statistics of the United States, 1789–1945* (Washington, D.C., 1949), pp. 231–32; and Simon Kuznets, "Changes in the National Income of the United States of America since 1870," *Income and Wealth of the United States* (Cambridge: Bowes & Bowes, 1952), p. 55.

[2] *Vertical integration* is the control within one business of two or more levels of activity necessary to process a commodity from raw materials to the finished product. *Vertical integration backward* is the moving from basic production into production of materials or supplies necessary for the basic production. *Vertical integration forward* is the moving from basic production into the marketing and distribution of the product. *Horizontal integration* is the control within one business of two or more establishments or subsidiaries producing the same goods or providing the same service. See Ralph W. Hidy and Paul E. Cawein, ed., *The Challenge of Big Business* (Lexington, Mass.: D. C. Heath, 1968), p. 114.

which came to prominence in America in the middle nineteenth century, stood as the initial model for the new style of internal organization and administration because of the enormity of the scale of financing and administration required to establish and operate them. Consequently, in response to the financial demands, investment banking and the corporate style of business organization developed, and the increase in administrative functions was met by functional departments, line and staff organization, central offices, statistics gathering, and cost accountancy. The railroads also disrupted the regional and local production and distribution structures and stimulated the development of a national market for manufacturers. National firms began to appear which initially expanded in the anticipation of securing a large share of the market, but as other producers of like perception also expanded, overcapacity occurred in some industries, resulting in price-cutting and other forms of competition.[3]

One of the earliest responses to this situation came in the petroleum industry, where Standard Oil pursued a course of action to alleviate price competition through combination, consolidation, and integration. This set a model for restructuring enterprises and industries that faced overcapacity or ruinous competition.[4] A wave of

[3] Alfred D. Chandler, Jr., and Stephen Salsbury, "The Railroads: Innovators in Modern Business Administration," in *The Railroad and the Space Program*, ed. Bruce Mazlish (Cambridge, Mass.: M.I.T. Press, 1965).

[4] Ralph W. Hidy and Muriel E. Hidy, *Pioneering in Big Business, 1882–1911* (New York: Harper & Bros., 1955).

business combinations came in the late 1880s and early 1890s in the meat packing and agricultural implement industries and also among producers of rubber, whiskey, rope, cotton and linseed oil, and leather.[5] The general national movement, however, was interrupted by the depression of the mid-1890s, but at the end of the decade, the movement resumed and continued on an unprecedented scale into the early years of the twentieth century.

Economic and business historians have paid particular attention to the Corporate Revolution because of its importance to the entire national and international economy in the twentieth century. Of particular interest to historians have been the reasons and motivations behind the movement and the timing of its occurrence. While many factors have been enumerated and analyzed, including the tariff, the markets for industrial securities, the shifting overseas demand for American goods, the impact of new technology, and the avaricious motives of the "Robber Barons," recent studies have emphasized the emergence of the national and urban market and such related factors as the inability of factory owners to enforce and maintain cartels and the inability of the established wholesalers to respond to the new needs for marketing new products.[6] Yet, there was another factor related to the new market scale and the creation of national competition: the technological implications of new products and new methods of production, both of which were geared to an emerging mass market. In contrast to the older handicraft and small-scale mechanization of production of the earlier periods, the new methods of manufacture based on large-scale machine production could not be designed precisely to suit the existing market; they were introduced in anticipation of ever larger markets, and the resulting "lumpiness" sometimes contributed to the development of overcapacity.[7]

One problem that American entrepreneurs, the politicians, and the public confronted was the traditional attitude toward large-scale enterprise. On the one hand, business leaders were faced with the unsettling problems and opportunities of the emerging national markets and highly competitive conditions, while on the other hand, established public policy was not prepared for the revolution. The result was an ambiguous and equivocal situation that left most business leaders in a dilemma: should they seize the economic opportunities, and run counter to sectors of public opinion, or await the public debate and lengthy court proceedings before moving into this new territory of business and industrial organization, and run the risk of letting someone else in their industry take the initiative? In

[5] Alfred D. Chandler, Jr., "The Coming of Big Business," in C. Vann Woodward, ed., *The Comparative Approach to American History* (New York: Basic Books, 1968).

[6] Ibid.; Glenn Porter and Harold C. Livesay, *Merchants and Manufacturers: Studies in the Changing Structure of Nineteenth-Century Marketing* (Baltimore: Johns Hopkins Press, 1971).

[7] "Lumpiness" in this context means that production capacity comes in large units that cannot be subdivided; hence, the market demand may best be met by 2¼ units, but a producer may produce 3 units or 2 units, not 2¼ units.

the face of the legal ambiguity, most entrepreneurs chose to test the legal limits.

During the 1870s and 1880s, numerous businesses sought relief from price competition through pooling agreements and trade associations, but because such agreements were not enforceable in courts under earlier common law precedent, they proved to be only temporary measures. The ineffectiveness of the pooling agreement strategy led the Standard Oil Company to organize a trust in 1879, thereby, developing a new strategy of industrial organization.[8] Legally speaking, trusts were an old device to separate ownership from control, used chiefly in the interest of minors and widows. The effectiveness of the application of this legal device to the organization of industrial enterprises by the Standard Oil Company created increasingly unfavorable public reactions during the 1880s and culminated in the passage of antitrust legislation by many states in the late eighties and early nineties and in the passage by Congress in 1890 of the Sherman Act, a federal antitrust law. Nevertheless, in the late 1880s, when the trust form of organization was under its greatest attack, the state of New Jersey, in pursuit of its long-standing policy of attracting corporate business, revised its general laws to permit a chartered company to hold stock in another corporation. This modification of New Jersey incorporation law opened a new alternative to business firms either anticipating consolidation or already under attack as trusts.[9] As other states gradually relaxed the restrictions that limited the size and privileges of corporations, industrialists and financiers promoted consolidations in the belief that they were immune from the federal antitrust prohibitions.[10]

Early Supreme Court interpretations of the Sherman Act did little to restrain those entrepreneurs who were skeptical of the validity of the legal restraints upon their consolidation efforts. Furthermore, in 1899, when the Court, ruling in the Addyston Pipe Case, decided that it was illegal for a loose association of manufacturers to fix prices, it appeared that outright mergers into holding companies was the only refuge for industries that faced severe price competition.[11] In light of the legal history of the first ten years of the Sherman Act, it is no wonder that the decade of the 1890s witnessed an enormous revolution in organization and structure of business.

However, the Court gradually began to shift its position. As prices started to rise during the last years of the century and as journalists and the public became increasingly concerned with the social and economic implications of the new, giant corporations, political and legal actions became more intense. In 1904, in the Northern Securities Case, the Court revived the moribund Sherman Act by ruling that holding companies were not automatically exempt from the

[8] Hidy and Hidy, *Pioneering in Big Business.*

[9] Alfred S. Eichner, *The Emergence of Oligopoly: Sugar Refining as a Case Study* (Baltimore: Johns Hopkins Press, 1969), pp. 140–50; and George W. Stocking and Myron W. Watkins, *Monopoly and Free Enterprise* (New York: Twentieth-Century Fund, 1951), pp. 257–60.

[10] Stocking and Watkins, *Monopoly and Free Enterprise,* p. 32.

[11] *Addyston Pipe & Steel Company* v. *U.S.,* 175 *U.S.* 211 (1899).

federal antitrust laws.[12] In 1903, in response to popular agitation, the Bureau of Corporations was established to compile data on corporate practices. Investigations conducted by the bureau provided data for the Justice Department cases against Standard Oil Company of New Jersey, the American Tobacco Company, and E. I. Du Pont de Nemours & Company. In 1911 these three corporations were commanded by judicial order to dissolve.[13]

Just as there was ambiguity and uncertainty regarding what organizational and industrial structures were acceptable, various new pricing and marketing practices that had been introduced with the emergence of the new national and urban mass market were also shrouded in legal uncertainty. From the beginning, the courts refused to enforce contracts with resale price maintenance clauses, but in the early years of the twentieth century three lower court decisions were made in which price maintenance was upheld in the case of the sale of patented articles. Then, in a reversal, the Supreme Court in the case of *Bobbs-Merrill Company* v. *Strauss* overruled the three cases, condemning the practice as a violation of the antitrust laws.[14] Other new practices included utilization of exclusive dealerships and price discrimination among dealers. While no court decisions dealt directly with these practices, the Clayton Act, passed by Congress in 1914, specifically forbade both exclusive dealerships and price discrimination.[15]

During this period, the most successful entrepreneurs proceeded to test the legality and social acceptance of their practices, and even though many of them were taken to court and forced to sign consent decrees that called for the dissolution of their corporate creations, they already commanded a dominating position in their industries and, therefore, successfully survived the political and legal attacks. This entrepreneurial daring helped define the limits of the new corporate structure that has come to dominate the American economy during the twentieth century.

While the American industrial community was undergoing a structural and legal revolution, the two major technological revolutions in the American photographic industry (gelatin emulsions and roll film system) opened the industry to the new mass market, permitting a new scale of organization to emerge. New entrepreneurial leadership, not committed to either the technological or business mind-set of the traditional leadership of the industry, attempted various organizational and structural techniques to solve the competitive problems the industry faced. After the ineffective pooling agreements and trade associations in the dry plate and paper sectors during 1894 and 1895, volume producers of durable goods—amateur

[12] *Northern Securities Co.* v. *U.S.*, 193 *U.S.* 197 (1904).

[13] *Standard Oil Co.* v. *U.S.*, 221 *U.S.* 1 (1911); *U.S.* v. *American Tobacco Co.*, 221 *U.S.* 106 (1911); and *U.S.* v. *E. I. Du Pont de Nemours & Co.* et al, 188 *Fed. Rep.* 127 (1911).

[14] *Bobbs-Merrill Co.* v. *Strauss*, 210 *U.S.* 339 (1908); confirmed in: *Dr. Miles Medical Co.* v. *Park*, 220 *U.S.* 373 (1911); *Bauer* v. *O'Donnell*, 229 *U.S.* 1 (1913); and *Strauss* v. *Victor Co.*, 243 *U.S.* 490 (1917).

[15] Sections 2 and 3 of the Clayton Act. See Stocking and Watkins, *Monopoly and Free Enterprise*, pp. 359–60.

cameras—and of perishable goods—plates, film, and paper—initiated vertical integration during the late 1890s and the early part of the first decade of the new century. Almost simultaneously, these sectors of the photographic industry also integrated horizontally. In the case of the photographic paper sector, the horizontal integration succeeded largely because of the backward vertical integration, while in plates and cameras the horizontal integration succeeded largely as a consequence of the pressures brought by forward vertical integration.

With the breaking of the technological set of the preceding fifty years and the dissolution of the leadership group that had dominated the industry for the past thirty years, a whole new business-technological worldview emerged. This new worldview inculcated the new leadership in the industry with many of the assumptions underlying the new corporate order in American "Big Business" at the time. Of particular importance was the international perspective that increasingly came to dominate the American photographic industry.

Market Growth: 1895–1909

With the introduction of the roll film system, the mass market for photographic materials and apparatus was opened between 1889 and 1894. From that time until 1909, the markets for photographic goods expanded dramatically, far exceeding the average growth rate of the rapidly expanding American industrial economy. From 1889 to 1909, industrial production in the United States grew at an annual rate of about 4.7 percent. On the other hand, production of photographic materials and apparatus grew at an annual rate of about 11 percent.

At the same time that this explosion occurred in the market, the Eastman Kodak Company assumed a position of dominance in the industry through its leadership in product innovation, production technology, and marketing strategies. From 1889 to 1909 the company's domestic sales grew at an annual rate of nearly 17.5 percent, while during the period 1894 to 1909, its total worldwide sales grew at an annual rate of about 20.8 percent. The roll film proved to be the most profitable item of production, and therefore, considerable attention was directed to the successful marketing of this product. While a good profit was made on roll film cameras, their cost was lowered dramatically during the 1890s in order to get them into the hands of the amateur who then would consume Eastman Kodak film.

Also of importance was the production of photographic paper, the major source of profits to the Eastman company when it was struggling to develop the roll film system. As the horde of novice photographers became consumers of photographic film, they likewise became major consumers of photographic paper. With the failure of the courts to uphold the paper production patents of Eastman during the mid-1890s, Eastman Kodak's control over the production of

Table 8.1 Growth of Eastman Kodak Sales Compared to Growth of U.S. Photographic Industry, All U.S. Manufacturing, and U.S. Population, 1889–1909

	1889	*1894*	*1899*	*1904*	*1909*	*Average Annual % of Increase*[a]
Eastman Kodak sales (millions of $)	.5	.6	2.3	5.1	9.7	17.5
U.S. photographic industry Total sales (millions of $)	2.7		7.8	13.0	22.6	11.0
Eastman Kodak sales as % of all U.S. photographic sales	16.4		29.2	39.4	42.9	
Frickey index of U.S. manufacturing production (1899 = 100)	66		100	121	166	4.7
U.S. population (millions of persons)	61.8		74.8	82.2	90.5	1.9

SOURCES: U.S., Department of Commerce, Bureau of the Census, *Abstract of the Census of Manufacture, 1914,* p. 672; *U.S.* v. *EKCo.,* pp. 2565–69; Edwin Frickey, *Production in the United States, 1860–1914* (Cambridge: Harvard University Press, 1947), series P13.

[a] Simple, not compound, rate of increase.

photographic paper evaporated, and numerous new small companies entered upon specialized production of paper. In the last years of the century, a few of these companies became dominant forces in the photographic paper sector of the industry. Although the major growth areas were associated with roll film, roll film cameras, and photographic paper, there also was a much slower overall growth in production of dry plates and dry plate cameras. Unable to compete with roll film cameras because of the patent barrier, numerous inventors developed improvements in dry plate cameras and, because of ease of entry, entered that market. By the first years of the twentieth century the dry plate camera market was glutted with small manufacturers, all of whom were in difficulty because of the low profits brought on by fierce price competition.

While the period from 1895 to 1909 witnessed vast increases of domestic sales in amateur and cine film materials and apparatus, it also witnessed a vast expansion of influence abroad by the American photographic industry. Without question, the Eastman Kodak Company represented the major force in the American photographic industry abroad. Prior to the 1890s the export of American goods had been so small as not to warrant disaggregation in export records, but the growth of sales of amateur goods abroad during the 1890s reflected the growing marketing network that extended throughout Europe, Asia, Australia, and Africa. Exports skyrocketed with a hundred-fold increase from 1894 to 1899. From 1899 to 1912 exports increased another ten-fold. The export figures, as given in table 8.2, do not provide the whole story, however. Not all products for foreign sale were produced in the United States. Major manufacturing facilities were erected in England, France, Germany, and Canada, and the timing of the opening of these facilities accounts for the fluctuations in export figures. Also, the English branch of Eastman

Table 8.2 U.S. Exports of Photographic Goods, 1894–1912

Year (ending June 30)	*Europe*	*N. America*	*S. America*	*Asia*	*Africa*	*Oceania*	*Total*
1894	$ 9,872	$ 2,001	$	$	$	$	$ 11,873
1899	1,074,421	50,198	27,198	1,585	1,508	9,555	1,164,465
1904	28,082	19,298	27,552	32,038	1,968	18,563	127,501
1909	3,373,530	531,958	124,365	76,150	895	77,818	4,184,716
1912	9,706,972	1,265,166	327,836	153,650	3,231	219,512	11,676,367

SOURCE: U.S., Department of Commerce, *Commerce and Navigation of the United States, 1894–1920.*
NOTE: Illustrative of the dominant role of Eastman Kodak in world markets, the American company's shipments to its Harrow facilities alone accounted for more than 80 percent of all U.S. photographic exports to England in 1912. Also, the effects of the rapid expansion of world photographic markets and of the discontinuous expansion of Eastman Kodak production facilities in Europe accounted for the wide year-to-year fluctuations in U.S. photographic exports. See *U.S.* v. *EKCo.*, pp. 2572–73.

Kodak handled nearly all of the foreign sales except for the Western Hemisphere. Therefore, although the figures on American exports can provide only a clue to the real expansion of the influence of the American photographic industry abroad, that clue indicates how rapidly the Americans assumed worldwide dominance.

Thus, the period from 1895 to 1909 was a period of change of leadership in the American industry. The new leaders who came to power in this revolutionary setting brought with them not only a new set of attitudes and strategies with regard to business organization which reflected contemporary trends in business enterprise but also a new technological configuration and new attitudes toward technological change.

Institutionalization of Technological Innovation: Eastman Kodak

The newly evolving technology policy at Eastman Kodak may be generally characterized as the institutionalization of technological innovation. During the mid-1890s George Eastman's style of entrepreneurship shifted from its strong focus on the technological functions to a broader effort directed at reorganizing the American industry and the Eastman Kodak Company. A concommitant of that shift was a growing personal interest in the acquisition and the construction of new production facilities and in finance. As a consequence of this shift, Eastman began to institutionalize many of the functions he himself had earlier personally performed. Clearly, this process had already been under way during the previous decade, but the production and technological innovative functions were among the last of the original duties that Eastman relinquished, an indication of the importance he accorded those functions for the successful development and maintenance of the company.

Another area of major concern to Eastman during the 1890s and the 1900s was the production of photosensitive materials. The man-

Fig. 8.1
Darragh de Lancey. Courtesy of Eastman Kodak Company.

agement of materials production at Kodak Park had been largely assumed by Henry Reichenbach in the early 1890s, but with his departure, Eastman once again had to assume considerable responsibility at the Park. Darragh de Lancey, a mechanical engineer and Eastman's first employee to graduate from the Massachusetts Institute of Technology, was hastily pressed into service upon Reichenbach's departure and proved to be an excellent supervisor, production organizer, and innovator (fig. 8.1). George Eastman, much to his own satisfaction, soon discovered that he could turn the responsibilities at Kodak Park over to de Lancey. Many years later Eastman credited de Lancey with "having switched Kodak Park from the empirical to the scientific path."[16] De Lancey sought to organize and operate the growing chemical plants according to the principles of engineering he had learned at M.I.T. and to place responsibility in the hands of young, scientifically trained personnel.

Eastman acquired a growing respect for the graduates of the recently developed technical schools of the United States. Despite the unpleasant experience with Reichenbach and Passavant, both of whom were technically trained college men, he maintained his confidence that the success of the photosensitive materials business depended upon such persons. During the 1890s he actively sought young technical graduates, especially in chemistry and chemical engineering, from M.I.T., Columbia, Purdue, Yale, Rose Polytechnic, and Rochester, and many of the positions of major responsibility in the rapidly growing company went to these people.[17] De Lancey himself frequently turned to M.I.T. for people needed for top technical and supervisory positions.[18]

Despite the growing number of college graduates at Kodak Park, Eastman fully supported and rewarded key craftsmen like Stuber. However, Stuber's methods of trial and error and, perhaps, his defensiveness toward the college graduates, led to some minor conflicts with de Lancey and Frank Lovejoy.[19] Eastman's respect for the talents of both the craftsman—for emulsion making was but an art at that time—and the scientifically trained engineer led to the amicable settlement of the conflicts. George Eastman, always frank and often outspoken with his employees, clearly defined the lines of authority and instilled confidence in his supervisors.

Eastman's laudatory remarks about de Lancey's management of Kodak Park were prompted by the revolution in the method of producing photographic film instituted under de Lancey's direction. While the Eastman companies had been world pioneers in introducing and developing the continuous flow method of coating photosensitive paper, they had remained with a static method of casting, coating, and drying celluloid photographic film on long glass tables.

[16] GE to Darragh de Lancey, 29 January 1920 (GEC).

[17] GE to Prof. T. M. Drown (Boston), 3 November 1891; GE to Dr. C. F. Chandler (Columbia College), 25 January 1892; and GE to Yale, Columbia, M.I.T., Purdue, Rose Polytechnic, 5 June 1899 (GEC).

[18] "Darragh de Lancey Recalls 10 Years Service with Eastman," *Waterbury [Conn.] Republican*, 12 April 1936.

[19] GE to de Lancey, 25 and 27 March 1899 (GEC). Similar problems had arisen earlier between George Monroe and de Lancey.

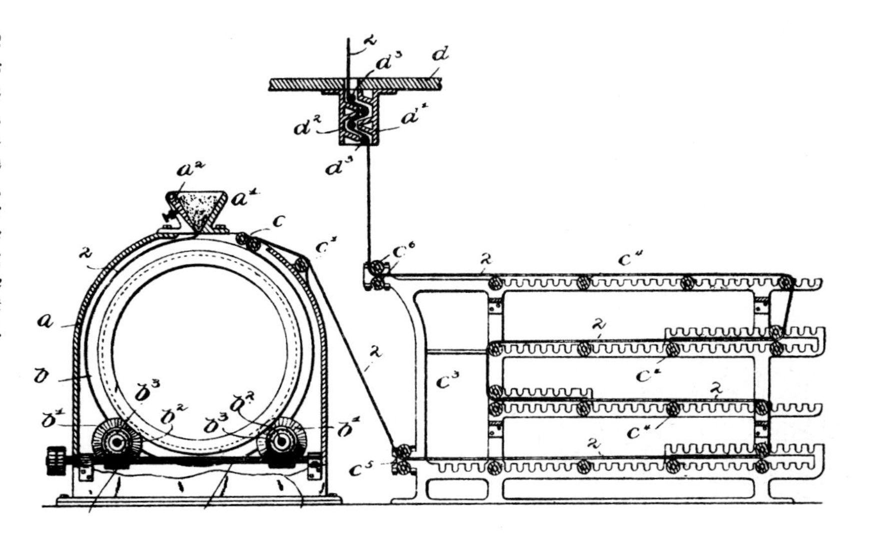

Fig. 8.2
Blair-Waterman apparatus for continuous production of sensitized film. Note: a^2, ***dope trough from which continuous stream is drawn forming base of film;*** c, ***drying-seasoning area for film; and*** d, ***zigzag entrance to dark room for applying sensitive emulsion to the film strip. From U.S. Patent # 588,790.***

The acquisition of the American Camera (1898) and Blair Camera (1899) companies served to remind Eastman of the continuous flow methods of coating celluloid film. The Eastman company's more integrated operation—producing and coating their celluloid film, rather than purchasing and coating it, as Blair and American Camera did—served to delay the adoption of continuous flow methods of production. Since 1890 the Celluloid Company had been producing rollable celluloid film by continuous casting machine for use by both Blair Camera and American Camera Company.[20] The New England camera companies instituted the continuous flow method of coating the film in 1896, about the time the Blair-Waterman patent was issued (fig. 8.2). Yet it was cheaper, at least at low output levels, for Eastman to leave the film support on the glass tables for coating than to strip it from the table, roll it, and coat it by machine. Furthermore, the wheel casting method of the Celluloid Company required complex and expensive machinery (capital intensive) and development of a film formula that would allow for rapid drying. In the early 1890s the Eastman company was not in the financial position to invest in such expensive machinery, and with the numerous difficulties that the company encountered in the production of the film, Eastman was reluctant to add additional variables by further altering the formula and method of production. However, during the mid-1890s, as the demand for film increased dramatically, Eastman and de Lancey began to lay plans for changing from the static table to a continuous flow method of production in order to reap production savings that could be important both in increasing profits and in allowing, if necessary, a lowering of price against competition from potential new entrants to the industry.

Prior to de Lancey's production revolution at Kodak Park, a series of experiments with the celluloid film during 1896 and 1897 had been directed toward developing a faster drying formula through the

[20] Marshall C. Lefferts testimony, *Goodwin* v. *EKCo.*, p. 2453; the patents were U.S. Patents #573,928 and #600,824.

Fig. 8.3
Frank W. Lovejoy. Courtesy of Eastman Kodak Company.

utilization of more volatile solvents.[21] The acquisition of the American Camera Manufacturing Company additionally gave the firm possession of the Blair-Waterman patent on continuous flow coating of celluloid film. By the summer of 1898 the development work on the new method of production was well under way. A new film-making building was designed to accommodate at least six of the new casting machines.[22] An important patent related to the new film-making machine, that of Bostwick and Harrison, was purchased, and the testing of the formula brought it to what was believed to be a satisfactory mix for the dope.[23]

At this time de Lancey, following his policy of acquiring technically trained men for supervisory and technical positions at Kodak Park, successfully wooed Frank William Lovejoy, who graduated in 1894 from M.I.T.'s new program in chemical engineering (fig. 8.3). De Lancey placed Lovejoy, despite his lack of experience with either celluloid or photographic chemistry, at the heart of the production revolution by appointing him superintendent of the crucial film support department. Lovejoy was soon in the midst of the planning and design of the new system of production.[24]

While de Lancey and Lovejoy directed the developmental work at Kodak Park, George Eastman personally supervised their progress and assisted and advised in the program. Aware of the problems associated with producing the large twelve-foot casting drums, which required nickel-plated surfaces, he directed the company's attorneys in New York to seek information regarding the source of the Celluloid Company's casting wheel.[25] From the New York attorneys Eastman gleaned additional information regarding the Celluloid Company's operations, including the production layout, details of the construction of the casting machine, and the speed of operation of the machine. Eastman relayed this information to de Lancey with various recommendations. One was that de Lancey hire a man to supervise the making of the machine and that this man then be placed in charge of the casting department once the machines were in regular operation.[26]

In response, de Lancey hired Perley Smith Wilcox, a mechanical engineer from Cornell University (class of 1897) who initially joined Swift and Company in Chicago as a draftsman and later served as an assistant to the mechanical engineer in Swift's St. Joseph, Missouri, plant. No doubt this experience exposed Wilcox to the continuous flow methods of operation then common in the meat packing industry. Upon joining Eastman Kodak, he was assigned to

[21] De Lancey testimony, *Goodwin* v. *EKCo.*, pp. 2325–29 and 2338; "Addendum to Main Brief and Argument for Appellant," ibid., pp. 244–46.

[22] GE to H. Senier, 22 April 1899 (GEC).

[23] GE to de Lancey, 1898 (GEC).

[24] *M.I.T. Centennial Alumni Register*, 11th ed. (Cambridge, Mass.: M.I.T. Alumni Ass'n., 1961), p. 6; and *F. W. Lovejoy: The Story of a Practical Idealist* (Rochester: Eastman Kodak Co., 1947), pp. 15–16.

[25] GE to Moritz B. Philipp, 7 July 1898 (GEC); de Lancey, Lovejoy, and Eastman already had some information regarding the Celluloid Company's operations from the testimony of the officers in the company in various patent suits.

[26] GE to de Lancey, 18 and 19 August 1898 (GEC).

work under Frank Lovejoy, supervising the design and construction of the new machinery.[27]

By the spring of 1899 the casting machinery had been installed, and the first trial runs, to the delight of everyone, were an immense success. At once de Lancey and Eastman planned to build additional machines with larger drums.[28] Attention then shifted to the design of the coating machinery. The men at Kodak Park drew upon the company's experience with the continuous coating of photosensitive paper,[29] and upon the information from the American Camera and Blair Camera Companies, although the latter was limited, as Eastman's appraisal indicated: "The reel system that has been in vogue at Northboro and Pawtucket is only a make-shift. . . ."[30]

One complication in the change in production method was the departure of de Lancey, who had a nervous breakdown and was urged by his physician to retire.[31]

Lovejoy assumed his position on a trial basis, having to accept the responsibilities of learning a new position and completing the installation of the new production machinery. With the assistance of Wilcox, Lovejoy successfully completed the program, and commercial production began early in 1900, with Wilcox supervising the Film Support Department.[32]

Hence, by the end of the century, Eastman's college-educated staff had successfully revolutionized the production of photographic film and provided Kodak Park with the most modern photographic production facilities in the world. The change came none too soon, because the wild popularity of the cinema, in addition to the growing popularity of amateur film photography, soon taxed even the new method of production. Throughout the next decade, expansion of film production facilities was very rapid. Yet, despite the cost advantages of the new system of production and the better competitive position in which it placed the Eastman Kodak Company, it did extract at least one substantial cost. The new method of production required more volatile dope, the manufacture of which necessitated a substantial deviation from the company-controlled Reichenbach formula for celluloid and opened the company to a later charge of infringement of the Goodwin celluloid patent.[33]

The contrast between the chemical materials production located at Kodak Park and the mechanical apparatus production on State Street was sharp. While Eastman and his protégés were developing a scientifically trained staff at Kodak Park, Eastman allowed the Camera Works to be dominated by a very capable but practically

[27] P. S. Wilcox testimony, *U.S.* v. *EKCo.*, pp. 446–47.

[28] GE to Henry A. Strong, 8 May 1899 (GECP); GE to Philipp, 22 April 1899; and GE to Senier, 22 April 1899 (GEC).

[29] The machine developed was similar to that used to coat Solio paper, U.S. Patent #358,848; de Lancey testimony, *U.S.* v. *EKCo.*, p. 415.

[30] GE to Strong, 1 July 1899 (GECP).

[31] "Darragh de Lancey Recalls," *Waterbury Republican*, 12 April 1936.

[32] GE to Strong, 24 May 1899 (GECP); GE to F. W. Lovejoy, 4 June 1900 (GEC); Lovejoy testimony, *Goodwin* v. *EKCo.*, p. 2360; and Wilcox testimony, *U.S.* v. *EKCo.*, p. 447.

[33] GE to George Davison, 27 March 1899 (GEC); and *Goodwin* v. *EKCo.*, pp. 244–46.

trained leadership. Brownell, of course, owned and operated the Camera Works until early in the new century, when he gladly sold the plant to Eastman Kodak, thereby relieving himself of what to him were burdensome cares of supervision and management. Thus freed, Brownell spent all his time designing cameras and apparatus. Eastman was delighted, for he regarded Brownell as "the greatest camera designer that ever lived," and he soon came to regard the area of product innovation as the new life line of the company.[34]

The increasing attention to continuous production development and innovation reflected a more general change from nearly exclusive reliance on the original set of roll film system patents to reliance on a multifaceted strategy that instigated the first stage of the horizontal integration of the photographic industry. This new and broader approach emphasized three strategies: (1) patents on products and production; (2) production efficiencies; and (3) continuous innovation. While Eastman realized that some of the film patents were not very strong, he maintained: "The patents which cover our present methods of manufacturing film should be kept up whether they are very strong or not, because of their moral effect on competitors. The film business has the greatest possibilities of profit of any branch of photography, and we must try to cover every avenue that leads to it. . . ."[35]
In the effort to "cover every avenue," the company acquired patents, some of which required the purchase of companies producing roll film cameras. In this way, Eastman Kodak entered into the first stages of horizontal integration. In addition, the shift from an empirical to a theoretical approach in production implemented by de Lancey not only provided efficiencies at Kodak Park that were reflected in the ever-important profit picture but also provided a capacity to produce at a cost well below that of any small potential entrant into the film industry.

With regard to the third strategy, which was directed to the apparatus sector of the industry, the role of patents came to play a diminished role, while continuous innovation became a more effective strategy, not only for the domestic market but for the European market as well. Eastman summarized the problem and the strategy:

I have come to think that the maintenance of a lead in the apparatus trade will depend greatly upon a rapid succession of changes and improvements, and with that aim in view, I propose to organize the Experimental Department in the Camera Works and raise it to a high degree of efficiency. If we can get out improved goods every year nobody will be able to follow us and compete with us. The only way to compete with us will be to get out original goods the same as we do.[36]

Eastman reminded Brownell constantly of the importance of newly designed cameras as a stimulus to business and as the key to main-

[34] GE to Davison, 27 September 1902 (GEC); and GE testimony, *U.S.* v. *EKCo.*, p. 239.
[35] GE to Eastman Photo Materials Co. Ltd., 23 April 1896 (GEC).
[36] Ibid.

taining the company's leading position in the industry throughout the world.

I enclose herewith copy of letter received from American Consul Blayney from Heidelberg. I send this to you because it bears upon the point that I have been trying to impress upon you lately; that is, that we have every reason to fear that the Germans will make very serious inroads upon our business if we conduct it as we have been conducting it the past year or two. It appears from this letter that they are not only copying us but that they are proposing to lead us in styles of our own cameras. When this thing occurs, your shop will be very much in need of work. The situation is wholly inexcusable, because we have and have had for years, better facilities than any of the Germans, and the only reason we have not kept a long way in advance of them is the lack of energy in your experimental department. If the Germans get out a 4 x 5 folding pocket camera before we get out the 3½ x 5½ which we have been working on so long, it would be simply a disgrace to us.[37]

In a more optimistic frame of mind a year and a half later, Eastman commented:

There is no question but what our cameras have always been better made than the German imitations. The only defect that they have had has been that they were for a time not kept up to date with improvements. Now . . . we are again at the head of the line and we ought to make quality our fighting argument.[38]

Thus, the mid-1890s represented a period of significant policy shift from dependence upon patents and litigation as the principal method of protection of an established market position to substantial dependence upon the institutionalization of continuous technological innovation, especially in the apparatus sector.

Another area of institutionalization of technical functions was the creation of the Experimental and Testing Department at Kodak Park. This laboratory, which was instituted with the opening of the Kodak Park plant in 1890, was initially not a research laboratory but was primarily a testing laboratory responsible for monitoring the quality of raw materials. The origins of the Experimental and Testing Department lay with the critical character and high chemical sensitivity of photographic chemistry. The normal industrial sources of raw chemicals could not insure the necessary chemical purity and, therefore, the Eastman company had to monitor nearly all of its supplies and also carefully test its own products in process and at the end of production.[39] Therefore, college-trained chemists were hired to carry out the routine analytical functions of the lab. In its earliest days Reichenbach and Passavant were associated with this laboratory, and in the mid-1890s Harriet Gallup worked there. After Lovejoy's arrival in Rochester, this laboratory became the center for much of the development work connected with Eastman Kodak's movement into the production of its own chemical raw materials. In

[37] GE to Brownell, 9 June 1902 (GEC).
[38] GE to Davison, 30 January 1904 (GEC).
[39] De Lancey testimony, *U.S.* v. *EKCo.*, pp. 408 and 416.

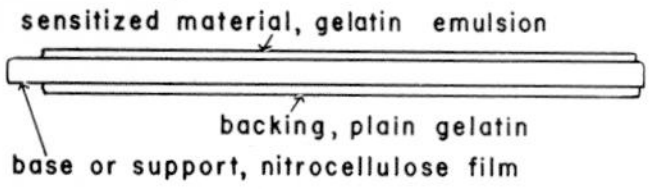

sensitized material, gelatin emulsion
backing, plain gelatin
base or support, nitrocellulose film

Fig. 8.4
Noncurling celluloid film.

addition, important developments in the photographic film were initiated in the laboratory. One important innovation was noncurling film. Ordinary uncoated celluloid film did not curl, but with the gelatin photosensitive coating on the film, the gelatin would swell or shrink as the atmospheric humidity varied. Consequently, the film tended to curl. Work on this problem, which became increasingly important as the company's products reached into the varied climates of the world, intensified during 1903. The problem was partially solved by placing a coating of gelatin on the back of the film, but because that coating did not contain the photosensitive emulsion, there was a difference in the hardness of the gelatin; the back coat was softer and more vulnerable to the developing solutions (fig. 8.4). This problem was finally solved by adding chrome alum to the gelatin backing.[40] Therefore, like the chemical laboratories of the German dye manufacturers,[41] this laboratory was initially an analytic lab, but as the search for new products and processes quickened, some research and development work emerged.

While George Eastman, during the mid-1890s, increasingly institutionalized the technical functions he had previously dominated, he recognized the vital strategic function of technology to the photographic business and, therefore, he continued to maintain a very close relationship with the leaders of his technical staff, to keep very well informed, and to initiate and participate actively in policy matters in the area of technology. Examples of his direct influence include his constant reminders to Brownell of the importance of product innovations in cameras, his participation in the change in film production method at Kodak Park, and his active interest in the development of color photography. Another example reflects interindustry diffusion of technology. Taking note in an engineering magazine of the description of the machines developed by Duane Church to produce watch mechanisms at Waltham Watch Company, Eastman thought such machines could be also used in the manufacture of shutter mechanisms. Within a short time he was in contact with Church, and Eastman Kodak became the first company to obtain a license under his patents.[42] In addition, Eastman regarded all secret formulae in the company to be in his safekeeping. No emulsion formula or other trade secret formula could pass from one entrusted employee to another, *regardless of who he was*, without the transmission first being authorized in writing by George Eastman. In this important way, he maintained a tight rein on the technological affairs of the company.[43]

The decade of the 1890s witnessed the growth of Eastman's respect for technically trained men and his desire for the company's name to be identified with science. While he unsuccessfully sought

[40] GE testimony, *EKCo.* v. *Blackmore*, pp. 1024–25; and GE to Davison, 13 June and 24 August 1903 (GEC).

[41] John Beer, *The Emergence of the German Dye Industry* (Urbana: University of Illinois Press, 1959), pp. 77–80.

[42] GE to Strong, 24 December 1901 (GECP); and GE to Duane R. Church, 28 February 1902 (GEC).

[43] GE to Senier, 5 October 1901; and GE to H. L. Quigley, 20 October 1905 (GEC).

Fig. 8.5 Seated, *Lord and Lady Kelvin;* standing, *Henry Lomb and George Eastman.* *Courtesy of Eastman Kodak Company.*

the services of Dr. Leo Baekeland,[44] at about the same time he successfully recruited one of the most prominent names in science of that day: William Thomson, or Lord Kelvin (fig. 8.5). While the prominent British physicist was at an advanced age and did not take an active part in any research in the company, early in the new century he lent his name and considerable prestige to the company by becoming vice-chairman of the board of Kodak, Ltd. Moreover, he made a much heralded tour of the Eastman Kodak facilities in America during the spring of 1902.[45]

Thus, during the mid-1890s Eastman personally withdrew from many of the technical functions of the company and institutionalized them. This process signaled a widespread shift of responsibility for technical affairs into the hands of professional engineers and scientists, and it betrayed Eastman's conscious effort to associate the company in the public eye with the prestige of science. What better way to develop a corporate image associated with technical quality than to gain the participation and endorsement of some of the world's leading scientists! This strategy was directed at both the business-financial community and at the multitude of amateur photographers who represented the newly created mass market. Eastman's early sensitivity to the role of science and technological change in business strategy persisted and deepened as the Eastman Kodak Company exploded in size and as Eastman's original functions became institutionalized; yet, he only very slowly and reluctantly loosened his grip on the reins of technological innovation and technical policy as the opportunities for industrial reorganization through horizontal integration presented themselves.

[44] GE to Davison, 26 October 1899 (GEC).

[45] GE to Lord Kelvin, 16 January 1902; and GE to D. O. Mills, 8 April 1902 (GEC).

9 / Horizontal Integration: Eastman Kodak

The period from 1895 to 1909 witnessed a major upheaval in the structure of the American photographic industry, an upheaval that reflected internal organizational changes within the constituent firms and changes in their operating strategies in response to the new mass markets opened by technological change: amateur roll film photography and cinematography. The key to the structural changes in the photographic industry was the horizontal integration (see chap. 8, n. 2) that occurred under the auspices of both Eastman Kodak and the Anthony and Scovill Companies, the latter largely responding to the strategy of the former. Therefore, to understand the forces stimulating the horizontal integration associated with Eastman Kodak is to understand, for the most part, the process throughout the industry. This process may be understood as having occurred in four stages.

Stage 1: Roll Film System

The first stage of horizontal integration in the American photographic industry evolved from the conscious patent policy pursued by the Eastman Kodak Company with regard to the roll film system. As an extension of his perception of the importance of control over technological innovation as a method of rationalizing the market place, George Eastman had advocated and successfully persuaded his associates to purchase or control patents with a direct bearing upon the company's product line. In continued pursuit of this patent acquisition policy, the company became interested in several patents that posed, in one form or another, a threat to the protected position of the company's roll film system, and in the process of acquiring these patents the company also acquired several small firms that owned or controlled them.

The motives behind these acquisitions were complex. First, the acquisition of patents would strengthen and temporarily extend the patent barrier around the system. Also, litigation could be avoided through purchase of patents and companies. Infringement suits, either by a patent holder against Eastman Kodak or by Eastman Kodak against an infringer of its patents, were expensive. Furthermore, the charges and countercharges of competing companies in the marketplace regarding contested patents could hurt the public reception of the new roll film system of photography. However, as George Eastman himself pointed out, the acquisition of competing enterprises could not be separated from the acquisition of patents. The owners of small companies often preferred to sell their companies rather than sell the valuable patents and be left with capital assets of little worth. Also, Eastman Kodak recognized that by purchasing a competing company, it was able to maintain the dominant position in the amateur market it had created. Three companies were important in the integration of roll film photography.

The first company, Boston Camera Manufacturing, was purchased by Eastman Kodak because of the combination of patents it owned that allowed it effectively to circumvent the Eastman roll film system. Since the Turner daylight-loading cartridge patent had just been issued and, therefore, still had nearly the full seventeen years to run, Eastman perceived that acquisition of that patent could prove highly significant in the continued long-term control of the roll film system.[1] While Eastman Kodak thus did acquire ownership of the Turner daylight-loading cartridge patent, it did not obtain control of the Houston patents on the front-roll system of roll film cameras.[2] The desire to acquire these important patents led to the acquisition of the American Camera Manufacturing Company.

The second company, American Camera Manufacturing, Thomas Blair established in 1896 in Northboro, Massachusetts, in order to provide a source of roll film cameras for his European company following Turner's sale of Boston Camera Manufacturing.[3] The company, which produced both cameras and film, acquired an exclusive license under Houston's front-roll system patents.[4] Also, it purchased raw film stock from the Celluloid Company and introduced the continuous rotary process of emulsion coating, a process patented by Blair and an associate.[5] When the company's cameras reached the market through Anthony, its exclusive trade agents, Eastman Kodak filed an infringement suit under its roll holder patents.[6] Blair, in turn, filed suit against the Rochester firm, charging infringement of the Houston front-roll patents.[7]

Early in 1896 William H. Walker in London recommended to George Eastman that he buy the company in order to acquire a competitor, but Eastman was not interested.[8] Yet, as the sales of the small company increased slowly during 1896 and 1897, Eastman recognized that the Houston system represented a major threat to his own roll film system of photography and that even if the courts upheld his claims and not those of Houston, a long and costly court battle would ensue. Fortuitously, Eastman and Blair returned from Europe in December of 1897 on the same steamer. During the voyage negotiations culminated in the agreements whereby the Eastman Kodak Company acquired control of the Houston patents as well as the other patents and business assets of the American Camera Manufacturing Company.[9]

The third company, Blair Camera, had sought unsuccessfully in the early 1890s to sell its assets to Eastman Kodak, but in the late 1890s, following the purchase of American Camera Manufacturing,

[1] EKCo. Board Minutes, 17 August and 30 October 1895, *U.S.* v. *EKCo.*, pp. 2607–11; and GE testimony, *EKCo.* v. *Blackmore*, pp. 1135–36.

[2] Houston's U.S. Patents #526,445 and #526,446; see memorandum of agreement between Turner and Houston, *U.S.* v. *EKCo.*, pp. 2026–29.

[3] Thomas H. Blair testimony, *U.S.* v. *EKCo.*, pp. 9–10, 17–19, and 24; and Blair to GE, 11 January 1898 (GEC).

[4] Memorandum of agreement between Houston and Blair, *U.S.* v. *EKCo.*, pp. 2029–41.

[5] Blair testimony, *U.S.* v. *EKCo.*, p. 10; and U.S. Patent #588,790 (24 August 1897).

[6] GE testimony, *U.S.* v. *EKCO.*, p. 230.

[7] Bills of complaint, *U.S.* v. *EKCo.*, pp. 3334–42; and GE testimony, in ibid., p. 223.

[8] EKCo. Board Minutes, Annual Meeting, January 1896, in ibid., pp. 2609–11; and GE to William H. Walker, 3 February 1896 (GECP).

[9] At a cost of about $50,000. GE testimony, *U.S.* v. *EKCo.*, p. 243; Government Exhibits 1 and 294a, in ibid., pp. 2001–15 and 3221–22; and "American Camera Mfg. Co." (GEN).

Eastman's interest quickened—because the Blair company was continuing to infringe on the Houston-Eastman-Walker roll holder patents; because the Eastman company had inherited, with the acquisition of American Camera Manufacturing, a suit against the Blair company over the Houston patents on the front-roll system; and because Blair Camera possessed patents on a new perforated-film system that represented a potential threat to the protected patent position of the Eastman Kodak Company.[10] Goff, too, was interested in a union of interests because he had invested a substantial amount of capital ($\sim$\$140,000) in the Blair company over the previous fifteen years, with no profits for over seven years and no immediate prospects of any dividends, and because the company had operation difficulties.[11]

Between the summer of 1898 and the spring of 1899 Goff succeeded in gaining control of the stock held by Blair, Anthony, and the more than thirty other shareholders and sold the company to Eastman Kodak.[12] Soon the Blair Camera production facilities and those of American Camera Manufacturing were moved to the Photo Materials building in Rochester, where both companies were operated for several years as separate companies, marketing under their own names the same products they had made under independent ownership.[13]

Stage 2: Photographic Papers

The paper sector of the photographic industry had undergone substantial change during the late 1880s and early 1890s, climaxing with the firms in the industry waging a vigorous price war during the mid-1890s. This competitive condition reflected the slackening of market demand during the early years of the depression. During that period, price cutting drove several of the smaller firms from the market, leaving by 1896 only a handful of producers, most of whom were moderate or large in size. Hence, as demand for photographic paper recovered during late 1895 and 1896 and began to expand rapidly as the amateur sector exploded during the late 1890s, some stability returned to the market place; but it was a tentative and uncertain stability that left prices fairly low and the leaders of the industry searching for some more permanent method of securing order and stability in this sector. Since Eastman Kodak was the only diversified firm in this sector, it is not surprising that the initial leadership for this effort came from the ranks of the nondiversified, single-product companies whose profit accounts were more directly affected by the price-cutting tactics.

[10] GE testimony, *U.S.* v. *EKCo.*, pp. 225–26; EKCo. Board Minutes, 7 December 1898, in ibid., p. 2621; GE testimony, *EKCo.* v. *Blackmore*, p. 21; and GE to Henry A. Strong, 13 March 1899 (GECP).

[11] GE to Strong, 8 April 1912 (GECP); and GE testimony, *U.S.* v. *EKCo.*, pp. 225–26.

[12] "Certificates of Condition," Blair Camera Co., 1897–98, Corporation Records, Commonwealth of Mass., Boston; and Goff-Eastman Agreement, 12 April 1899, Government Exhibit 98, *U.S.* v. *EKCo.*, pp. 2179–86.

[13] GE to George Davison, 17 April 1899 (GEC).

Understanding the origin and control of technological change provides a major clue to the market and industrial structure in this sector of the photographic industry. Whereas a single firm dominated the photosensitive film sector through control of a broad spectrum of key patents, in the photographic paper sector there were few significant patents, and no single firm controlled a crucial set of them. Furthermore, in contrast to the photographic film or film camera sectors, the most significant product innovations originated outside the industry and were introduced by new entrants into the field.

The failure of the Eastman Kodak paper patents in 1896 signaled a rethinking of appropriate alternative strategies to those of patent control and constant product innovation. From 1894 to 1910 the industry leaders developed a series of strategies some of which were soon discarded and some of which were retained and refined but all of which aimed to rationalize the marketplace in order to avoid price competition and to defend the dominant position of the leading firms in the industry. The sequence of three basic strategies marked a movement away from reliance on technological innovation to reliance on marketing and supply innovations, in imitation of strategies pursued by other large-scale enterprises at that time. As will be seen, even these strategies were not fully successful and required numerous modifications and adjustments.

Strategy 1

The first major strategy organized by the photographic paper sector of the industry after the general breakdown of the Eastman Kodak patent strategy (1894) but before the final court decisions on the Kodak patents (1896) consisted of a pooling arrangement operated through the Merchant Board of Trade. The board enlisted the cooperation of the major jobbing houses in New York City and the major producers of photographic papers and dry plates to enforce a standard price level. Like most such pools in the United States, this one failed, largely because the board failed to gain the support of the Stanley Dry Plate Company and the Photo Materials Company, major producers in the dry plate and photographic paper sectors, respectively. The leaders of the paper industry responded to the failure by marketing cheaper brands of paper and by offering "special seconds."

Following a vigorous price war in 1895 and 1896 among the leaders of the photographic paper sector, the aristotype papers emerged as the most popular, and the American Aristotype Company became the leader of the photographic paper sector of the industry. The only other notable producers of Aristotype papers, the New Jersey Aristotype Company and the United States Aristotype Company, both of Bloomfield, New Jersey, produced only a small quantity of paper. The Eastman Kodak Company dominated the production of gelatin printing-out paper (POP) with its Solio paper, though the Photo Materials Company of Rochester posed a major competitor with its Kloro paper. Other firms that produced only small quantities

Collodion POP	Gelatin POP	DOP
1. American Aristotype (Jamestown, N.Y.)	1. Eastman Kodak (Rochester, N.Y.)	1. Eastman Kodak (Rochester, N.Y.)
2. New Jersey Aristotype (Bloomfield, N.J.)	2. Photo Materials (Rochester, N.Y.)	2. Nepera Chemical (Yonkers, N.Y.)
3. United States Aristotype (Bloomfield, N.J.)	3. New Jersey Aristotype (Bloomfield, N.J.)	
	4. Kirkland Lithium Paper (Denver, Colo.)	
	5. Palmer & Croughton (Rochester, N.Y.)	

Fig. 9.1
Photographic paper companies by product sector.

Table 9.1 U.S. Photographic Paper Sales, 1895–99

Company	*1895*	*1896*	*1897*	*1898*	*1899*
American Aristotype				$647,119	$807,528
Eastman Kodak					
Solio	$ 94,465	$178,344	$241,089	316,124	492,480
Other Papers	204,270	94,541	23,225	14,949	26,033
Total	298,735	272,885	264,314	331,073	518,513
Nepera Chemical					
(Jan. 1 to Dec. 31)				236,942	
(Jan. 1 to Nov. 1)				159,706	311,004

SOURCE: *U.S.* v. *EKCo.*, pp. 2438, 2565–69, and 2652.

Fig. 9.2
Leo Hendrik Baekeland. Courtesy of Carl Kaufmann, Wilmington, Delaware.

of gelatin POP included New Jersey Aristotype Company, Kirkland Lithium Paper Company of Denver, and Palmer & Croughton of Rochester, New York. In the rapidly declining bromide developing-out paper (DOP) sector, the Eastman Kodak Company and the Nepera Chemical Company of Yonkers, New York, were the only producers of any note.

Just as prices began to stabilize in 1896 and 1897, Nepera Chemical introduced a unique photographic paper, Velox, which began to make inroads on the market positions held by American Aristotype and Eastman Kodak (see table 9.1). Leo Baekeland, with his unusually rich technical and academic background, investigated and tested in Nepera's laboratory facilities a large number of chloride and bromide emulsions for photographic papers (fig. 9.2). He tested many photographic qualities, varying chemical and physical conditions such as composition, temperature, and period of ripening for the emulsions.[14]

Baekeland worked on one paper in particular, Velox. In contrast to the goal of most paper manufacturers throughout the world,

[14] Excerpts from Baekeland's experimental notebook and "Notes on Formulae Used by Nepera Chemical Company," in Carl B. Kaufmann, *Grand Duke, Wizard, and Bohemian: A Biographical Profile of Leo Hendrik Baekeland* (privately printed, 1968), pp. 39–42 and appendix 1; and Leo H. Baekeland, "Photochemical Industry of the United States," *International Congress of Applied Chemistry, Fifth,* 4 vol. (Berlin: Deutscher Verlag, 1904), 4: 373.

Fig. 9.3
Nepera Chemical Company trademark.

Baekeland did not seek a "faster" paper but a DOP so slow that it could be handled and developed in subdued artificial light ("gaslight") but could also be exposed effectively when brought in close proximity to the same light. Working from a gelatin chloride formula which Eder and Pizzighelli had originally introduced in the early 1880s and which he altered himself shortly thereafter, Baekeland made two important changes that made his "gaslight" paper a significantly new paper. By not washing the emulsion, which ordinarily removed the excess silver nitrate, and by not heating the emulsion for a long period of time ("ripening"), he produced a paper with the remarkable Velox properties.[15]

When Velox paper was first introduced by Nepera Chemical, it was not given a very enthusiastic reception by the typically conservative professional photographers, but by the end of the century, sales soared, especially among amateurs, and the Nepera Chemical Company became an important force in the photographic paper sector of the industry (see table 9.1). By early 1898, Nepera Chemical's new product had begun to affect the leaders of the industry. In response, Eastman Kodak introduced in 1898 Dekko, a paper to compete with Velox, and at about the same time the Photo Materials Company introduced its Azo paper.[16]

Photo Materials had had variable success since its founding in 1892. The introduction of Kloro paper as a competitor of Solio paper had brought considerable pressure on Eastman Kodak during the mid-1890s. The Photo Materials Company remained outside the paper and plate pool organized at that time and increased its market share by cutting prices. This strategy paid off in the short run but was devastating in the long run. The only year in which the company made a profit was 1895. Eastman Kodak and other companies had met the lower prices with even lower-priced, cheaper brands, so that soon the profit margin evaporated for the small Rochester firm. Furthermore, while the company had expended considerable capital for experimental and research work, its leadership was fundamentally unfamiliar with the technical side of the business. The lack of coordinated leadership eventually led to Reichenbach's departure of 1897.[17]

By early 1898 most of the stockholders in the company were irritated by the deficit position of the firm, and they saw little prospect of improvement. Following the urgent pleas of major stockholders in Photo Materials, Eastman Kodak agreed to purchase the firm in July 1898.[18] Eastman Kodak made the purchase in order to

[15] Baekeland testimony, *U.S.* v. *EKCo.*, pp. 38–39; Kaufmann, *Grand Duke*, pp. 55–62 and appendix 1; and Erich Stenger, *History of Photography: Its Relation to Civilization and Practice*, trans. Edward Epstean (Easton, Pa.: Mack Printing Co., 1939), p. 50.

[16] Government Exhibit 210, *U.S.* v. *EKCo.*, p. 2565; and testimony of Leo Baekeland, Arthur B. Enos, Rudolph Speth, and John S. Cummings, in ibid., pp. 38, 451, 278, and 115.

[17] Max Brickner testimony, in ibid., pp. 435–38.

[18] Abram J. Katz testimony, in ibid., pp. 432–34; and EKCo. Board Minutes, Special Meeting, 19 July 1898, and Monthly Meeting, 10 August 1898, in ibid., p. 2616.

Fig. 9.4 Photo Materials Company building; later the headquarters of the Blair Camera Company Division of Eastman Kodak Company. Courtesy of Eastman Kodak Company.

gain the building and the production machinery at a bargain (fig. 9.4). The building provided space for the continued production of the Photo Materials paper products and production facilities for the newly acquired American Camera and Blair Camera companies.

Despite the purchase of the Photo Materials Company and the introduction of Dekko by Eastman Kodak, Eastman Kodak and American Aristotype found themselves threatened in the market place by the new Velox paper. Furthermore, although by informal agreement the price war had ceased, prices had not returned to prewar levels. Moreover, with the decisions then being made in the courts regarding the antitrust laws, the two companies could not make any formal agreements. The combined threat posed by the Nepera Chemical Company and the status of the antitrust laws stimulated a search by the leadership of the industry for a new strategy to insure their position without further erosion of profits.[19]

Strategy 2

During the first half of 1898, there occurred in Europe a business development that suggested to the leadership of the paper sector of the American photographic industry an opportunity for an entirely new strategy, one built largely on the example of the recent activities of the photographic paper industry in Germany. For nearly forty years the producers of photographic paper throughout the world had depended upon two paper mills for their stock of raw paper. This limited source was due, of course, to the special requirements of paper suitable for photosensitive coating, i.e., the paper had to be mineral free, the rags and pulp free of foreign substances, the plant very clean, and the sizing compatible with the photo-

[19] GE to Directors, Kodak Ltd., 30 March 1899 (GEC).

graphic toning, developing, and fixing baths. The two paper mills were located at sources of virtually mineral-free water, and they were careful to insure that the process of manufacturing the paper did not introduce contamination. During the last two decades of the nineteenth century, as the demand and the popularity of photography grew, these two companies became more aware of the unique position they held in the industry.

Early in 1898 the management of the two firms agreed to consolidate their interests in the production of raw paper for photographic papers, forming a new company, the General Paper Company of Brussels, Belgium. This company, in cooperation with the Dresden photographic paper combine, dictated the price and annual supply of raw paper to the baryta and emulsion coaters of Europe.[20] In this way, they restricted the entry of new photographic paper producers into the industry, sharply increased the price of raw paper (from two and one-half francs to six francs per kilo), and maintained the retail price of photographic paper. Curiously, French, English, and American producers obtained the raw paper from General Paper for a substantially lower price than that paid by German producers. While the latter still had substantial profits from their sales to the German domestic market, their export capacity was hindered.[21]

By the late spring of 1898, as the effect of the price increase became known to American photographic paper producers,[22] Charles S. Abbott of American Aristotype and George Eastman of Eastman Kodak came to see in the union an opportunity for potential control of the American photographic paper industry (fig. 9.5). They envisioned a strategy for rationalizing the American market through control of the crucial foreign supply channels while not running aground on the American antitrust laws.[23]

Following several overtures to the General Paper Company, Abbott and Eastman set sail for London for negotiations in the fall of 1898. The General Paper Company and the cooperating baryta coating companies had a number of customers in the United States with whom they had personal connections and contracts through their representatives in New York. These agents and even the officials of the General Paper Company recognized how difficult the position of these smaller companies in the United States would be once they had agreed to sell exclusively to Eastman and Abbott.[24] Likewise, Abbott and Eastman recognized the sensitivity of the issue of "cut-

[20] Most photographic raw stock was coated with a barium mixture consisting principally of barium chloride. This *baryta* coating (1) prevented the emulsion from sinking into the fibers of the paper; (2) protected the emulsion against impurities in the basestock; and (3) reflected light, thereby increasing the potential density of the print and improving the definition of the image.

[21] Willy Kühn, *Die photographische Industrie Deutschlands wirtschaftswissenschaftlich gesehen in ihrer Entwicklung und ihrem Aufbau* (Schweidnitz, 1928), pp. 63–64; Fritz Wentzel, *Memoirs of a Photochemist* (Philadelphia: American Museum of Photography, 1960), pp. 52 and 57; and F. Hansen, *Industrie photographischer Bedarfsartikel* (Berlin, 1901), p. 12.

[22] *Photo-Miniature* 1 (October 1899): 84; and EKCo. Circular Letter, 9 December 1898, *U.S.* v. *EKCo.*, p. 2055.

[23] GE testimony, *EKCo.* v. *Blackmore*, pp. 1038–39.

[24] Baekeland testimony, *U.S.* v. *EKCo.*, pp. 40–43.

Fig. 9.5
George Eastman and Charles S. Abbott. Courtesy of Eastman Kodak Company.

ting-off" former customers of General Paper Company. Eastman approached the General Paper Company with the idea that efforts would be made to purchase or merge the American paper companies, thereby alleviating the potential hardship these companies might otherwise face.[25] Following a long series of negotiating sessions a contract was signed by the three parties which provided that Eastman Kodak and American Aristotype would pay a set price for the paper they purchased; would use exclusively the output of the General Paper Company; and would not experiment in developing other raw stock suitable for photographic purposes. In return, the General Paper Company would grant the two American companies the exclusive agency for their output of raw paper for North America and would provide a discount on all paper used (up to a limit of 25 percent of total consumption) for production of so-called fighting brands, low-priced papers used to undersell competition in the American market. In the final agreements, restrictions were included that insured a supply of raw paper to the New Jersey Aristotype Company.[26]

Also in late December, following the public announcement of the contract in the United States, Leo Baekeland sailed to Brussels, where he met with the manager of General Paper Company and sought to insure a continued supply of paper from the traditional sources. This brief negotiation ended abruptly when Baekeland would not consider an ultimate ceiling on expansion of the Nepera Chemical output of 10 percent or 15 percent above the current level.[27] Nonetheless, the agreements were made with the tacit understanding that the American Aristotype and Eastman Kodak interests would try to acquire the other American firms, including the Nepera Chemical Company.[28]

[25] L. Goffard to GE, 9 June 1906, reprinted in ibid., pp. 251–52.

[26] Agreement, 20 October 1898, Government Exhibit 14, *U.S.* v. *EKCo.*, pp. 2046–51.

[27] Baekeland testimony, in ibid., pp. 40–42.

[28] Goffard to GE, 9 June 1906, in ibid., pp. 251–52.

Eastman and Abbott returned to the United States, where they and their associates drafted a new sales plan for the American market. In December they had announced to the trade the agreement with the German suppliers, and they solicited the opinion of dealers regarding the desirability of introducing a "fair trade" policy. At the same time they announced a general increase in the wholesale price of photographic papers, reflecting for the first time the price increase imposed on American producers by the General Paper Company in May 1898.[29] Late in December they announced the results of the poll of dealers and, accordingly, instituted a new sales policy, one that reflected a commitment to a general policy of price maintenance by the leadership of the photographic paper manufacturers and reflected the principal components of the sales policy introduced by the Merchant Board of Trade in 1894: dealers in photographic paper were to receive a direct discount of 15 percent, instead of the previous 25 percent, and an additional 12 percent upon compliance with the maintenance of list prices. In addition, it was announced that the European raw stock was to be sold only to coaters who agreed to operate according to these terms of sale. Companies producing gelatin papers that continued to obtain the European stock through Eastman Kodak included the New Jersey Aristotype Company, the Photo Materials Company, and Palmer & Croughton. The last two were owned and operated by Eastman Kodak.[30]

Some of the new, small companies were immediately and directly affected by the new control of the supply of raw paper stock from Europe. For example, the American Self-Toning Paper Company of Newark, New Jersey, organized in 1896, introduced a self-toning paper that eliminated the toning bath following fixing. Eastman made several offers to purchase the firm, but they were refused. Shortly thereafter, the company began to find it difficult to procure the vital raw paper stock. Furthermore, the rebate system instituted by Eastman Kodak also substantially cut orders from dealers. In March of 1900 American Self-Toning Paper went into receivership.[31]

At about the same time that the new sales policy was announced, problems began to arise with the enforcement by the General Paper Company of the restrictions on the shipment of raw paper to the United States. American companies scrambled for new sources of supply, and soon rumors of new or alternate sources of raw paper both in Europe and in the United States were rife. Such rumors were not without foundation, because numerous small paper mills in Europe, induced by the high price being charged by the General Paper Company, began to turn their attention to production of raw stock for photographic papers.[32] Such rumors only served to stir the

[29] EKCo. Circular Letter, 9 December 1898, in ibid., p. 2055.

[30] Ibid., 31 December 1898, pp. 2055–56.

[31] George Nelson Waite testimony, *U.S.* v. *EKCo.*, pp. 49–52; James Blackmore testimony, in ibid., pp. 125–26; and *EKCo.* v. *Blackmore*, pp. 855–64.

[32] William B. Dailey testimony and Baekeland testimony, in ibid., pp. 58–61 and 39–41; and Kühn, *Die photographische Industrie Deutschlands*, pp. 63–64.

concern of Abbott and Eastman, who promptly relayed them to the General Paper Company, where investigations were undertaken.[33]

The merger of the raw paper interests into the General Paper Company, then, opened an opportunity for the leadership of the American photographic industry to imitate the strategy of the European photographic paper industry, i.e., to regulate prices and access to the marketplace by controlling the raw paper supply. However, because of the close personal and business relations of the European paper suppliers with many of the leaders of the American coating companies, Abbott and Eastman secured control of the raw paper supply only with the understanding that the companies that were left out of the agreement would be purchased when they could be obtained at a reasonable price. Hence, the strategy of controlling the raw paper supply led directly to the third strategy, the consolidation of the American photographic paper companies through the creation of a holding company.

Strategy 3

While the Eastman Kodak Company had previously acquired two photographic paper companies, the Western Collodion Company in 1894, and the Photo Materials Company in 1898, the acquisition of these companies had been in response either to a desire to acquire a new technology or to an opportunity to acquire plant facilities at a bargain price. Later in 1898, however, Eastman Kodak acquired the first paper company, Palmer & Croughton, in response to the new understanding with the General Paper Company.[34] The small company traced its ancestry to the Hovey albumen paper company in Rochester.[35] In 1898 the firm operated on a very small scale, producing primarily gelatin POP, but it was one of the firms General Paper was concerned about.[36] An agreement for the sale was reached late in 1898, and soon the Palmer & Croughton production facilities were moved to the Photo Materials Building.[37]

Of much greater importance was the consideration of the amalgamation of the major photographic paper manufacturers. Abbott of American Aristotype initially pressed the issue. He was concerned about the continuing leaks in the European paper supply, the rising sales of Nepera Chemical's Velox paper, and the increasingly vulnerable position of a single-product firm, such as his own, in the industry.[38] Eastman was initially indifferent to the idea of consolidation. He had just completed the reorganization of the Eastman companies into an English holding company, Kodak Limited, and he

[33] GE to Herman Lipps, 2 March 1899; and Goffard to GE, 8 April 1899, in *U.S. v. EKCo.*, pp. 3278 and 3282–84.

[34] GE to Directors, Kodak Ltd., 21 April 1899 (GEC).

[35] Douglas Hovey (albumen paper, 1866); American Albumenizing Company (Hovey & Brown, 1880); Peerless Albumen Paper Works (Brown, 1881); A. M. Brown (albumen paper, 1887); Brown & Palmer (May 1892); and Palmer & Croughton (August 1897).

[36] GE testimony, *U.S.* v. *EKCo.*, p. 246.

[37] "Palmer & Croughton" (GEN).

[38] GE to Strong, 13 March 1899 (GECP).

Table 9.2 Profits and Acquisition Price of Companies Combined in General Aristo Company, 1898

Company	*1898 Profits*	*General Aristo Acquisition Price*
Eastman Kodak Company paper divisions, Rochester, N.Y.	$152,764	$1,727,000
(Photo Materials Co. and Palmer & Croughton)	19,860	
American Aristotype Company Jamestown, N.Y.	160,084	750,000
Nepera Chemical Company Yonkers, N.Y.	130,000[a]	750,000
New Jersey Aristotype Company Bloomfield, N.J.	27,134	69,448
Kirkland Lithium Paper Company Denver, Colo.		12,500
Total	489,842	3,421,448

SOURCE: *U.S.* v. *EKCo.*, pp. 2058–69; 2373a–j; 2648–49.

[a] This figure is an estimate based on the following considerations. In general in the photographic paper industry at this time, sales during the first half of the year were substantially below those during the second half (reflecting Christmas sales). Hence, despite the growth trend in sales with Nepera Chemical, the profits during the first half of 1899 were undoubtedly less than those of the second half of 1898. Consequently one might estimate total 1898 profits at somewhat less than twice the profits of the first half of 1899.

complained to Strong: "I do not feel like undertaking any great additional burdens; unless a very good thing can be figured out, I shall not encourage it."[39] Moreover, the diversified Eastman Kodak, enjoying immense success with its amateur roll film system, was under no particular pressure to act quickly.

During the spring of 1899, however, Eastman warmed to the idea. He became increasingly concerned about Abbott's worries and feared that Abbott, as a means of self-protection, might amalgamate with other paper or dry plate manufacturers that were antagonistic to Eastman Kodak interests.[40] Also, he liked Abbott's relatively low valuation of American Aristotype for the purposes of merger (table 9.2). Furthermore, Eastman believed that eventually some means of price maintenance would have to be arranged in order to insure profits for the manufacturers and quality for the consumer;[41] yet, recent court decisions prohibited price agreements among independent producers, while permitting combinations under the holding company form of organization.[42] After receiving advice from his New York attorneys, who indicated in their opinion that an amalgamation of paper producers would be legal, Eastman decided to respond to Abbott's urgings.[43]

In developing a plan of consolidation, Eastman decided against

[39] Ibid.

[40] GE to Directors, Kodak Ltd., 30 March 1899 (GECP).

[41] GE to Strong, 29 March 1899 (GECP).

[42] E. C. Knight (1895); Trans-Missouri Freight Association (1897); and Addyston Pipe (1899).

[43] GE to Strong, 13 March 1899 (GECP).

Fig. 9.6 American Aristotype Company offices, Jamestown, N.Y. Courtesy of Eastman Kodak Company.

having either the New York or English Eastman companies acquire the other companies, because of internal financial and policy considerations. Instead, he favored the formation of an outside company, ultimately controlled by Eastman Kodak stockholders and officers, which would acquire the major paper producers, including the paper division of Eastman Kodak.[44] After careful preparation, Eastman presented his plan to the American and English Board of Directors and received their approval, with the provision that the Eastman company have the option to purchase the new company within three years.[45] Serving as both promoter and underwriter, Eastman acquired, during the summer of 1899, options on American Aristotype, Nepera Chemical, New Jersey Aristo, Kirkland Lithium, and the Paper Division of Eastman Kodak; in August he founded the new holding company, General Aristo (see table 9.2).[46] In so doing, Eastman conducted a major refinancing and reorganization of the industry without recourse to outside banking and investment interests. Of course, Eastman, and his close associates who assumed the risk and underwrote the stock issue, profited handsomely. George Eastman clearly demonstrated his successful transformation from a primarily technological innovator into an organizational and financial innovator.

The General Aristo Company, unlike many of the combinations created at this time in other industries, was essentially founded and operated by one dominating firm (see table 9.2). While Charles Abbott of American Aristotype had initially promoted the combine, George Eastman and Eastman Kodak controlled nearly two-thirds of its capital stock. Abbott was the only member of the Board of Directors not associated with Eastman Kodak. Similarly, Eastman Kodak personnel dominated the management and supervision of the General Aristo Company. Again, Charles Abbott of American Aristotype, who, through his previous relations with George Eastman and Eastman Kodak, had quickly gained Eastman's confidence and trust and soon joined Eastman's inner circle of intimate friends and colleagues, was the only non-Eastman Kodak manager to play any major role in the company.[47] Reflecting the influence of American Aristotype and Eastman Kodak, production facilities of the constituent firms were gradually concentrated in Rochester and Jamestown, New York (figs. 9.6 and 9.7).

However, despite the dominant position that General Aristo held in the photographic paper market, foreign and domestic raw paper

[44] GE to Directors, Kodak Ltd., 30 March 1899 (GECP).

[45] EKCo. Board Minutes, 18 May and 3 June 1899, *U.S.* v. *EKCo.*, pp. 2622–24.

[46] Option Agreement: GE, Sheldon, Abbott, Sheldon, 8 June 1899, in ibid., pp. 2373a–d; Option Agreement: GE, Jacobi, Baekeland, and Hahn, 8 July 1899, in ibid., pp. 2061–69; Option Agreement: GE, DeJonge, and Franc, 22 June 1899, in ibid., pp. 2373e–j; Option Agreement: GE, Kirkland, Colwell, Kirkland, Colwell, July 1899, in ibid., pp. 2058–61; and "General Aristo Company," New York Corporations Abstract of State Records, [File No. 8/0–48.] Bureau of Corporations, Dept. of Commerce and Labor, National Archives, Washington, D.C.

[47] GE to Strong, 4 March 1905; and GE to H. M. Fell, 13 March 1905 (GECP).

Fig. 9.7
American Aristotype Company factory no. 1, Jamestown, N.Y. From Vernelle A. Hatch, ed., Illustrated History of Jamestown *(Jamestown, N.Y.: C. E. Burk, 1900).*

continued to flow to a handful of small independent companies. General Aristo, therefore, became increasingly concerned that these new sources of raw paper would circumvent the advantageous position created through the control of the General Paper Company supply to North America. Upon learning that the Nepera Chemical Company had been using a substantial amount of American raw paper,[48] both General Paper and General Aristo became concerned lest the new American supply undermine the position of both companies. Therefore, early in 1900 the General Paper Company successfully negotiated an exclusive contract with the American Photographic Paper Company of Boston, the company which had been supplying Nepera Chemical.[49] A second threat to the position of General Paper and General Aristo was a European baryta coating firm that threatened to break the American paper barrier. Eastman went to Europe in the spring of 1901 and successfully negotiated the purchase of the Benecke firm on behalf of General Paper and General Aristo.[50]

General Aristo experienced a substantial increase in sales and profits. With the additional stimulus of a rapidly growing amateur sector of the industry, sales of photographic paper increased more than 18 percent from 1898 to 1899.[51] During the next three years total sales for General Aristo increased substantially. It was estimated that in 1901 General Aristo produced 95 percent of all photographic paper sold in the United States.[52] In addition, dividends on common stock approached 20 percent.[53] In view of the company's success, Eastman Kodak exercised in 1902 its option to purchase General Aristo.

[48] GE testimony, *U.S.* v. *EKCo.*, p. 255.

[49] Agreement between American Photographic Paper Co. and General Paper Co., 10 April 1900, in ibid., pp. 3015–25.

[50] General Aristo Co. Board Minutes, 8 May 1901, in ibid., p. 2657.

[51] After taking account of the general price increases instituted early in 1899 ("President's Address," Annual Meeting of General Aristo Co., 22 January 1900, in ibid., pp. 2653–54).

[52] 226 *Fed. Rep.* 74 (1915).

[53] "Report of Vice President," Annual Meeting of General Aristo Co., 21 January 1901, *U.S.* v. *EKCo.*, p. 2656.

From the time of its merger with General Aristo until the end of the decade the position of Eastman Kodak in the photographic paper market declined—although the American firm continued to dominate the North American market. Two general thrusts characterized the company's policy during this period. One was a major move into the European paper market, and the other was a set of responses to the rise of new competition in the domestic market. Eastman learned that the leading European collodion paper manufacturer, Vereinigte Fabriken Photographischer Papiere, of Dresden, had its weakest markets in those European areas that Eastman Kodak was most desirous of entering with its American aristo collodion paper: Britain, France, Portugal, and Spain. Because of the contracts American Aristotype had signed in the early 1890s with Carl Christensen, this paper could not be marketed in Europe.[54] The Christensen contracts were now owned by the Vereinigte, which had acquired most of the major collodion paper companies of Europe.[55] Early in 1903 Vereinigte and Eastman Kodak, the two largest producers of photographic paper in the world and the two exclusive customers of General Paper Company, reached an agreement, according to which Vereinigte conceded its western European collodion paper markets to Eastman Kodak in exchange for a sizable amount of the American company's stock. Although the European thrust was executed during the next decade with considerable energy, collodion paper sales fell precipitously, following an international pattern. Consequently, Eastman Kodak did not realize its expectations in the deal.[56]

The second thrust of Eastman Kodak policy during the decade from 1900 to 1909 was in the form of defensive responses to the development of limited but important competition in the domestic market. Originally, Eastman Kodak had envisioned that the control of the raw paper supply from General Paper would certainly diminish the prospect of domestic competition, but the limitation of the supply of raw stock simply prompted American and German paper companies to develop new kinds of raw stock for photographic papers. Soon American coating companies obtained very satisfactory paper from Europe. The ability to obtain the raw stock did not, however, automatically insure success to new entrants in the American market, especially in light of the Eastman Kodak terms of sale, which were in effect during this period. Yet, product innovation did prove to be an effective strategy of entry into this market. During the first decade of the twentieth century, Eastman Kodak pursued the course of follower rather than that of innovator in photographic

[54] Option Agreement, 8 June 1899, *U.S.* v. *EKCo.*, p. 2373a; GE testimony, in ibid., p. 276; and "American Aristotype Co." and "Vereinigte Fabriken Photo Papiere" (GEN).

[55] "Vereinigte Fabriken Photo Papiere" (GEN); Wentzel, *Memoirs of a Photochemist*, p. 69; and Kühn, *Die photographische Industrie Deutschlands*, pp. 114–15.

[56] Kühn, *Die photographische Industrie Deutschlands*, pp. 64–65; *Photog. Korresp.*, 37 (1900), 105; "Vereinigte Fabriken Photo Papiere" (GEN); GE testimony, *U.S.* v. *EKCo.*, p. 276; and EKCo. of N.J. Board of Directors Meetings, 16 July 1903 and 28 May 1904, in ibid., pp. 2643–46.

Fig. 9.8
Cyko trademark.

papers. This course, in sharp contrast to the Eastman Kodak innovative strategy in the production of cameras and apparatus, perhaps reflected the differences in investment and in cost of development and introduction of new photographic materials as opposed to new pieces of apparatus.

Two areas of vulnerability in the Eastman Kodak product line, platinum paper and new types of DOP, presented opportunities for new entrants into the industry (see table 9.3). Platinum paper, one of the earliest prepared papers marketed in the United States, was introduced by the Platinotype Company of London through Willis & Clements of Philadelphia in the late 1870s.[57] The Philadelphia manufacturer of this expensive paper, which appealed to high-class studio and art photographers, dominated the American market until early in the twentieth century. Since General Aristo had no platinum paper manufacturing capacity, Eastman Kodak sought to develop a popular platinum paper; however, when developmental efforts had failed by 1906, Eastman Kodak initiated negotiations with Willis & Clements, seeking to purchase the firm.[58] When those efforts failed Joseph Di Nunzio of Boston was approached. A former employee of American Aristotype, Di Nunzio had developed and successfully marketed an excellent platinum paper, Angelo. Therefore, in order to add platinum paper to its product line, Eastman Kodak purchased the Di Nunzio firm in 1906.[59]

Of greater significance, the introduction of several new DOPs in the early years of the century produced substantial erosion of the Eastman Kodak market share and, consequently, significantly changed the structure of the photographic paper market from 1900 to 1909 (tables 9.3 and 9.4). In 1900 General Aristo held a 95 percent market share, while Eastman Kodak held only a 69 percent market share by 1908.[60] New kinds of photographic paper, most notably gelatin DOP introduced by Defender, Ansco, and Artura, began to achieve success.

Most of the new gelatin DOPs were introduced by new paper companies that were founded in the late 1890s or the early twentieth century. Defender Photo Supply Company of Rochester, founded in 1899 by former employees from Photo Materials and New Jersey Aristotype, introduced Argo paper, which captured a small part of the professional market.[61] By 1908 the company had secured about 10 percent of the total photographic paper market and held the position of America's second largest photographic paper manufacturer. Ansco, the result of a merger of the two established firms of Anthony and Scovill, successfully introduced a new DOP, Cyko, which late in the first decade of the century attracted a growing

[57] Alfred Clements testimony, in ibid., p. 314.

[58] GE to Strong, 14 May 1906 (GECP).

[59] "Joseph Di Nunzio" (GEN); Joseph Di Nunzio Balance Sheet, 6 July 1906, *U.S.* v. *EKCo.*, pp. 2466–67; and GE to S. V. Haus, 29 June 1906 (GEC).

[60] 226 *Fed. Rep.* 74 (1915).

[61] John S. Cummings testimony, *U.S.* v. *EKCo.*, p. 434; Frank W. Wilmot testimony, in ibid., p. 191; "Defender Joins Du Pont," *Du Pont Magazine,* August 1945, p. 19; and "Fifty Years of Progress," *Defender Spotlight,* May 1945, p. 2.

Table 9.3 U.S. Photographic Paper Sales, 1902–11

Company	*1902*	*1903*	*1904*	*1905*	*1906*	*1907*	*1908*	*1909*	*1910*	*1911*
Artura	$ 8,644	$ 14,327	$ 24,573	$ 46,695	$ 98,890	$ 228,980	$ 353,516	$ 329,272[a]	$ 662,580[b]	
Eastman Kodak:										
DOP	437,612	493,506	633,587	835,927	1,135,964	1,528,137	1,763,170	1,958,966	2,066,783	
Artura								[a]	662,580	
POP	523,924	457,951	395,645	354,753	333,045	300,933	268,366	243,820	223,968	
Bromide	94,494	103,719	112,871	131,800	159,750	180,807	166,694	171,763	174,145	
Angelo					35,639	105,198	91,402	89,419	111,533	
American Aristotype	1,201,196	1,251,007	1,189,264	1,229,107	1,199,333	1,010,751	717,347	491,553	329,279	
Others	43,600	46,487	44,908	40,092	64,363	59,980	51,079	41,801	30,990	
EKCo. Total	2,300,826	2,352,670	2,376,275	2,591,679	2,928,094	3,185,806	3,058,058	2,997,322	3,599,278	
Defender[c]							422,083	426,910	409,472	
Haloid										$ 85,699
Ansco							200,000		585,478	773,210[d]
Kilborn										88,378
Willis & Clements					273,715	241,982	205,643	166,822	141,568	112,847
U.S. Aristotype										13,210
Photo Products										61,731

SOURCES: *U.S.* v. *EKCo.*, pp. 314–5, 346, 358, 385A, 403, 420, 2073, 3466–70 and 3472; and Ansco Company, *Journal*, p. 483, located at Smithsonian Institution, Washington, D.C.

[a] Does not include three months of EKCo. ownership.

[b] Owned by EKCo.

[c] These figures are 91 percent of total sales; for justification of this assumption, see *U.S.* v. *EKCo.*, p. 420.

[d] Estimated from 1910 data.

Table 9.4 Sales and Market Share of Photographic Paper Manufacturers, 1908

Company	*U.S. Paper Sales (in millions of $)*	*% of Market*	*U.S. DOP Sales (in millions of $)*	*% of Market*
Eastman Kodak	3.06	69	1.76	60
Defender	.42	10	.46	16
Artura	.35	8	.35	12
Willis & Clements	.21	5	—	—
Ansco	.20	5	.20	7
Haloid / Kilborn / Photo Products	.16[a]	3	.16[a]	5
Total U.S. Paper Sales (approx.)	4.40	100	2.93	100

SOURCE: Derived from data contained in table 9.3.
[a] Estimate based on half of 1911 sales.

Table 9.5 Comparison of Artura and American Aristoype Paper Sales, 1904–9

	Artura		*American Aristotype*	
Year	*Net Sales*	*% of Increase over Prior Year*	*Net Sales*	*% of Increase/ (Decrease) over Prior Year*
1904	$ 24,573		$1,189,264	
1905	46,695	90	1,229,107	3
1906	98,890	112	1,199,333	(3)
1907	228,980	132	1,010,751	(16)
1908	353,516	54	717,347	(29)
1909	480,000[a]	36	491,553	(31)

SOURCE: *U.S.* v. *EKCo.*, pp. 2173 and 3472.
[a] Estimated.

market among the professional trade.[62] The company that introduced the most successful paper, Artura Photo Paper Company of Columbus, Ohio, developed a noncurling, nonblistering, and nonabrasion DOP which in the middle and last years of the decade drew many professional photographers from American Aristotype's POP when the Jamestown plant was encountering a period of difficulty with quality control.[63] By 1909 Artura had secured nearly a fifty percent share of the professional market (table 9.5).[64] These and other companies had to turn to new sources of raw paper in Europe and develop new emulsions for the new grades of raw paper employed.

The strategy exercised by Abbott and Eastman in securing control of the European raw paper supply reflected assumptions regarding the photographic paper industry that prevailed during the last third

[62] Gustave Gennert testimony, *U.S.* v. *EKCo.*, pp. 188–89.
[63] GE to Strong, 8 March 1907; and GE to Moritz B. Philipp, 2 December 1907 (GECP).
[64] T. E. Minshall testimony, *U.S.* v. *EKCo.*, p. 67.

of the century. It was assumed (1) that only one kind of paper could be used to carry photographic emulsions and, therefore, (2) that the consumer of photographic materials would be brand-conscious with regard to the raw paper stock. As a generation of professional photographers emerged during the 1890s that did not coat their own paper, they became less loyal to the traditional European raw papers.[65] Moreover, both European and American coaters, under the stimulus of having to use new sources of paper, succeeded in developing and introducing new forms of paper, most notably gelatin DOP. The strategy of the General Paper Company and its clients failed in the long-term goal of controlling photographic paper manufacture, but it succeeded in allowing substantial profits for a decade. The General Paper Company contracts actually placed Eastman Kodak at a decided disadvantage in trying to meet the new domestic competition because the contracts prohibited it from purchasing paper from any other source and from experimenting and producing any paper for itself.[66] Consequently, when the General Paper Company sought to produce new paper so that Eastman Kodak could introduce a gelatin DOP to compete with Artura and, after two years, failed, Eastman Kodak had very little strategic room for maneuvering.[67]

As the contract between Eastman Kodak and General Paper approached the expiration date of December 1907, George Eastman seriously considered not renewing the arrangement;[68] however, when negotiations began, General Paper spokesmen, faced with the threat of other European raw stock producers, made clear that if Eastman failed to renew, the company would consider entering the American photographic paper market by purchasing or making an arrangement with one of the new gelatin DOP companies. The threat represented a very real one for Eastman Kodak. Aware of the major inroads being made into its professional market, and restricted in the original contract against experimenting or developing a new source of paper, Eastman Kodak found itself in a vulnerable position. If it did not sign the contract, it would lose its own raw paper supply, and no other firm in the world could provide the needed quantity of paper.[69] The company learned a valuable lesson in the importance of personally controlling its own sources of supply. George Eastman hurriedly scouted some potential American sources and hired a paper chemist to explore the possibility of Eastman Kodak's developing its own production facility at Kodak Park; however, the company had little alternative but to renew the contract.[70]

[65] GE testimony, *EKCo.* v. *Blackmore*, pp. 1046 and 1051.

[66] Agreement between General Paper Co. and General Aristo Co., 17 March 1900, *U.S.* v. *EKCo.*, pp. 2390–93.

[67] GE testimony, *EKCo.* v. *Blackmore*, pp. 1052 and 1063.

[68] GE to Strong, 7 June 1906 (GECP); and GE to C. S. Abbott, 15 April 1904 (GEC).

[69] GE testimony, *EKCo.* v. *Blackmore*, pp. 1052–53.

[70] GE to Strong, 7 June 1906 (GECP); GE to W. H. Oglesby Paper Co., 3 August 1906 (GEC); and David E. Reid obituary, *Rochester Democrat & Chronicle*, 16 April 1934.

In response to the new competing gelatin DOP, Eastman Kodak sought to develop a similar paper. Finally, in 1908, Eastman Kodak marketed such a paper under the brand name Nepera, but it failed to meet the standards of studio photographers and, therefore, to stem the trend of professionals to Artura paper.[71] With their own efforts at product innovation a failure, the leaders at the Rochester firm had to devise an alternative strategy. With a large income flow and handsome reserves, Eastman Kodak once again, as it had in the past decade and a half, chose the route of outright purchase of a company in order to obtain a proven and successful product. From among the competing DOP companies, Eastman Kodak leaders had little difficulty in making a choice. Defender's Argo paper, a cheap brand of lower quality DOP addressed to the professional, would not halt the trend to the Artura paper. Ansco's Cyko had, at this time, proven itself only to amateurs. Moreover, the acquisition of Ansco at this time would have required the purchase of substantial assets that would have been of little value to Eastman Kodak. Only the Artura paper had proven itself in the market place.

The Ohio company willingly entertained overtures for purchase because of the personal situation of its three principal stockholders.[72] Early in 1909 Eastman Kodak purchased Artura and within a year transferred the production facilities to Kodak Park in Rochester.

Once again, having failed to keep pace with important product innovations in the photographic paper sector, Eastman Kodak found it necessary, in order to retain its position of supremacy in this sector of the industry, to purchase an entire technology.

In 1910, to the mutual satisfaction of both companies, Eastman Kodak terminated the raw paper contract with General Paper Company. As General Paper watched its profits dwindle because of large sums spent fighting competition, it became willing to terminate the agreement amicably. At the same time, Eastman Kodak initiated experiments and development work that led to the establishment of its own paper mill at Kodak Park.[73]

Stage 3: Plate Cameras and Optical Accessories

Period 1: 1895–99

In the middle of the 1890s two forces combined to strengthen the popularity of plate cameras: (1) the difficulty Eastman Kodak experienced with its roll film during the period 1893–94, which did considerable damage to the popularity of roll film photography; and (2) the emergence of the country from the nadir of the depression.

[71] GE to Baker, 26 October 1909 (GEC).

[72] Ibid.; and Minshall testimony, *U.S.* v. *EKCo.*, pp. 66–70.

[73] GE to Goffard, 13 December 1909; Goffard to GE, 5 January 1910 and 24 March 1910 (GEC); and General Paper Co. to EKCo., 15 March 1910, *U.S.* v. *EKCo.*, pp. 3309–15. See also GE testimony, *EKCo.* v. *Blackmore*, pp. 1054 and 1063.

Fig. 9.9
Irving-Clay camera of the Scovill & Adams Company. From American Annual of Photography, *1897.*

At the middle of the decade small folding cameras, such as the Henry Clay camera of the Scovill & Adams Company, caught the popular fancy and encouraged sales. Yet, growing camera sales, popular interest in photography in general, and the relative ease of entry into the plate camera sector enticed many new entrants into the industry, resulting in a period of overcapacity, fierce price competition, and diminished profits. The patent barrier that Eastman Kodak had erected around the popular Kodak roll film system served to steer would-be competitors into the plate camera sector, where the general simplicity and traditional design served to make production relatively easy and unrestricted. While there were numerous patents on plate cameras, none was sufficiently comprehensive to prevent successful imitation and circumvention. Consequently, the number of companies in the plate camera sector of the industry doubled from 1894 to 1899. Major new companies that entered during this period included C. P. Goerz American Optical, Western Camera, Ray Camera, Monroe Camera, Multiscope & Film, Ernest Gundlach, and Reflex Camera.

However, accompanying a sharp increase in the popularity of Kodak cameras at the end of the decade (see table 9.6)—a reflection of restored confidence in the roll film system of photography—the sales of plate cameras began to fall from their peak in 1898. A year later the number of plate camera producers reached its maximum as the industry became plagued with sharp price competition and sharp declines in profit margins.[74] The sharp decline in Kodak roll film camera prices, partially in response to the lowering of plate camera prices, further exacerbated the problems of the industry. By the end of the century, the various firms in the plate camera sector faced a situation similar to the one that the dry plate and photographic paper producers had faced in 1894, and therefore, their leaders began to consider methods of improving their position, including combination, price stabilization, and product innovation.

Period 2: 1899–1903

The period from 1899 to 1903 witnessed several major and minor combinations in this sector of the industry, reflecting the general trend toward merger and combination throughout American business. A number of stimuli prompted the creation of a plate camera combine. First, the plate camera sector lent itself to combination because, with the exception of Anthony, the combining firms were all fundamentally single-product producers, thereby simplifying the merger and the evaluation of the assets of the participants. Second, the general business climate both nationally and locally suggested and encouraged the move to combination. Court decisions that gave tacit approval to combination as a legal means of dealing with the problems of price competition had already stimulated numerous mergers in many industries across the nation. In Rochester, the suc-

[74] "Committee Report to Stockholders of Rochester Optical & Camera Co.," 19 June 1893, *U.S.* v. *EKCo.*, p. 3343.

Table 9.6 Eastman Kodak Roll Film Camera Sales, 1892–1900

Name of Camera	*1892*	*1893*	*1894*	*1895*	*1896*	*1897*	*1898*	*1899*	*1900*
Kodak	$116,967	$59,609	$38,999	$ 34,637	$ 26,360	$ 22,887	$ 2,693		
ABC Kodak	81,631	28,324	19,075	17,838	7,169				
Pocket Kodak				104,849	108,737	70,870	14,299	$ 8,640	
Bullet & Bulls-Eye				9,897	166,995	236,997	255,548	207,361	$131,039
Folding Pocket Kodak							104,435	155,554	
Cartridge						76,722	98,493	90,310	112,520
Eureka						10,604	18,223	20,865	
Kodet					24,700	17,831	1,853		
Falcon					3,743	37,172	37,771	17,634	
Brownie									35,047
Flexo									32,495
Kodako									232,326
Total	198,598	87,933	58,074	167,221	337,704	473,083	533,315	500,364	543,427

SOURCE: *U.S.* v. *EKCo.*, pp. 2565–69.

cessful achievement in the summer of 1899 of a combination of the photographic paper manufacturers into General Aristotype provided a local model for successful combination in the photographic industry. Third, Rochester, because of the example of the phenomenal success of Eastman Kodak, was emerging as a very good market for photographic stock.[75]

In the spring of 1899 the leaders of Rochester Optical and Rochester Camera and Supply initiated the first and most significant merger. Under the leadership of William F. Carlton, Rochester Optical had built a national reputation for good quality amateur and professional cameras and had introduced during the 1890s a popular series of plate cameras and accessories under the trade name Premo. Under the leadership of Harvey B. Carlton, Rochester Camera, founded in the early 1890s, developed a successful line of cameras sold under the trade mark Poco. With the support and active participation of several leading banks and bankers in Rochester, the two Carltons organized a combination of their firms with other major plate camera makers in Rochester, Chicago, and New York. Included in the combine from Rochester were two firms founded in the late 1890s, Monroe Camera and Ray Camera. Two firms from outside Rochester joined the combine: Western Camera and E. & H. T. Anthony & Company. The former, started in Chicago about 1895, had sought to woo the amateur trade from roll film cameras with its Cyclone line of magazine cameras, priced as low as six dollars. In 1899, as the merger proceeded, Western Camera moved to Rochester.[76]

Just prior to organization of the combine, the Carltons sought to insure both adequate quantities of crucial optical supplies and access to the market place by drawing contracts with the Bausch & Lomb Optical Company of Rochester, thereby assuring a supply of lenses and shutters, and by making Eastman Kodak the sole trade agent for the new company in the United States.[77] While George Eastman refused to participate financially in the plate camera combine, he did secure the agreement for the trade agency because his company could obtain a small profit on the sales of a sector in which it did not manufacture and because it could also exercise some influence on the sector that represented the major source of price competition for the Kodak camera.[78]

[75] GE to Davison, 15 August 1899 (GEC).

[76] GE to Strong, 18 March 1899 (GECP); Cossitt testimony, *U.S.* v. *EKCo.*, pp. 504–9; Mosher testimony, in ibid., p. 102; *Rochester City Directory*, 1892, 1895, 1898, and 1899; *City Directory of Chicago*, 1896, 1897, and 1899; Prospectus of Rochester Optical and Camera Company, [4 December 1899], Rochester Historical Society, Access #10,746, on loan to Rochester Public Library; "Catalog and Price List of the Rochester Camera Co.," May 1898; "Catalog," Monroe Camera Co., April 1897; "Ray Cameras, 1897," Mutschler, Robertson & Co.; "Ray Cameras, 1899," Ray Camera Co.; "Catalog . . . ," Western Camera Manufacturing Co., 1897; "Cyclone Cameras and Photographic Supplies," 1898; and "Cyclone Cameras . . ." [1899], all in the Eastman Kodak Co. Patent Museum.

[77] Agreement between Rochester Optical and Camera Company and EKCo., 4 December 1899, *U.S.* v. *EKCo.*, pp. 2131–35.

[78] GE to E. W. Peet, 1 November 1899; and GE to Davison, 10 October 1899 (GECP).

The outlook for Rochester Optical and Camera, organized in the fall of 1899, was favorable. The plate camera sales had grown steadily from the early years of the 1880s, and the six combining firms accounted for approximately 85 percent of the nation's sales of plate cameras in 1899. Among the larger companies not included in the combine were Scovill & Adams of Waterbury, Connecticut; Gundlach Optical and Seneca Camera, both of Rochester; Imperial Camera of LaCrosse, Wisconsin; Conley Camera of Rochester, Minnesota, the production plant for Sears & Roebuck Company; and many small shops across the nation whose sales were largely confined to local and specialized markets.[79]

While the Rochester Optical and Camera Company started business with a number of assets, many of them soon evaporated. First, early in 1900, problems developed between Rochester Optical and Camera Company and Eastman Kodak, resulting in termination of the sales agreement.[80] Second, the merger occurred during the year immediately following the peak of sales in the plate camera sector, and thereafter the popularity of the roll film system brought substantial reductions in plate camera sales. Because the Rochester Optical and Camera Company had secured substantial production facilities and staff predicated on the continued growth of the plate camera business, the company found itself faced with overcapacity.[81] Late in 1900 the dominant banking interests in the firm became dissatisfied with the management and replaced the Carltons with John Robertson, formerly of Ray Camera, who emerged as the principal manufacturing officer. The banking interests, however, continued to dominate the board, and no strong figure with technical expertise emerged to provide leadership. From 1901 to 1903, as the market and profits shrank, the company not only did not make a profit but lost money at the rate of nearly $10,000 per month.[82]

Under the circumstances, the leadership began to search for new products. Based on J. E. Thornton's English tin can pack for sheet film, Robertson and associates developed a new style of film pack, which was then patented.[83] After securing an agreement with the Seed Dry Plate Company to furnish cut sensitized film for the packs, the new system appeared on the market in the spring of 1903 and met with some success.[84]

The internal problems in management at Rochester Optical and Camera during the first year of operation prompted the departure of five employees who organized in Rochester their own plate camera company, the Century Camera Company. The small company quickly developed an impressive line of folding plate cameras and

[79] Prospectus of Rochester Optical and Camera Company. See David N. Sterling, ". . . The Conley Story," *Photographic Collectors' Newsletter*, August 1975.

[80] Agreement Canceling Agreement of 4 December 1899, *U.S.* v. *EKCo.*, pp. 2135–36.

[81] "Committee Report to Stockholders of Rochester Optical and Camera Company," 19 June 1903, in ibid., pp. 3342–46.

[82] "Rochester Optical Company" (GEN); and GE to Davison, 11 October 1902 (GEC).

[83] U.S. Patent #728,718 (19 May 1903).

[84] John Robertson, quoted in "Rochester Optical Company" (GEN).

Fig. 9.10
Century Camera Company building in Rochester, N.Y. Courtesy of Eastman Kodak Company.

secured a national reputation for quality. Despite the falling market for plate cameras, the company's sales displayed impressive rapid growth, securing approximately 25 percent of the plate camera market during 1902.[85] Nevertheless, with overcapacity in the industry, no company could succeed if the prices were reduced to near the cost level. In January of 1903, Century Camera and three other camera manufacturing companies in Rochester agreed to a fixed price list for their cameras. As an indemnity against cheating, each party deposited $5,000 with a disinterested party who had the power to act in the event of departure from the agreement.[86] Because of its position in the market, Rochester Optical and Camera likely was a participant.

Although the camera combination did not prove very successful, other photographic companies followed this strategy. In the spring of 1903, the Century Camera Company purchased the Imperial Camera Company of LaCrosse, Wisconsin, and moved its operations to Rochester. This small company was one of the few camera producers outside Rochester still making folding plate cameras.[87] Early in 1902 the two old-line companies in the American photographic industry, E. & H. T. Anthony & Company and Scovill & Adams Company, joined forces; however, this merger proved to be of little significance to the plate camera business because the combining firms held only a small segment of the plate camera market. More

[85] Century Camera Company Sales in *1901*: 4 mos., $21,530; 2d half, $57,004; in *1902*: 1st half, $78,002; 2d half, $85,197; in *1903*: 5 mos., $91,349 (Price-Waterhouse & Co. Audit Report of Century Camera Co., 15 July 1903, *U.S.* v. *EKCo.*, pp. 2447, 2451, and 2454).

[86] Ibid., p. 2447.

[87] Ibid., pp. 2451–53. There also was an Imperial Camera Company in Grand Rapids, Michigan.

significantly, the Gundlach Optical Company of Rochester and the Manhattan Optical Company of New York and New Jersey combined in 1902, bringing together two of the leading optical companies in the country. Gundlach Optical, founded in the late 1880s by a former optical designer from Bausch & Lomb and later dominated by Henry H. Turner and John C. Reich, produced cameras, shutters, and a wide range of optical goods.[88] Manhattan Optical developed a reputation for high quality optical and camera production and boasted during the middle 1890s the technical leadership of Dr. Ludwig Hugo Schroeder, previously the famous director of England's renowned Ross & Company optical firm. The firm, with its combined production facilities in Rochester, ranked third in sales of plate cameras but also did a fine business in photographic objectives and large lenses for telescopes.[89]

Besides Gundlach-Manhattan, other major firms that supplied much of the demand for photographic optical goods and also some of the needs for plate cameras included Bausch & Lomb, C. P. Goerz American Optical Company, and Wollensak. While the leadership in optical design clearly belonged to Germany, the Bausch & Lomb Optical Company of Rochester emerged as the largest optical manufacturer and most dominant optical firm in the Western Hemisphere. Founded as a small optical shop in Rochester in the early 1850s by the German immigrants John J. Bausch and Henry Lomb, it grew as a consequence of its introduction of microscope production in the 1870s and of photographic lens production for Eastman in the late 1880s. The company expanded rapidly as the demand from Eastman Kodak skyrocketed. By the early twentieth century Bausch & Lomb had secured a position of major leadership in photographic optics and shutters, scientific instruments, and spectacles.[90] The C. P. Goerz American Optical Company, an important link between the American and the German optical and photographic industries, was founded in the mid-1890s, following a triumphant showing of its parent firm's (C. P. Goerz) anastigmat at the Chicago World's Fair. It supplied many American camera manufacturers with high quality lenses for their top-of-the-line cameras, counting among its customers the Eastman Kodak Company. In the early twentieth century the American company dropped its own production facilities and became principally an American sales agency for the German firm.[91] Another important optical supply firm, Wollensak, was founded in 1899 by Andrew Wollensak, a German immigrant who was a shutter designer at Bausch & Lomb.

[88] Moritz von Rohr, *Theorie und Geschichte des Photographischen Objektivs* (Berlin: Julius Springer, 1899), pp. 244–45.

[89] GE to Davison, 11 October 1902 (GEC).

[90] Blake McKelvey, *Rochester: The Flower City, 1855–1890* (Cambridge: Harvard University Press, 1949), p. 243; *Dictionary of American Biography*, s.v. "Edward Bausch"; and Ronald J. Nimmer, "Edward Bausch; William Bausch," (University of Wisconsin Library School, 1963), MS at Rochester Public Library.

[91] Josef M. Eder, *History of Photography*, trans. Edward Epstean (New York: Columbia University Press, 1945), pp. 409–11; and testimony of Otto Goerz, L. J. R. Holst, and Charles F. Schmid, *U.S.* v. *EKCo.*, pp. 136–38, 522–27, 360, and 553.

Within a decade the company developed a major market for its shutters, soon becoming the world's largest producer of shutters.[92]

By the early part of the twentieth century, Rochester, New York, was the photographic capital of the world. With the exception of the dry plate industry, which was concentrated in St. Louis, the largest companies in the camera, optical, and photographic materials industries were located in Rochester. The question arises as to why such a concentration of the industry should have occurred in Rochester. Rochester had no outstanding natural resource other than a plentiful supply of pure water and air, commodities that were not uncommon throughout the United States in the late nineteenth century. Rochester did, however, have a substantial German immigrant population, some of whom possessed skills of importance to optical and instrument production. The German population may also have attracted later German immigrants, so important to the optical industry. The fortuitous location in Rochester of the Bausch & Lomb Company and the Eastman Kodak Company was, of course, highly significant. These two companies reinforced each other's growth. Eastman Kodak represented a major market for Bausch & Lomb's photographic lens department, while the close proximity of Bausch & Lomb was convenient for Eastman Kodak. With two large and growing companies in a relatively new industry already located in one small city, Rochester became the location where employees who were prompted to leave these firms and start their own businesses were already known and most likely to find credit and partners. A further factor is that size begets size. There was a regenerative effect, as was illustrated during the merger movement of the period 1895 to 1909. The larger and more successful companies, already located in Rochester, had the capital to acquire smaller firms that were then moved for consolidation to Rochester. Once Rochester had become an important center for the photographic industry, it tended to draw talent from across the country and from Europe, thus further concentrating both companies and talent. That concentration of the industry in Rochester then encouraged and facilitated a cross-fertilization among the companies as employees of various levels of skill and research and executive ability changed employers.

Period 3: 1901–9

From the end of the 1880s to the close of the 1890s, George Eastman and the Eastman Kodak Company focused their attention almost exclusively on the amateur market; however, with the creation of General Aristo, a new era emerged, with the company paying increasing attention to marketing and the creation of exclusive dealerships. In order to insure full cooperation of the dealer, Eastman Kodak needed a full line of photographic materials and supplies, including materials and apparatus for professional and commercial

[92] *Encyclopedia of Biography of New York*, 2 vols. (New York: American Historical Society, 1916), 2: 141. Wollensak founded Rauber & Wollensak Optical Company in 1899; the company became Wollensak Optical Company in 1902.

photographers. Motivated by this increased attention to marketing and the appearance of certain opportunities, the leadership gradually moved Eastman Kodak into the plate camera sector of the photographic industry.

The initial interest in nonroll film apparatus and accessories arose in the spring of 1901 when the company purchased the stock of Warnica & Company of Rochester, a very small plate holder company. Then, the company entered upon the production of plate cameras after the termination of the abbreviated sales contract with Rochester Optical and Camera. Yet, the sales remained quite small.[93] In the spring of 1903, Eastman's interest in the plate camera business quickened with the opportunity to purchase the Rochester Optical and Camera Company for a substantial discount on its assets. The stockholders of the debt-ridden firm voted to negotiate the sale of the company with Eastman Kodak. The combination of the bargain price and the company's possession of the important film pack patents and licenses motivated the Eastman Kodak purchase.[94]

Once Eastman Kodak took possession, it reincorporated the company and reduced the scale of production closer to the level of demand at the time. Charles Ames, formerly the manager of the Blair Camera Company, assumed responsibilities as treasurer and sales manager, while John Robertson served as production superintendent and product designer. Considering the poor profit performance of the combine throughout its short life and considering the weakening of the plate camera market, George Eastman held little hope of making the operation very profitable,[95] but in its first year under Eastman Kodak ownership, Ames and Robertson produced a profit between $80,000 and $90,000. Eastman confided to Abbott: "I do not think I have ever known a case where a good business was more wantonly wrecked than this one."[96] The performance demonstrated the caliber of the Eastman Kodak management skills and the importance of its marketing operations (table 9.7).

Once Eastman had decided to enter whole-heartedly into the plate camera sector, his attention immediately turned to the most successful company in the business, the infant Century Camera Company of Rochester. Eastman made overtures to purchase the company, and its management responded favorably, leading to its acquisition by Eastman Kodak in the summer of 1903 (table 9.7).[97] Completing the series of plate camera acquisitions in 1903, Eastman Kodak also purchased a small camera company, the Multiscope &

[93] Plate camera sales in 1900, $8,274; in 1901, $13,466; in 1902, $27,196 (*U.S.* v. *EKCo.*, p. 2566).

[94] "Committee Report to Stockholders of Rochester Optical and Camera Company," ibid., pp. 3342–46; "Rochester Optical Company" (GEN); GE to Davison, 11 October 1902 (GEC); and GE testimony, *EKCo.* v. *Blackmore*, p. 1070.

[95] GE to Strong, 1 March 1904 (GECP).

[96] GE to Abbott, 26 July 1904 (GEC).

[97] Mosher testimony, *U.S.* v. *EKCo.*, p. 103; and "Century Camera Company" (GEN).

Table 9.7 Sales by Century Camera, Folmer & Schwing, and Rochester Optical Divisions Eastman Kodak Company, 1903–9

Division	*1903*	*1904*	*1905*	*1906*	*1907*	*1908*	*1909*
Century Camera	$107,203	$128,849	$140,053	$166,658	$111,266	$146,786	$135,149[a]
Folmer & Schwing		70,190	80,000[b]	87,781	169,099	134,443	139,264[c]
Rochester Optical	153,495[d]	441,994	451,083	616,781	682,669		

SOURCES: *U.S.* v. *EKCo.*, pp. 2570–71; 3123–24.

[a] Breakdown: $65,741 amateur/ $66,493 professional/ $2,915 sundries and repairs; this figure does not include the Century share of $68,349 for sales of lenses and shutters previously attributed to Century and Folmer & Schwing separately but aggregated in 1909.

[b] George Eastman's estimate of annual sales, GE to Davison, 22 May 1905 (GEC); actual sales for the nine month period 5 April to 31 December 1905 were $45,916.

[c] Breakdown: $121,811 amateur/ $12,707 professional/ $4,746 sundries and repairs; this figure does not include sales of lenses and shutters. See note *a* above.

[d] Includes only the sales for the five month period 1 August through December 1903.

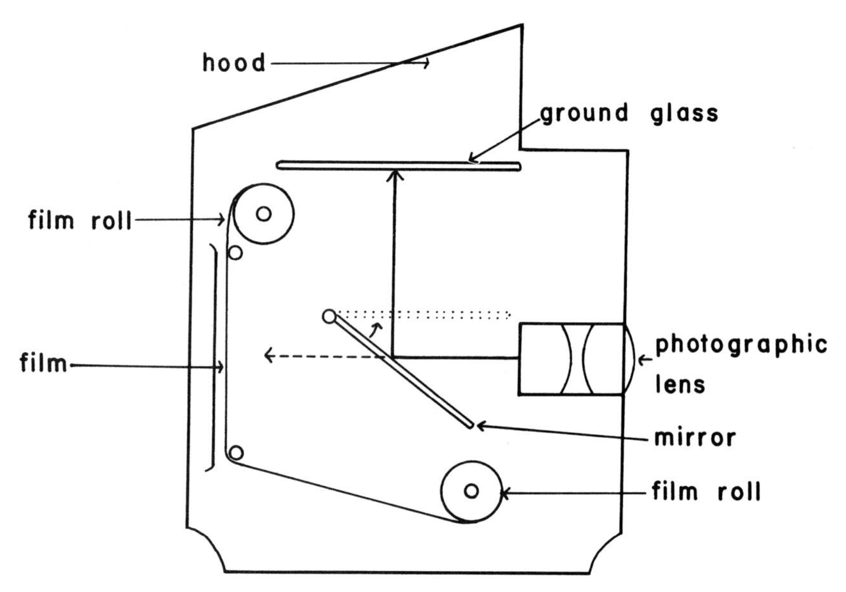

Fig. 9.11
Single-lens reflex camera.

Film Company of Burlington, Wisconsin. The plate camera company did a relatively small business but did have some important patents.[98] Typical of Eastman's strategy, when the Eastman Kodak Company entered a new sector of the industry, the acquisition of a substantial market share and of important patents was paramount. Although the plate camera sector could not be effectively controlled through patents, as was the roll film camera sector, nevertheless, patents were still significant enough that their acquisition, even through outright purchase of companies, was deemed desirable. While Eastman had modified his patent strategy over the years, its core still remained intact.

Eastman Kodak's entry into the plate camera sector had come quickly and decisively during the spring and early summer of 1903, and then for two years, acquisition of businesses in that sector ceased. During that time, however, a new type of plate camera grew in importance: the reflex camera. The single-lens reflex camera consisted of an ordinary large camera box and lens with a mirror behind the lens that cast the image onto a horizontal ground glass screen, providing a much better view of the object to be photographed than did the ordinary finder (fig. 9.11). Before exposing the film, the photographer moved the reflecting surface to one side and made the exposure in the ordinary manner, with a focal plane shutter. While such cameras had been on the European market since the 1880s, no American company produced reflex cameras until the Reflex Camera Company of Yonkers, New York, introduced them in 1899. Founded by a Dutch engineer, Lodewyk J. R. Holst, who had produced reflecting cameras in Amsterdam and later managed production for C. P. Goerz American Optical, Reflex Camera soon developed a mar-

[98] Sales in 1899, $18,000; in 1900, $26,000; in 1901, $38,500; in 1902, $38,600; in "Multiscope & Film Company" (GEN).

ket, especially among photojournalists and professional and serious amateur photographers.[99]

Early in the twentieth century the Reflex Camera Company suddenly faced competition from an older company, the Folmer & Schwing Manufacturing Company of New York. Folmer & Schwing Manufacturing entered the photographic business about 1891, at first selling cameras and supplies and later assembling a product line directed to the high quality professional and serious amateur plate camera market. Early in the twentieth century the company developed the Graflex series of single-lens reflex cameras and high quality folding plate cameras for professionals and serious amateurs. The high quality of its product made it a tough competitor for Reflex Camera. Also, Folmer & Schwing made arrangements with Eastman Kodak for its products to be marketed through Eastman Kodak dealers. In so doing, they came under the terms of sales of Eastman Kodak, thereby ensuring themselves a minimum fair trade price in the retail market and a place in the stock of Eastman Kodak dealers. The rapid growth of sales of a style of camera not in the Eastman Kodak product line attracted George Eastman's attention. In the spring of 1905 Eastman completed a series of negotiations that resulted in the transfer in ownership of the New York company to Eastman Kodak (table 9.7).[100]

Thus, the movement to combine firms in the plate camera business markedly centralized camera production in Rochester, New York, a city clearly identified in the public mind as the camera capital of the country. A most significant by-product was the eventual bringing together into one organization of important camera draftsmen and designers (see Appendix). In these later phases of combination, in which Eastman Kodak participated, it reversed the policy it had pursued in the consolidation of the photographic paper sector, i.e., in general not to retain the leadership or technical personnel of acquired companies. Lack of experience in the production of plate cameras led Eastman Kodak to retain a substantial number of important camera craftsmen and designers as a part of its supervisory, production, and experimental staff, and this staff continued to provide a flow of technical innovations.[101] The camera factories served as laboratories for design and experimental work. Although attention has always been focused on the accomplishments of the Eastman Kodak Industrial Research Laboratory, the skilled craftsmen and designers in the camera factories produced a continuous stream of product innovations and production modifications that were most instrumental in the maintenance of Eastman Kodak's strong position in the world camera market.

[99] Holst testimony, *U.S.* v. *EKCo.*, pp. 516–19 and 524.

[100] Ibid., pp. 520–22; W. F. Folmer to Rudolph Speth, 3 December 1912, copy in "Folmer & Schwing Mfg. Co." (GEN); and Agreement between W. F. Folmer and EKCo., 13 May 1905, ibid., pp. 2186–96. Sale price was about $67,000.

[101] Eastman Patent Employees, *U.S. v. EKCo.*, pp. 3455–56.

Dry Plate Sector: 1895–1909

Although production facilities increased substantially in size and ownership became more concentrated in most sectors of the photographic industry during the period 1895 to 1909, the degree of concentration was comparatively less in the dry plate sector. Like other manufacturers in the industry, the dry plate producers felt the pressure of declining sales and profits during the depression of the 1890s, and the ensuing competitive struggle resulted in the emergence of an oligopolistic structure in the industry. The character of competition and the structure of the industry from 1895 to 1910 fell into three phases.

Market Struggle: 1895–99

During the latter part of the 1890s there emerged an oligopolistic structure the shape of which was influenced by three major factors: (1) failure of price maintenance agreements; (2) regional concentration; and (3) the origins and diffusion of technical knowledge. The independent position taken by the Stanley brothers of the Stanley Dry Plate Company of Newton, Massachusetts, contributed significantly to the failure of the price consortium of 1894. The Stanleys decided to resist the price association even though they held only a small fraction of the total dry plate market. They reasoned that the manufacturers' advance in plate prices and profits under the protection of the association in the face of decreasing material costs would simply encourage the entry of new plate makers: "the business would be split up, and the greater competition that would follow would soon make the business worth nothing. In trying to grasp too much the trust would lose all."[102] While the other plate and paper manufacturers certainly recognized the cogency of this point, they continued to assume that the traditional power of the jobbers was sufficient to prevent "outside" manufacturers from successfully marketing their products, even at a lower price. However, the Stanleys failed to take into account certain factors inherent in the dry plate industry that limited its member firms in number.

When the Stanleys remained independent of the association, the principal jobbers and many of the principal dealers refused to handle the Stanley plate any longer. Prepared for this boycott, the Stanley Company moved at once to develop its own marketing department, which began selling directly to professional photographers and amateurs, thereby lowering the price of the Stanley plate further because of the elimination of the jobber and dealer profits.[103] Just as the Eastman Dry Plate & Film Company a decade earlier had decisively shown the independence of the manufacturer from the jobber, once again the newly emergent manufacturing sector demonstrated the weakness of the jobber in this new industrial en-

[102] "A Boycott Declared against the Stanley Dry Plate Company," *The Newton Graphic* (Newton, Mass.), 7 September 1894, p. 1.
[103] Francis E. Stanley testimony, *U.S. v. EKCo.*, p. 78.

vironment. The scale of manufacturing operation, even for a single product such as dry plates, now warranted the manufacturer's operating his own marketing system. During the long struggle the dealers continued to boycott the Stanley plate, fearing the loss of their supply of other brands of dry plates; however, contrary to what the association intended, the sales of Stanley plates skyrocketed. In the second half of the decade, the Stanleys doubled the output of their Newton factory and opened a factory in Montreal to serve their Canadian market.[104]

Although the Stanley brothers defeated the dry plate association, they did not, of course, demonstrate the validity of their criticism, i.e., that new producers could enter and successfully defeat the association. Despite Stanley's long-standing reluctance to use advertising,[105] the Stanley plate was an established, well-known, and popular product. Therefore, its success did not demonstrate that unknown, new producers could successfully compete against the association. Fundamentally, the industry had moved into a mature phase wherein certain plate manufacturers, having developed and nurtured an established national reputation and brand identification, attracted a loyal following. Furthermore, the development of fast and reliable emulsions was a cumulative process, knowledge of which was a zealously guarded secret. Therefore, by the late 1890s, not only did a potential new entrant have difficulty producing a competitive product and attracting a market, but even the older producers who had catered almost exclusively to local and regional markets found their sales shrinking. Consequently, most of the smaller producers left the industry during the mid or late 1890s, and about a dozen plate manufacturers emerged to dominate the American dry plate market.

This emerging structure of the dry plate sector was also characterized by regional concentration. The location of virtually all of the dry plate producers, large or small, either on or east of the Mississippi River reflected the regional distribution of population and also the concentration of skilled labor and entrepreneurial and capital resources. St. Louis, with its initial leadership in dry plate production (Cramer & Norden) was home to the three dominant companies in the industry: Seed, Cramer, and Hammer. By the late 1890s these three firms were producing about three-quarters of all the plates used in the Western Hemisphere.[106] The development and improvement of the emulsions in the American industry centered around key technical personnel working in these three companies. The centralized location of these companies, with their ready access to the eastern and western markets through the rail network and to foreign markets through the Mississippi ports, served to provide

[104] Ibid., pp. 78–79; and "Killed in Tip-Over," *The Newton Circuit* (Newton, Mass.), 2 August 1918, p. 5.

[105] Interview with Raymond Stanley of York Harbor, Maine, 18 August 1970. Raymond Stanley is the son of Francis E. Stanley.

[106] William Hyde and Howard L. Conrad, *Encyclopedia of History of St. Louis: A Compendium of History and Biography for Ready Reference*, 4 vols. (St. Louis: Southern History Co., 1899), 3: 1729.

Fig. 9.12 New York Dry Plate Company advertising. From American Annual of Photography, *1897.*

advantages that simply complemented the advantages they already had, such as superior emulsion quality and established national reputations. St. Louis was to the dry plate industry what Rochester was to roll film photography and what Dresden was to the European photographic paper industry.

However, a notable pattern of location change emerged during the 1890s, reflecting the increased competition among dry plate manufacturers and the efforts of certain producers to relocate nearer markets and major sources of supply and transportation. In 1891 the Stanley company moved from Lewiston, Maine, to Newton, Massachusetts, near Boston. Wuestner moved from St. Louis to Jersey City, near New York City. In the mid-1890s Record and Damon of Harvard Dry Plate moved from Cambridge, Massachusetts, to Guttenberg, New Jersey, near New York City, and renamed the company the New York Dry Plate Company. Lovell Dry Plate moved in the late 1890s from Portland, Maine, to New Rochelle, New York, a suburb of New York City. The attraction of New York is obvious. Despite these moves, only the Stanley company rose to national prominence; the other companies failed to survive for any sustained period. Clearly, the price difference due to transportation costs was secondary to the quality and reliability of the plate, to the national reputation of the brand, and to the marketing and administrative capacity of the company's leadership.

The decline of the small regional producer of dry plates reflected the manufacturer's increased critical dependence upon technological knowledge. With the emergence of a dozen producers and the growing dominance of the three St. Louis companies, the dry plate market began to exhibit some of the characteristics predicted by oligopolistic competition, i.e., each producer took "account of his *total* influence upon the price, indirect as well as direct."[107] Yet, with producers such as the Stanley company in the market, even oligopolistic behavior departed at times from theoretical prediction. That dry plates were not standardized commodities but differed in quality and price complicated market behavior.

The key to emulsion quality was continual improvement, the sources of which included research by master emulsion makers and academically trained assistants and acquisition of formulae for improved emulsions from European and other American producers. The large, leading American plate producers could afford to pursue a variety of combinations of these avenues to emulsion improvement, moving further ahead of their small competitors in plate quality. The larger plate and film producers could finance experiments by master emulsion makers, whereas in the smaller plants the emulsion maker, who often had also to supervise the entire plant's production, had limited research time. During the 1890s some of the larger manufacturers began to hire chemists and other technically trained personnel who not only directed their technical talent to improvements in the production processes but, equally important, devoted their talent to product innovation as well. Trained chemists who played major roles in photosensitive materials manufacture included Milton B. Punnett (Case) at Seed Dry Plate, Charles Force Hutchison (Rochester) at Eastman Kodak, Albert Sellner (Stuttgart Polytechnic) at Cramer, Leo Baekeland (Ghent) at Nepera Chemical, and Albert Hahn (Cornell) at Nepera Chemical.

American dry plate manufacturers also drew upon the experience and innovations that originated in Europe. As indicated earlier, Cramer sought to import European orthochromatic emulsions from Europe, first through Edward V. Boisonnas, the Swiss photochemist, and later through B. J. Edwards and his son, Austin Edwards, both British emulsion makers.[108] Eastman's master emulsion maker, William G. Stuber, had spent some time in Zurich, Switzerland, studying under the well-known emulsion maker, Dr. J. Hugo Smith.[109] Wuestner introduced orthochromatic nonhalation emulsions based upon the formulae of the English plate maker Sandell.[110] Punnett added to his understanding and knowledge of photochemistry generally and dye sensitization specifically from his studies in Berlin and Vienna with Dr. Vogel and Dr. Eder. Along with these known examples of the direct dependence of American emulsion technology upon European photochemistry and practical innovation is the stream of chemical and mechanical information drawn from European sources that filled the numerous photographic journals, both European and American, at the time. The indirect influence of this

[107] Edward Chamberlin, *The Theory of Monopolistic Competition* (Cambridge: Harvard University Press, 1933), p. 46; see also Joan Robinson, *The Economics of Imperfect Competition* (London: Macmillan & Co., 1933).

[108] Rudolph Schiller, *Early Photography in St. Louis* (St. Louis: W. Schiller & Co., [n.d.]), [p. 10]; and Wentzel, *Memoirs of a Photochemist*, p. 19.

[109] *William G. Stuber: A Biography* (Rochester: 1951); Carl W. Ackerman, *George Eastman* (London: Constable, 1930), pp. 107–8; and Stuber testimony, *U.S. v. EKCo.*, pp. 430–32.

[110] *Photog. T.* 24 (1894): 146 and 291.

Fig. 9.13
Excelsior Dry Plate Company trademark.

information is difficult to measure, but the prevalence of the literature and isolated references to it by Cramer and Eastman indicate its considerable influence on American photographic technology.

The mobility of master emulsion makers provided one of the chief means of diffusing the emulsion technology that contributed to the increased reliability and sensitivity of dry plates (see appendix). Between 1880 and 1900, several emulsion makers moved from one company to another, carrying their secrets with them and learning the techniques of others at their new locations. In Portland, Maine, during the 1890s, Leonard F. Libby and Charles O. Lovell founded a number of different dry plate companies. Eventually, Lovell moved his company to New Rochelle, New York.

In Rochester during the 1880s knowledge of emulsions passed quickly among several small firms as emulsion makers moved from one firm to another. Prominent in these moves were Charles Forbes, William Walker, James Inglis, and Arthur Van Voorhis. The Eastman Dry Plate & Film Company benefited most from the mobility, acquiring perhaps some knowledge of formulae used by Forbes, Inglis, and Walker. Another Rochester emulsion maker, George Monroe, reaped only limited success during the 1880s from his own company. Late in the decade he worked in St. Louis as an emulsion maker for Cramer Dry Plate. When Althans and Hutchings left Cramer to join Hammer in the founding of the Hammer-Althans company, Monroe joined the new company as one of its emulsion makers. Therefore, the Hammer company began with a knowledge of the Cramer formulae and some knowledge of certain Seed formulae that Monroe had obtained. When, early in 1892, George Eastman persuaded Monroe to join the Eastman company, the Rochester firm gained substantial knowledge of the Cramer and Hammer formulae and some knowledge of the Seed formulae.[111] After Monroe lost his position, he joined the Columbian Dry Plate Company of Jamestown, New York.

With the loss of Monroe, Hammer turned to one of Hutchings's former partners, Frank Pratt, an emulsion maker with three dry plate companies in Iowa City, Iowa, in the early 1880s who had joined the Excelsior Dry Plate of Rockford, Illinois, in the early 1890s. Pratt joined Hammer in November of 1892. With the departure of Pratt from Excelsior, that firm, with some knowledge of Pratt's formulae, combined with the Rockford Dry Plate, continuing under the Excelsior name but with B. F. Green as emulsion maker. Thus, by the mid-1890s there had been a diffusion of emulsion knowledge, however imperfect, that interlinked Cramer, Eastman, Hammer, Excelsior, Seed, and Columbian (see appendix). In the East, by the end of the century, there was at least some interlink between Libby and Lovell. While this imperfect interlink had some bearing on the general levels of quality of the leading plates, it was during the next half decade, with the movement of certain key emulsion makers among the

[111] Monroe knew of Cramer's boiling formula and that it would cover 20 percent more glass than any ammonia formula. Monroe also knew that Seed took a week for ripening of its emulsion (GE to Walker, 9 February 1892 [GEC]).

Fig. 9.14 Seed Dry Plate Company trademark.

leading dry plate firms, that the network of plate makers tightened and the channels of information improved.

Attempts at Combination: 1899–1904

In keeping with the general business trend of the period, the major dry plate companies began to consider in the 1890s the possibility of forming a combination. This strategy reflected the industry's recognition of the failure of pools and trade associations to maintain prices. These manufacturers recognized the holding company as a legal means of enforcing price stabilization. That serious talk of combination circulated among the dry plate manufacturers early in 1899 indicates that their idea of combination preceded that of the plate camera manufacturers and was coeval with that of the leaders in the photgraphic paper industry.[112]

Although the industry was moving toward oligopolistic behavior during the late 1890s, plate prices continued under a downward pressure. During 1900, in response to the growing competition of Eastern plates, especially those of Stanley, Hammer increased its discount to dealers. Even the leader of the industry, M. A. Seed, carefully considered following the price leadership of Hammer but finally decided that the Seed quality would warrant a 5 percent price differential. The Board of Directors of Seed Dry Plate sought to avoid a direct price confrontation with Hammer, fearing that it could only severely curtail profits. The actions of the Seed board clearly reflected policy formulation that approached theoretical oligopolistic behavior. The board also instituted a sales policy that included the requirement that dealers sell Seed plates at or above a company-established price, thereby following in broad outline the sales policy instituted during the previous few years by Eastman Kodak in the roll film system market. As the pressure of the lower Hammer prices mounted during the first months of 1901, the Seed company again considered altering its discounts and prices, but it deferred a decision pending the outcome of negotiations to sell the business to another corporation that planned to combine all the dry plate companies.[113]

During the period from 1899 to 1901, various groups, both within and outside the dry plate industry, attempted to create a combination. George Eastman and Charles Abbott encouraged the formation of a combine of the major dry plate companies, offering financial and organizational assistance with the understanding that Eastman Kodak would act as sole sales agent for the combine. However, Eastman's expectations were limited: "I do not expect that this proposition will work immediately but it will be put to soak."[114] At a meeting with Eastman and Abbott in Buffalo in late July 1899, Cramer reacted favorably; he returned to St. Louis to talk with Seed and Hammer, and he broached the subject to the Stanleys.[115] The

[112] GE to Directors of Kodak Ltd., 30 March 1899 (GECP).
[113] "M. A. Seed Dry Plate Company" (GEN).
[114] GE to Strong, 26 July 1899 (GECP).
[115] GE to Philipp, 29 July 1899 (GECP).

Fig. 9.15 Milton B. Punnett. Courtesy of Case Institute of Technology Alumni Association.

plate companies considered the idea during 1899 and 1900 but did not act favorably upon it; however, by 1901, with the failure of the independent St. Louis promoter, W. Davies Pittman, and the pressure of an advance in the price of Belgian glass for the plates, the manufacturers began to reconsider the possibilities of combination.[116] F. Ernest Cramer, son of Gustav Cramer, initiated negotiations with George Eastman and encouraged him to assume the leadership in combining Seed, Cramer, Hammer, and Eastman. Although Cramer possessed no purchase options, he had discussed the matter informally with the other St. Louis manufacturers.[117] In a meeting at the offices of Cramer Dry Plate in July 1901, attended by the leadership of the major St. Louis companies, the Rochester leaders proposed a formula for evaluation of the businesses entering the combination: five times the average of three years' annual profits for goodwill, plus the value of live assets at inventory cost.[118] The proposition received a mixed reception; yet, negotiations continued. Though Salzgeber, representing Hammer, stated after the meeting, "nothing doing," Cramer took the proposition to his Board of Directors.[119]

In August the tempo of action quickened. In Boston, Abbott and Eastman discussed the proposition with the Stanley brothers. Later in the month Eastman entertained Miles Seed, Gustav Cramer, and their families in Rochester while negotiations continued, but Hammer and Cramer could not reach an accord on valuation of their businesses. At this point, Eastman recognized that only outright purchase of the plate companies by Eastman Kodak could consummate the consolidation. The old-time plate makers could not come to terms among themselves, even with the leadership of Abbott and Eastman. Yet, Eastman Kodak, inundated with the details of establishing Eastman Kodak of New Jersey, a holding company designed to assume the ownership of Eastman Kodak of New York, General Aristotype, Kodak Ltd., and several other related domestic and foreign companies, was not in the financial position to purchase these companies at the generous evaluations the owners demanded. Therefore, Eastman postponed further negotiations with the plate companies, noting that "nothing further can be done about the dry plate consolidation until after we get the New Jersey Company incorporated and complete its purchase of Kodak and Aristo shares."[120]

At the same time the power and influence of knowledge of emulsion secrets manifested itself in the marked increase in sales by an old dry plate company, Standard Dry Plate Company (tables 9.8 and 9.9). Milton B. Punnett, the trained chemist employed by the M. A. Seed Company, left the St. Louis firm and soon accepted an offer from the Standard Dry Plate Company of Lewiston, Maine, to attempt to revive the faltering enterprise (fig. 9.15). In 1897 inter-

[116] GE to Abbott, 3 August 1901 (GECP).
[117] Cramer testimony, *U.S.* v. *EKCo.*, p. 75.
[118] Salzgeber testimony, in ibid., pp. 83–84.
[119] Ibid., p. 84; and Cramer testimony, pp. 72–73.
[120] GE to E. O. Sage, 21 August 1901 (GECP).

Table 9.8 Professional Photographers Using Various Brands of Dry Plates, 1900–1905 (by percent)

Company	*1900*	*1901*	*1902*	*1903*	*1904*	*1905*
Seed	44.0%	46.0%	50.0%	54.7%	54.7%	48.7%
Cramer	20.0	17.0	15.6	12.3	9.3	9.5
Stanley	17.0	19.5	17.6	12.8	11.6	10.1
Hammer	13.0	14.7	13.3	11.9	11.5	11.1
Eastman Kodak	6.0	10.9	8.0	2.9	3.5	2.3
Standard	2.7	4.0	6.5	11.6	21.6	34.6
Rockford	1.1	0.4	0.7	0.3	0.0	0.0
American	0.9	0.5	0.4	0.2	0.5	0.4
Record	0.0	0.0	0.0	0.3	0.2	0.4
Others	8.0	7.0	8.4	1.4	1.5	1.5

SOURCE: *U.S.* v. *EKCo.*, p. 2604.
Note: Some photographers used more than one brand. Therefore, the totals represent more than 100 percent.

ests from the Auburn Box Company of Lewiston had acquired a controlling share in the nearly moribund firm and had sought to breathe life into it by appointing Fred L. Leavitt, a local dentist, as manager and supervisor. Leavitt's efforts did not meet the expectations of the new management, so they sought Punnett, who not only knew the production methods of the nation's leading dry plate maker but had at his command its emulsion formulae as well. Within a short time after his arrival in Lewiston, Punnett succeeded in raising the quality of the Standard plate, and its low price and high quality brought it into serious competition with the national dry plate brands.[121] Despite this success internal dissension developed, with Punnett as the focal point. Punnett made an offer to Eastman for the sale of both the company and his services. The offer arrived at a propitious moment, because the new Eastman Kodak holding company had just completed formal organization and included in its treasury a generous portion of shares reserved for the acquisition of dry plate companies. Eastman Kodak consummated a purchase agreement with Standard Dry Plate and Milton Punnett in the spring of 1902, acquiring an important new emulsion formula and the services of one of the most knowledgeable photochemists and emulsion makers in the industry.[122] Much to the delight of George Eastman and other leaders at Eastman Kodak, the Standard emulsion formula was adapted to film with a substantial reduction in the quantity of requisite silver nitrate, netting savings of double the purchase cost in the first year. In addition to the economy of the formula, it also possessed twice the rapidity of the Eastman formula.[123]

[121] M. B. Punnett to Mrs. Borsch, 14 May 1941, in office of Case Alumni Assn., Case Western Reserve University, Cleveland, Ohio; Freeman Gillice, interview by Catherine McIntosh, 25 October 1956, at Office of Corporate Information, Eastman Kodak Company, Rochester, N.Y.; and "Standard Dry Plate Company" (GEN).

[122] Agreements: M. B. Punnett and EKCo., and Nathaniel J. Jordon and EKCo., *U.S.* v. *EKCo.*, pp. 2208–16.

[123] GE to Davison, 30 June 1902 (GEC).

Table 9.9 Net Sales of Leading American Dry Plate Manufacturers, 1902–10

Company	*1902*	*1903*	*1904*	*1905*	*1906*	*1907*	*1908*	*1909*	*1910*	*% of Eastman Kodak Professional Customers Using Plates in 1910*
Seed	$600,000[a]	$700,000[a]	$730,000[a]	$670,000[a]	$742,057	$738,917	$686,235	$619,972	$539,370	30.6%
Cramer	500,000[b]				[c]	[d]	[d]	[d]	[d]	11.1
Stanley	400,000[e]		234,512	185,884	292,108	339,608	401,227	460,818	460,193	27.7
Hammer	375,000 to 400,000									11.7
Standard	108,900	144,187	251,929	401,410	343,823	325,654	321,729	374,228	372,229	33.4
Eastman	87,814	72,336	61,018	49,794	38,534	30,618	25,759	22,592	14,468	100.0

SOURCE: U.S. v. EKCo., pp. 76–78, 83, and 2136.

[a] Estimated U.S. sales from total sales figures.

[b] Cramer had sales between those of Seed and Hammer (Salzgeber testimony, p. 83).

[c] Decline in sales in 1906 (F. Ernest Cramer testimony, p. 76).

[d] Steady increase in sales each year (pp. 76–77).

[e] Stanley estimated sales @$685,000 at retail prices. At a 45% discount, that would represent sales of about $400,000 (p. 78).

During June and July George Eastman conducted negotiations with the M. A. Seed Dry Plate Company and with the Stanley Dry Plate Company, coming to provisional agreements with both regarding Eastman Kodak's purchase of the two companies.[124] Also as a part of the purchase agreement, several of the technical and managerial personnel contracted to continue in the employ of Eastman Kodak. Charles Hutchison, Eastman Kodak's second ranking emulsion man, later observed that "it was when Kodak bought Seed that it forged to the front in the plate field."[125]

After examining the accounts of the Stanley company, the Eastman company decided not to proceed with that purchase. Having acquired Standard and Seed, Eastman may have thought it a considerable expense and little technical advantage to include the Stanley company. Undoubtedly, the unsophisticated character of the company's production and accounting systems did little to entice the conservative leaders from Rochester.[126] Hence, by late summer of 1902 Eastman, temporarily satisfied with the two acquisitions, expected that the sales of the other companies would soon fall, and then they could be acquired for less money than in mid-1902. In this expectation Eastman was to be disappointed.

Beyond his general interest in consolidating the industry in order to protect the market, Eastman consciously sought to strengthen the company's emulsion technology. There loomed on the horizon the French film company, Lumière, which threatened to initiate production of photographic film in America. In connection with the dry plate acquisitions, George Eastman observed at the time that "by the time the Lumière people get to running we hope to be able to put up quite a good front."[127] Later Eastman indicated that he pursued the purchase of the Seed company principally to acquire its emulsions for the Eastman Kodak film.[128] Possibly Eastman expected that the high quality of Seed and Standard emulsions would be a technological advantage parallel to the advantage that control of the raw paper supply had provided in the consolidation of the photographic paper industry. The market pressure of the high quality emulsion in the hands of the Eastman Kodak Company did not, however, encourage the other leading dry plate producers to sell to Eastman Kodak. While high quality emulsions were a trade secret, both Cramer and Hammer possessed very good formulae and were able to maintain and improve their market positions.[129]

In the middle of 1903 Eastman Kodak procured a small English plate and paper firm, Cadett & Neall of Ashtead, Surrey, which held a geographical market structure that somewhat paralleled that of the acquired Vereinigte. As it had done with the Seed company, Eastman Kodak maintained the Cadett & Neall production facilities

124 Ibid.; and GE to Philipp, 4 June 1902 (GEC).

125 Charles Hutchison, interview by Jean Ennis, 17 October 1950, Office of Corporate Information, Eastman Kodak Co., Rochester, N.Y.

126 GE to Dennis G. Brussel, 20 October 1902 (GEC); and GE to Abbott, 7 January 1904 (GECP).

127 GE to de Lancey, 6 August 1902 (GECP).

128 GE testimony, *EKCo.* v. *Blackmore*, p. 1112.

129 Cramer testimony, *U.S.* v. *EKCo.*, p. 76.

Fig. 9.16
Cadett & Neall Limited, Ashtead, England. Courtesy of Eastman Kodak Company.

separate from its Harrow plant for several years following acquisition and also maintained the brand name identity (fig. 9.16). American emulsion makers from Standard and Seed were sent to Ashtead to assist in the improvement of the Cadett & Neall emulsions (see Appendix).[130]

Again, late in 1903, George Eastman's attention began to refocus on the American dry plate industry, as Cramer cut his prices in response to the increase in sales of Seed and Standard plates and as the Stanley dry plate continued to hold a strong position in the American market.[131] The combination of Francis Stanley's all-consuming interest in the steamer automobile business and Freelan Stanley's poor health quickened the twins' interest in divesting themselves of the dry plate business. Eastman's interest in the Massachusetts company was furthered when he learned that the Standard emulsion had been leaked to the Stanleys.[132]

Eastman Kodak purchased the company in January 1904 and moved the business to the new dry plate plant at Kodak Park in Rochester, where the Stanley plate continued to be produced as a separate brand.[133] At the same time, the Seed plant in St. Louis underwent remodeling and expansion, a reflection of Eastman's immediate concern for improved efficiency in production and his desire to provide sufficient room for production in the event of completion of the dry plate company consolidation with the purchase of Cramer and Hammer (fig. 9.17). Eastman argued, "It may seem a little cheeky to provide capacity for the business of our competitors but while we are building we might as well figure on taking them in as it is certainly liable to come to pass."[134] And, "I

[130] "Cadett & Neall, Limited" (GEN); GE to Punnett, 9 February 1905; and GE to H. L. Quigley, 20 October 1905 (GEC).

[131] GE to H. C. Reiner, 9 May 1903 (GEC).

[132] *Tribune* (Watertown, Mass.), 8 January 1903, p. 1; and GE to Abbott, 27 November 1903 (GECP).

[133] GE to Carlton F. Stanley, 11 January 1904 (GEC).

[134] GE to Abbott, 14 January 1904 (GECP).

Fig. 9.17
Seed Dry Plate Company building, St. Louis, Mo. Courtesy of Eastman Kodak Company.

think we can make more money by hard work in the field this year than we can by buying out either Hammer or Cramer."[135]

Having acquired mastery of the emulsion technology, Eastman Kodak was then to rely principally upon its marketing ability. Eastman initiated a piecemeal strategy of consolidation, but after 1904 the dry plate market and the general social-political-legal atmosphere gradually changed. Eastman Kodak made no further acquisitions in the dry plate industry and never acquired overwhelming dominance of the dry plate market (tables 9.8 and 9.9). Cramer, Hammer, and Eastman Kodak held a sufficient market share to insure that profits in the industry would not be raided by recurrent price wars. Furthermore, the dry plate industry did not continue as a major growth sector in the photographic business. Capital invested in other areas of the industry could reap better returns. Finally, the cream of the emulsion technology had been skimmed from the industry and successfully transferred to Eastman Kodak's all-important roll film. Thus, the company's use of emulsion technology as a strategic tool, interacting with other business, economic, and personal factors, played an important role in the shaping of the dry plate industry in the early twentieth century.

Eastman Kodak Dominance: 1905–9

Although significant outside competition remained in the dry plate industry during the second half of the first decade of the century, Eastman Kodak's plate companies held about two-thirds of the total sales of dry plates in the United States. The Eastman brand plate steadily declined in importance, but Seed, Stanley, and Standard maintained their strong positions. Cramer and Hammer retained their places as the two most important independent companies. Since Eastman Kodak maintained and enlarged the Seed production facilities in St. Louis, that city remained the center of the dry plate industry.

[135] Ibid., 25 January 1904.

Fig. 9.18
"Papa Cramer" illustrating some points in his new book to a number of his salesmen. From James F. Ryder, Voigtländer and I in Pursuit of Shadow Catching *(Cleveland: Cleveland Printing and Publishing Company, 1902).*

In 1905 and 1906 Eastman Kodak began to use its strong marketing position to put pressure on its two St. Louis rivals. At the beginning of 1906 the company announced the extension of its camera and paper sales policy to dry plates, i.e., a special discount to dealers who did not handle any brand dry plate other than those produced by Eastman Kodak.[136] This announcement made the independent dry plate producers anxious. Almost at once the Directors of the Cramer Dry Plate Company met to lay plans for the coming campaign. They decided to circumvent the photographic supply houses altogether and sell directly to professional photographers at the discount formerly given to dealers.[137] In effect, they substantially reduced their price to the professionals and assumed the financial burden of creating and operating their own direct marketing organization (fig. 9.18). The strategy proved effective, and Eastman Kodak did not persist in its campaign to extend its sales policy to dry plates. The independents' reduction in plate prices initiated a trend in the profession to these plates.

Another change during the latter years of the first decade brought attention to the pioneer dry plate firm of John Carbutt. Although Carbutt maintained an active policy of product innovation during the last two decades of the nineteenth century, he did not succeed in introducing competitive methods of production that would allow him to compete, pricewise, for the bulk of the professional trade. Therefore, over the last years of the century the company's business rapidly declined. When John Carbutt died in 1905 at age 73, his son sought, without success, to revive the business. About 1908 or 1909 the family sold the Carbutt Dry Plate and Film Corporation of Wayne Junction, Pennsylvania, to the Defender company of Rochester. After a period of considerable technical difficulty, Defender brought Rowland S. Potter, an emulsion maker, from England and

[136] *Eastman Kodak Company Trade Circular,* 7 January 1906, pp. 1–4, Government Exhibit #27, *U.S.* v. *EKCo.,* between pp. 2080–81.
[137] Cramer testimony, in ibid., pp. 73 and 77.

placed him in charge of the operation. In 1911 Defender transferred the production facilities to Rochester, where they were consolidated with the existing Defender plant. Potter played an important role in introducing improvements in the company's plates and paper, and soon the Defender X-ray plates attained a national reputation (see sales figures for Defender sales of dry plates in text table below).[138]

1908	$22,901
1909	20,926
1910	34,656

The character and products of the dry plate industry changed significantly during the period from 1895 to 1909. Numerous improvements were introduced, and the quality and reliability of dry plates increased. These changes were linked to the mobility of master emulsion makers and the diffusion of emulsion secrets through the acquisition of one company by another. George Eastman succinctly indicated the effect of consolidation upon emulsion technology: "In the last few years we have developed a scheme for making dry plates, by combining all of the improvements that we have found in the different factories we have bought, that is well nigh perfect."[139]

The structure of the industry reflected the general trend of consolidation in American business. Whereas in 1895 local and regional markets sustained a couple of dozen fairly large dry plate producers, by 1909 three companies, Eastman Kodak, Cramer, and Hammer, dominated an oligopolistic dry plate market. As in the case of photographic paper and plate cameras, the substantial profits from the new roll film amateur business provided the capital for the acquisition of important firms in the industry and, in the case of dry plates, the partial consolidation of the industry.

With the acquisition of a number of companies and the growth in scale and complexity of the Eastman Kodak business, the company underwent substantial organizational change in the fifteen years from the middle of the 1890s. In the first fifteen years of the company's operations, George Eastman had directly supervised and in many cases conducted most of the company's major business and production functions. Until the late 1880s he had directed the emulsion making, and he had directly supervised photographic materials production until the opening of Kodak Park in the early 1890s. Slowly, however, he transferred the research-invention-development and production functions to trusted employees. He then directed his attention largely to the formulation of long-term policy and strategy, to negotiations in the purchase of companies, and to closing major contracts both at home and abroad. Eastman's financial background enabled him to deal with intricate financial matters independently

[138] Wentzel, *Memoirs of a Photochemist*, pp. 17–18 and 123; *Defender Spotlight*, May 1945, p. 3; and J. S. Friedman, *History of Color Photography* (Boston: American Photographic Publishers, 1947), p. 86. Defender sales figures from W. F. Worall testimony, *U.S.* v. *EKCo.*, p. 420.

[139] GE to Reiner, 10 October 1908 (GEC).

Fig. 9.19 George Eastman. Courtesy of Eastman Kodak Company.

of the large banks and without intermediate promoters and underwriters.

In the middle of the 1890s, on the eve of the company's rapid growth, the company's management and organizational structure was relatively simple. With the Board of Directors serving as the principal policy-making center, the company still operated largely as a partnership of Eastman and Strong. The two principals held more than a majority of the outstanding stock, and Eastman commanded enormous respect and authority among the board members who themselves represented the remaining major blocks of stock. The small board, five of whose members had been directors since the original incorporation, represented a cross section of socially, culturally, and financially prominent Rochester families. Consisting of local friends of Strong and Eastman, it did not represent technologically progressive industries, such as local firms prominent in the telegraph, telephone, optics, or tobacco machinery industries. Perhaps it was easier to attract new capital from less growth-oriented sectors, such as buggy whips, nurseries, staves, and shoes. During the next fifteen years, although the composition of the board changed, its character did not.

From the mid-1890s Eastman placed increasing emphasis on building a responsible and talented staff. While he continued to supervise carefully the production side of the business, he demanded detailed reports from his managers. Supervision of the crucially important photosensitive materials production at Kodak Park passed from de Lancey to Lovejoy to Haste, all graduates of M.I.T. De Lancey retired and Lovejoy, in the middle of the first decade of the century, moved to the corporate offices, where he at first assisted and later prepared to succeed to the responsibilities of Eastman.

From the 1890s Eastman gradually built a small staff: personal secretary, sales department, advertising department, and corporate and patent legal counsel. Foreign operations were directed through managers in London and Paris, who were in close contact with each other and in constant communication with Eastman. Late in the 1890s all of the Eastman companies were organized under an English holding company, Kodak Limited, but the combined difficulties in policy management and English tax law stimulated the organization in 1902 of Eastman Kodak of New Jersey, a holding company combining the complete international holdings of the company.

One of the key benefits of the merger and consolidation movement in the industry was the opportunity it presented for selection and development of especially talented and skilled managerial and technical personnel for the rapidly proliferating leadership positions within the Eastman Kodak organization. The integration of the industry permitted a cross fertilization and sharing of ideas among a large number of production and business organizations. Although George Eastman and his inner circle stamped their own larger viewpoints and strategies upon the policies of the consolidated company, it expanded so rapidly that most of the top echelon managers and supervisors developed considerable autonomy. Although the industry possessed its share of mediocre and unimaginative leaders between 1880 and 1900, Eastman, in integrating the industry, brought together not only the best of the industry's technology but also the most talented of its technical and supervisory personnel, building a leadership team committed largely to Eastman's goals and strategies and fully indoctrinated in the importance of quality and reliability in photosensitive products, of technological innovation as a key product strategy, of international diversification as a key marketing strategy, and of a no-debt financial strategy. George Eastman had developed these technological and financial strategies early in his career, but the marketing strategy and nuances of the other strategies gradually emerged from the leadership team Eastman built. As a part of these strategy modifications, the Eastman team also initiated and executed further organizational and industrial changes, through pursuit of vertical integration of the Eastman Kodak Company.

10 / Vertical Integration: Eastman Kodak

Although horizontal integration played an important role in Eastman Kodak's strategy during the 1890s and the first decade of the twentieth century, an equally or, in some respects, more important role was played by the strategy of vertical integration that was interwoven with it. The sequence of strategies of horizontal and vertical integration alternated as the company's product line moved from a nearly exclusive emphasis on products controlled by patents to materials and apparatus that were not controlled by patents: from roll film, roll film cameras, and bromide paper to a range of photographic papers, dry plates, and plate cameras.

Forward Vertical Integration

One of the most crucial elements in Eastman Kodak's successful exploitation of the roll film system was the creation of a system for distributing photographic products to the new mass amateur market. At first the photographic supply houses were the only establishments for the distribution of the popular Kodak camera and the roll film, but this old network proved to have too few retail outlets to meet the needs of the new market. During the 1890s the sales department at Eastman Kodak broadened its network to include thousands of drug, jewelry, optical, department, and hardware stores, which carried Kodak cameras for amateurs as a sideline.[1]

[1] GE testimony, *EKCo.* v. *Blackmore*, p. 1013.

Two major stimuli to the development of a nonprofessional network were (1) the introduction in the mid-1890s of the daylight loading cartridge, which freed the user or dealer from the delicate operation of removing the film from the camera in a dark room; and (2) the introduction in 1900 of the popular Brownie camera. The popularity, simplicity, and low cost ($1.00) of the Brownie camera prompted many merchants to add Kodak cameras as a sideline (fig. 10.1).

One of the problems that Eastman Kodak faced in the middle and late 1890s was that many of the dealers that carried Eastman Kodak products later added other competing lines of cameras and photographic materials. The top management of the sales department and George Eastman, not happy with creating a new sales network and then having other companies capture it, sought a solution to the problem. They adopted the general strategy employed by the unsuccessful plate and paper pool in the middle 1890s: allowance of special discounts to dealers who abided by the terms of sale, e.g., those who sold Eastman Kodak products exclusively.[2] At the turn of the century, as the company placed more emphasis on manufacturing products that were not protected by patents, it extended its terms of sale to all Eastman Kodak products. If the dealer handled even one product that competed with Eastman Kodak, he would then lose the extra discount on the entire Eastman line.[3] This method effectively created a network of exclusive Eastman dealer-

[2] EKCo. to The Trade, 10 February 1898, *U.S.* v. *EKCo.*, pp. 3037–38.
[3] Ibid., 20 December 1899, pp. 3045–47.

Fig. 10.1
The Brownie camera.
Courtesy of Don Ryon,
Eastman Kodak Company.

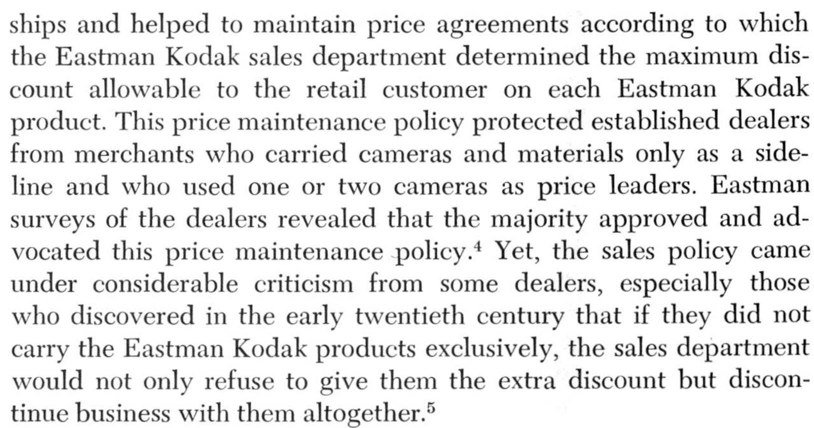

ships and helped to maintain price agreements according to which the Eastman Kodak sales department determined the maximum discount allowable to the retail customer on each Eastman Kodak product. This price maintenance policy protected established dealers from merchants who carried cameras and materials only as a sideline and who used one or two cameras as price leaders. Eastman surveys of the dealers revealed that the majority approved and advocated this price maintenance policy.[4] Yet, the sales policy came under considerable criticism from some dealers, especially those who discovered in the early twentieth century that if they did not carry the Eastman Kodak products exclusively, the sales department would not only refuse to give them the extra discount but discontinue business with them altogether.[5]

During the late years of the 1890s, as the amateur sales soared, the company reorganized the entire sales program. Since the mid-1880s, when Eastman had created the company's own independent marketing department, demonstrators and salesmen had blanketed the country, calling on dealers and photographers with whom they had a special relationship. As sales and the number of personnel increased, this informal structure began to falter. In a reorganization of the department in 1899 the company divided the domestic market into geographical regions with a district sales office established in each region and with demonstrators and salesmen assigned to regions and to specific customers. At the same time wholesale depots were established in New York, Chicago, and San Francisco to service the three basic regions of the country.[6] This decentralized sales system insured that each dealer or professional photographer had an Eastman Kodak man assigned to him, responsible for his needs; however, the system not only made each salesman and demonstrator personally accountable, it insured a steady flow of marketing information to Rochester that proved most valuable in marketing and production decisions.[7] By the beginning of the twentieth century a regiment of seventy traveling men visited nearly every professional photographer or dealer in the country from at least once to a dozen times per year.[8]

As Eastman Kodak also committed itself to professional markets, a different sales tactic was required. While small supply houses and merchants who carried photographic products as a sideline agreed in general to confine themselves to Eastman Kodak products, the large supply houses, located in major cities and catering to professional photographers, were much less inclined to accept the Eastman Kodak terms of sale. Although they handled the company's products, they did not promote them very vigorously because of the small profit margin that remained after loss of the special discount. And so, in 1901, George Eastman urged that his company open its

[4] Ibid., 8 June 1896, 10 February 1898, 6 January 1900, pp. 3032–33, 3036, and 3047.

[5] Decision, *EKCo.* v. *Blackmore*, 277 *Fed. Rep.* 694 (1921); and *Eastman Kodak Company* v. *Southern Photo Materials*, 295 *Fed. Rep.* 98 (1923).

[6] GE to Henry A. Strong, 1 July 1899 (GECP).

[7] Sales Correspondence and Reports, *U.S.* v. *EKCo.*, pp. 2660–3014.

[8] GE to [A German Chemical Co.], 3 February 1902 (GEC).

own store in Chicago, as it had in New York. At once, a number of considerations came under review: managerial responsibilities of a new store, advertising costs, and anticipated losses until a permanent trade could be established. Concluding that acquisition of an existing store might prove more economical, the company purchased Chicago's largest photographic supply house, Sweet & Wallach. This success prompted the implementation of a larger plan, one that originated with Charles Abbott.[9]

Abbott, because of his close relationship with Eastman and his intimate understanding of the details and general policy of the company, sought to resolve both the difficulties of the sales department in dealing with large supply houses and Eastman's difficulties in promoting a combination of dry plate companies. He suggested obtaining options on some of the largest supply houses and possibly even purchasing them, reasoning that perhaps the plate manufacturers would then prove more tractable when negotiations for sale opened again.[10] While the idea of creating a chain of company-owned retail stores was not new—Kodak Limited had adopted it when creating its sales network in Europe—Eastman had previously eschewed the suggestion of employing such a strategy in the United States. Now, however, he accepted it.[11]

Even though the dry plate combination was not fully achieved, the adoption of the plan did initiate a two-year period of company acquisition of supply houses, particularly houses in the upper Middle West: Wisconsin, Minnesota, Iowa, Nebraska, and Illinois. In 1902 Abbott, who assumed responsibility for negotiations and settlement of the purchases, acquired six houses in Chicago, Milwaukee, Minneapolis, St. Paul, and Duluth. In addition, he obtained two houses in Boston: Benjamin French & Company, one of the oldest in Boston, and Horgan, Robey & Company, the largest supply house in the East. In 1903 Abbott added five more houses, in Davenport, Omaha, Lincoln, Des Moines, and Quincy, Illinois. In early 1904 Eastman Kodak extended its network to Toronto and Montreal.[12] This purchase of supply houses dovetailed with the company's acquisition of dry plate and plate camera companies as a part of a new emphasis on production and distribution of materials and apparatus for professional photographers. In conducting this series of acquisitions, Charles Abbott clearly demonstrated his mastery of the art of negotiations.

A gradual shift in the Eastman Kodak domestic sales policy and general strategy occurred during the second half of the first decade. In January 1908, the company discontinued the rebate—or special discount—exclusive dealership, and price maintenance except on patented goods or goods produced by secret formulae. This change reflected personnel changes in the company's Sales Department and state court decisions upholding price maintenance on patented goods. The later overruling of those decisions in *Bobbs-Merrill*

[9] GE to Strong, 17 and 24 December 1901, and 30 January 1902 (GECP).
[10] Ibid., 11 February 1902.
[11] GE to George Davison, 1 October 1900 (GEC).
[12] Ibid., 30 June 1902; and GE to Strong, 1 March 1904 (GECP).

Company v. *Straus* (1908) then threatened the strategy that Eastman Kodak had employed since the mid-1890s. A related response to antitrust legislation and the court decisions of this period was the merger and consolidation of all the companies that Eastman Kodak had acquired and operated as separate firms.[13]

With social, legal, and political pressures mounting, Eastman Kodak returned to the forward vertical integration strategy at the end of the first decade. In 1908 the company purchased major supply houses in Philadelphia and Atlantic City. In 1910 five other retail stores were acquired in large cities strategically located across the country: Atlanta, New Orleans, Denver, Seattle, and Los Angeles.[14] These significant acquisitions completed the phase of forward vertical integration in the American market, with the company possessing a network of twenty retail photographic stores.

In Europe the Eastman companies faced some quite different problems. After completion of the early phase of American supply house acquisitions, Eastman asked Abbott to go to Europe as grand strategist to help develop the European markets for Kodak Limited. Abbott arrived in Europe in the fall of 1902, faced with a very difficult situation. In eight years the European sales strategy had changed from the establishment of sole agencies to (1) the creation of branch companies—there was one in Paris and one in Berlin; (2) the establishment of company-owned wholesale houses—in eleven European cities and in Melbourne; and (3) the opening of company owned retail stores—there were nineteen in Britain, Belgium, France, Russia, Germany, Austria, Italy, and Australia.[15] Much of the policy of the European operation was in accord with the price-cutting strategies of George Davison of London and H. M. Smith of Paris, who found themselves caught between increasing competition, on the one hand, and George Eastman's opposition to price-cutting, on the other. In general, these policies prevailed for a time, but in 1902 eight of the eighteen retail stores lost money. Eastman sent Abbott to Europe to examine the situation and develop a basic campaign directed toward restoring profitable prices.[16]

When Charles Abbott left the United States, he had in mind merging the leading European manufacturers of dry plates, paper, and film with Eastman Kodak. Attention was to be focused upon England at first, but because of the general flow of goods throughout Europe, the Continental manufacturers were also to be considered. During his first six months in Europe, Abbott succeeded in negotiating a plan of acquisition of England's largest photographic manufacturers (outside of Eastman Kodak), Ilford, and he approached, with initial success, Imperial Dry Plate, Cadett & Neall Dry Plate, and Lumière of Lyon. The overtures and negotiations, however, demonstrated to Abbott that the larger companies placed

[13] Eastman Kodak Terms of Sale, 15th edition, 1 January 1908, *U.S.* v. *EKCo.*, pp. 3103–06; and GE to R. C. Sheldon, 14 March 1907 (GEC).

[14] Stockhouse Sales, 1902–12, *U.S.* v. *EKCo.*, p. 3350.

[15] Prospectus of Eastman Kodak Co., New Jersey (1902), *U.S.* v. *EKCo.*, pp. 3260–62.

[16] GE to C. S. Abbott, 18 December 1902 and 11 December 1903 (GEC).

a much higher value on their assets than he could offer. When, much to George Eastman's delight, Ilford's stockholders refused to approve the agreement, negotiations with Acworth and Lumière were dropped, Abbott reasoning that the introduction of the recently acquired high-quality Seed plate into a British-owned firm could quickly bring the dry plate market to the Eastman interests, at which time the larger concerns might be less demanding in their price. In accordance with this new plan, in the spring of 1903 Eastman Kodak purchased the Cadett & Neall company and the Vereinigte Fabriken Photo Papiere Western European marketing network and initiated its campaign to introduce the Seed emulsion to England; but the company found the effort more difficult than anticipated. Thereafter, Abbott dismissed the horizontal integration strategy and focused more directly on forward integration through improving the management of existing company-owned retail stores, reestablishing where appropriate sole agencies with strong local companies, and extending stores to areas not previously represented.[17]

During the next decade Eastman's worldwide marketing network expanded rapidly. The Rochester office assumed responsibility for the Orient and focused attention particularly on the Japanese market. Kodak Limited extended wholesale branches to Switzerland, Denmark, Holland, Spain, Egypt, East Africa, South Africa, and India. New channels of distribution were also opened in Norway, Sweden, and Finland. Much of this expansion came after Abbott's return to America in late 1904.[18] The most troublesome branch of the foreign company early in the century was the one in Australia. The company had opened a wholesale and retail branch in Melbourne late in the century; but, like so many of the foreign stores, it faced fierce local and national competition. In 1902 the Melbourne branch, remote from supervision, acknowledged the largest loss of any Eastman Kodak store. The Eastman operation was sold to Australia's largest photographic mercantile company, Baker & Rouse, with the understanding that they would maintain a certain minimum level of orders with Eastman Kodak and that the London company would have the option at a later date to purchase Baker & Rouse. In 1908 when Baker wanted to retire, the Australian company successfully petitioned Eastman Kodak to exercise its option.[19]

Gradually during the latter half of the first decade, Eastman Kodak's international policy, motivated by high tariff barriers and local public relations considerations, moved to establish production facilities in major nations such as Canada, Britain, France, Germany, and Australia.[20] Although the Kodak trademark represented

[17] GE to Davison, 8 September 1902 (GEC); and GE to Abbott, 8 December 1902 and 13 and 28 April 1903 (GECP).

[18] GE to E. O. Sage, 23 August 1897; GE to Frank Babbott, 2 March 1906 (GECP); and Wyatt Brummitt, "Kodak and the World Market," Brummitt MS, Office of Corporate Information, Eastman Kodak Co., Rochester, N.Y., pp. 1–2.

[19] GE to Davison, 8 September 1902 and 17 October 1904; GE to Moritz B. Philipp, 2 December 1907; and GE to Baker, 22 June 1910 (GEC).

[20] GE to Strong, 8 April 1908 (GECP).

Fig. 10.2
James Haste. Courtesy of Eastman Kodak Company.

quality and reliability, the company sought to diminish its "Yankee-ness."[21] Serious overtures were again made by Lumière to sell to Eastman Kodak in 1907–8, but the opportunity no longer looked attractive to George Eastman, as the company was then pursuing a strategy that focused more on marketing than on the acquisition of manufacturing companies.[22]

Backward Vertical Integration

In the late 1890s, as Eastman Kodak's sales and profits soared, George Eastman began to consider developing the company's capacity to supply the basic raw materials necessary for its manufacture of photographic materials. Because the requisite raw materials were primarily chemicals that carried high profit margins, at an early point Eastman directed investment of company profits in this direction as a cost-saving measure. Further considerations included better control of the critical quality of these vital raw materials through the company's own production, and independence from other firms, which might raise prices, halt production because of labor problems, or leak formulae secrets to potential competitors.

As might be expected, Eastman Kodak first sought to produce critical materials necessary for the production of the nitrocellulose film base and for the emulsion. In 1897 Eastman gave consideration to the construction of a cellulose nitrating plant at Kodak Park and late that year began interviewing and hiring key personnel for the plant.[23] James Haste, an 1896 graduate in chemical engineering from M.I.T., planned the production facility, supervised the construction and installation of machinery during 1898, and early in 1899 directed the first commercial production of nitrocellulose at Kodak Park (figs. 10.2 and 10.3).[24] By mid-year, contracts with the traditional supplier, Charles Cooper & Company of Newark, were terminated and the Kodak Park Chemical Plant filled the company's nitrocellulose needs. The relative savings on this one item alone can be estimated by noting that in 1901 the market price for large quantities of nitrocellulose for film purposes was $1.50 per pound, while production cost at Kodak Park was $.42 per pound.[25]

The Kodak Park Chemical Plant also initiated production of other basic chemicals employed in production of the film base, including sulfuric acid, nitric acid, and fusel oil (a mixture of amyl alcohols). In 1898 the company began producing another major item, silver nitrate, at a considerable cost reduction. In the early 1900s Eastman Kodak established the Powder and Solution Department at Kodak Park, where fine chemicals for photographic purposes—developing

[21] GE to Abbott, 25 October 1902 (GECP).
[22] GE to H. M. Smith, 6 May 1907; and GE to Baker, 31 August 1910 (GECP).
[23] GE to G. Adelmann, 14 August 1897 (GEC).
[24] James H. Haste testimony, *Goodwin* v. *EKCo.*, p. 2364, and *U.S.* v. *EKCo.*, p. 421; and *Rochester Democrat and Chronicle*, 8 January 1929, p. 1.
[25] Frank W. Lovejoy testimony, *Goodwin* v. *EKCo.*, p. 2346; and GE to Abbott, 20 August 1901 (GECP).

Fig. 10.3
Interior of the nitrocellulose building, Kodak Park, c. 1902. Courtesy of Eastman Kodak Company.

and toning solutions—were either produced or purchased in bulk and then repackaged for amateur and professional use.[26] Young, academically trained chemists and chemical engineers were assigned the tasks of creating these new chemical production facilities at Kodak Park, insuring the employment of efficient, well-conceived, and tested methods of production based on chemical and engineering principles.

A second major area of supply control that George Eastman addressed in the late 1890s was, as indicated above, the raw paper supply. In this instance he sought not to produce the paper but to gain, through contracts, an insured supply from Europe. Abbott and Eastman, through their negotiations with the General Paper Company of Brussels in 1898, guaranteed themselves an adequate flow of the vital raw material. Eastman did, however, in the late 1890s establish a baryta coating plant at Kodak Park, thereby taking one step in the direction of production of photographic paper.[27] In 1906, as the renewal date for the contract with General Paper approached, Eastman hired a practical chemist with an interest in paper production to experiment with the production of raw photographic paper. After Eastman renewed the contract with General Paper, he assigned his chemist to work on the development of a noninflammable film base.[28]

As the Eastman Kodak Company acquired dry plate companies during the period 1902–4, George Eastman turned his attention to an adequate and high-quality supply of plate glass. Previously, the

[26] Lovejoy testimony, *Goodwin* v. *EKCo.*, p. 2352; GE testimony, *EKCo.* v. *Blackmore*, p. 1076; GE to James Pender, 6 July 1906; GE to Brown Bros., 11 August 1898; and G.E. to Charles W. Markus, 19 January 1904 (GEC).

[27] GE to Walker, 9 July 1897 (GECP).

[28] David E. Reid obituary, *Rochester Democrat and Chronicle*, 16 April 1934; and GE to W. H. Oglesby Paper Co., 3 August 1906 (GEC).

relatively low Eastman Kodak demand had been supplied by British and Belgian glass companies, but early in 1903 the company acquired a small but important company, Thatcher & Whittemore, which purchased, cut, and sorted glass for plates in Gilly, Belgium, and then shipped them to its depot in St. Louis for distribution to the various dry plate manufacturers in the United States. Eastman retained the two young St. Louis men to operate the company, but their speculation and mismanagement served to impede the flow of the crucial material. Consequently, when Eastman heard that the Belgian glass companies were considering forming a combine, he secured long-term contracts with a few of the producers of the best quality glass and, in the fall of 1904, terminated the Thatcher & Whittemore business.[29]

While basic chemicals, photographic paper, and glass were major supply needs for Eastman Kodak, the company also began to produce many of its other needs. From the 1890s the company produced its own boxes for packaging and shipping (fig. 10.4); in the early twentieth century it began producing some of its own shutters; and in the early 1900s it developed its own printing facilities for labels and directions.[30]

Perhaps as significant as the supplies that Eastman Kodak produced for its own needs were those it chose not to produce. In the areas of optical goods—lenses and shutters—its traditional close ties with Rochester's Bausch & Lomb Company no doubt had much to do with the company's continued dependence upon the old optical firm. The company also purchased some shutters from the new Rochester firm, Wollensak. The profits on these optical and mechanical elements were high, as Eastman knew, and he finally applied pressure to Bausch & Lomb to reduce prices on the lucrative Eastman Kodak orders.

Three other areas that Eastman Kodak had not entered by the end of the first decade of the century were photographic paper, gelatin, and certain chemicals, particularly fine chemicals. These very vital products, all of exacting quality for photochemical purposes, were supplied nearly exclusively by German industry.[31] Long-term contracts did much to insure dependable supplies of these raw materials, which were vital to the Rochester firm. Already, late in the first decade of the century, Eastman gave consideration to relieving this dependence upon European industry and planned to develop production capacities in Rochester. As the surplus account and the

[29] GE to Rudolph Speth, 18 March 1903; GE to C. W. Thatcher, 10 October 1903; GE to M. A. Seed, 24 February 1904; and GE to Abbott, 2 September 1904 (GEC).

[30] GE to Strong, 24 December 1901 (GECP); and GE to Davison, 29 August 1903 (GEC).

[31] Virtually all organic fine chemicals were purchased from Germany at this time. For example, developers, including the highly popular metol and hydroquinon, came from Germany and were repackaged under the Kodak label. Also, the Eastman company depended on Germany for fine chemicals for experimental purposes. Even certain chemicals employed on an industrial scale, such as acetic anhydride, were provided by German chemical firms (GE to Strong, 7 July 1908 [GECP], and GE to E. H. Janson, Mgr., Kodak Gesellschaft m.b.H., Berlin, 28 September 1914 [GEC]).

Fig. 10.4 Interior of box factory at Kodak Park, c. 1906. Courtesy of Eastman Kodak Company.

technical capacity and knowledge of the firm grew, the feasibility of the American firm's entering these critical production areas increased. Yet not until the years immediately prior to World War I did the company enter upon production of these supplies.

It is striking that Eastman Kodak did not begin to furnish its own raw material needs through the acquisition of existing supply companies but through the creation of its own manufacturing capacity, which was designed, installed, and supervised largely by young engineers and chemists from American technical institutes, particularly M.I.T. In traveling this route, Eastman Kodak not only gained the profits that its former suppliers had made but, by introducing the latest methods of chemical production, undoubtedly gained even further advantage by new cost-saving methods. The contrast between the backward vertical integration strategies pursued by creating new supply facilities and the forward vertical and horizontal integration strategies pursued by acquiring existing firms highlights the discriminating scrutiny that Eastman and his inner circle bestowed on the whole integration and consolidation strategy. Established supply houses and photographic paper and dry plate companies had important established reputations and good will that would be costly to duplicate. Furthermore, the manufacturers of the acquired paper and plates possessed unique, vital technological secrets, and the camera companies had highly skilled craftsmen and designers who could not be duplicated. The supply areas that Eastman chose at this time not to enter—most notably raw paper and gelatin production—were those which, like the photographic industry itself, were still dominated by secrets, trial-and-error methods, and skilled artisans. Eastman's sensitivity to the differing roles of the experienced artisan and the academically trained student of science or engineering and also to market considerations influenced the patterns of horizontal and vertical integration not only of Eastman Kodak but of its imitative competitors, such as Anthony and Scovill, as well.

11 / Horizontal and Vertical Integration: Anthony and Scovill

During the late 1890s it became increasingly clear that Eastman Kodak had emerged into the leadership position in the American photographic industry in both the amateur and professional market sectors. This change in leadership left the traditional leaders, Anthony and Scovill, in a defensive position. Their continued reliance into the 1890s on the jobbing function, which initially gave control of the industry and apparently assured profits, proved disastrous because their long-held conception of themselves as "photographic merchants" no longer permitted them to operate effectively as leaders in an industry that had moved rapidly between 1880 and 1900 from decentralized production at the local level, with centralized national distribution of producers' supplies and materials, to centralized national production and national mass distribution of consumer semidurable and consumable goods. The traditional national distributors of producers' supplies and materials failed to perceive and act upon the changing character of the industry. What production facilities these old-line companies had developed over the years were confined largely to producers' durables, and therefore, their production experience did not even lend itself to the new growth areas in production in the industry.

The older companies even failed to embrace an imitative posture once Eastman Kodak led the way. It is often not the revolutionary technological innovator but the more assiduous imitator who emerges as the dominator of an industry, but in the photographic industry at the turn of the twentieth century, the failure of the old-line companies to imitate the Eastman Kodak roll film system or to circumvent its patent barrier left them outmoded and with little strength to face the new industrial climate. Had the older companies consummated a few key mergers with dry plate and photographic paper companies, even as late as in 1896, they could have enchanced their chances of preventing Eastman Kodak's dominance of the professional materials and apparatus sectors. The Eastman company only reluctantly entered and consolidated plate camera and dry plate production. However, instead of seizing the initiative and developing a viable production capacity, the photographic merchants, especially the Anthony company, sought at that time to disengage from manufacturing.

Although the old-line companies did not eschew technological change or product innovation, their conception of the business potential of patented technological innovation remained tied to their older conception of the nature of the industry. Throughout the period from the 1870s to the beginning of the twentieth century, Anthony's, and to a lesser extent, Scovill & Adams' production employees developed and patented numerous improvements in cameras and ancillary products; but their conception of technological innovation saw these exclusively as "improvements" in established products. Furthermore, the old-line companies relied on the spontaneous inspiration of employees who "hit upon" something and then patented it. The leadership at Scovill or Anthony did not consciously seek to develop and introduce a *system* of improvements with interlocking patents that could serve to discourage imitation. Furthermore, the firms did not pursue a conscious policy of annual model changes. Here again, the conception of the market was im-

portant. These firms aimed their production at the professional photographer who sought a piece of quality equipment that represented for him a capital investment. The Eastman policy of annual model changes in cameras developed in the mid-1890s, in response to (1) a mass consumer market where the consumer made a purchase on a less rationally calculated basis than the professional photographer; and (2) an international market where patent protection proved less reliable than in the domestic market. With no conscious patent policy or strategy and only a limited perception of the potential interrelation of patents and marketing, the old-line companies acquired a miscellaneous collection of patents but did not consciously seek to direct their technological innovative activity.

Two events in the second half of the 1890s awakened these companies to the fatal weakness of their reliance on the jobbing function for control of and survival in the industry: (1) the failure in the mid-1890s of the Photographic Board of Trade to enforce through marketing restrictions the price agreements on dry plates and photographic paper; and (2) the loss in 1899 of the sales agency for American Aristotype when it merged with General Aristo and turned a major portion of its marketing over to Eastman Kodak. The removal of that lucrative contract represented a major blow to an already shaky firm.[1]

In the period of 1899 to 1901, Anthony belatedly sought to strengthen its position. Recognizing the weak market position of the once strong camera manufacturing department, the company sold to the Rochester camera combine in 1899 its plate camera business; yet, it retained the fledgling roll film camera capacity it had developed in the late 1890s through the acquisition of a patent licensing agreement from American Camera Manufacturing. The agreement contract continued in force under Eastman Kodak's ownership of American Camera. Of more fundamental importance were three additional moves made by the Scovill and Anthony interests. First, Scovill and Anthony joined in July 1901 in acquiring a major interest in the recently organized Goodwin Film & Camera Company, whose major assets consisted of the Goodwin celluloid film patent and a very small and primitive film production facility in Newark, New Jersey. Second, six weeks later Frederick Anthony approached Eastman and sought without success to interest him in buying the Goodwin company. Third, in less than six months the Anthony and Scovill interests brought several companies together to form a new consolidated firm, the Anthony and Scovill Company. The new business pressures and opportunities had begun to force the old-line companies to alter their traditional conceptions and strategies.

The key to these changes was the Goodwin patent. The Hannibal Goodwin patent application had remained in the U.S. Patent Office during most of the 1890s. The interference proceedings that had taken place in the Patent Office between the Goodwin application and the Reichenbach patent in 1893 had resulted, through Goodwin's failure to prosecute the final stages of the proceedings, in his

[1] It lost $62,000 in revenue. GE to Henry A. Strong, 8 May 1899 (GECP).

conceding to Reichenbach the claims for celluloid film made with a high (60 percent) camphor content. For the next two years, the Goodwin application lay dormant in the Patent Office. Then, just one day before the application would have lapsed, Goodwin and his attorneys submitted an amended specification of considerable length with new and much more sophisticated chemical and physical language, obviously drawing upon the expert testimony provided in the earlier interference proceedings. Again, as in every previous amendment, the examiner rejected the new amendments. From 1895 to 1898, efforts at amendment and successful issue of the patent quickened—but with little effect on the examiner, who maintained that the claims were anticipated both in earlier patents and in the general photographic literature.[2]

With considerable persistence, Goodwin's attorneys continued, finally taking the case early in 1898 on appeal to the Board of Examiners of the Patent Office. The breadth of Goodwin's claims stood in the way of the examiners' making a favorable decision. Furthermore, the examiner argued that the new language of the Goodwin claims introduced in the 1895 amendment failed to fall under the original claims of the application in 1887.[3] The Board of Examiners weighed the evidence of the two briefs, the contents of the Goodwin File Wrapper and Contents and the Goodwin-Reichenbach Interference, and decided that it was indeed a very complex case but that Goodwin's claims appeared to be novel and were consistent with the original application. They concluded: ". . . We deem it best to allow the claims, broad as they are, in the shape in which they are presented, in order that the applicant may, if he can, sustain them in the courts, since in that forum the many difficult questions of law and fact raised by this record . . . can best be settled. . . ."[4]

Therefore, the Patent Office issued the Goodwin patent on 13 September 1898, more than eleven years after the original application had been filed. Goodwin and his attorneys interpreted the patent as covering any process of producing a flexible roll celluloidal film except one that, like the one in Reichenbach's patent, employed large quantities of camphor. Moreover, they believed that Eastman Kodak had abandoned the use of camphor in its film base because of its deleterious effect on the photosensitive emulsion and, therefore, regarded the Goodwin patent as a key or controlling patent in the industry. Undaunted, Eastman casually remarked that "the Reverend Goodwin ghost [was] up again." The Goodwin patent was not taken seriously in Rochester or New York City at the office of the company's patent counsel.[5]

Before Goodwin's death in 1900, he formed the Goodwin Film & Camera Company to exploit the patent and to initiate small-scale

[2] "Goodwin File Wrapper and Contents," U.S. Patent Office, in *Goodwin* v. *EKCo.*, pp. 3575–3830.

[3] "Goodwin Appeal File Wrapper," in ibid., pp. 2682–778.

[4] Decision of Board of Examiners, "Goodwin Appeal File Wrapper," in ibid., pp. 2759–76, and quotation, p. 2776.

[5] GE to Moritz B. Philipp, 12 September 1898 (GEC).

production of celluloid film. Extensive efforts by Goodwin's attorney's eventually led to involving the Scovill and Anthony companies in the ownership of the Goodwin firm. Although the two old-line companies now had an opportunity to produce the most profitable item in the photographic industry without risking any capital beyond that required for initiating production, this was not part of the vision of those involved with the Goodwin patent.

The new interests in the Goodwin patent, still restrained by their marketing orientation, did not envision themselves as creating a company to rival the Rochester company. They entered into the Goodwin company for speculative purposes rather than to produce and market roll film. Within a month, Frederick Anthony called on George Eastman in Rochester and sought to sell Eastman Kodak the Goodwin company for half a million dollars in cash and an equal amount in Eastman Kodak stock. At that meeting Anthony admitted that he placed such a high valuation on the patent because he understood that Eastman Kodak had infringed it, and therefore, a large claim might be recovered from the Rochester company. Eastman relayed to his patent attorney his response to Anthony's veiled threat:

. . . We were only looking for somebody to sue us under that patent; that we had been hoping for years that somebody would muster enough courage to do it, and that now he and Richard Anthony and Mr. Adams had acquired control, we had renewed hopes. I told him that I had not consulted you to the point, but that I thought we would be even willing to aid him to such an extent as to admit exactly what we were using, and thus relieve him from the necessity of proving something that might be difficult . . . that if he would only sue us, there would be no hard feelings, it would be regarded as a friendly act.[6]

Frederick Anthony's speculative fever must have abated as a consequence of Eastman's attitude. Although Eastman was capable of playing a convincing role, many of his remarks here revealed his true feelings and understanding. He concluded this missive to his attorney thus: "I am thankful that we do not want to buy the Goodwin patent." Any idea the Anthony and Scovill people might have had of making a quick profit by selling the patent to Eastman Kodak evaporated that day. From that point, any return on the Goodwin patent lay along the rough roads of litigation and the production and marketing of film.

Consequently, the most significant aspect of Eastman's failure to purchase the patent was that the Goodwin patent served as the catalyst in the merger of the Anthony and Scovill companies. During the remaining months of 1901, the Anthony and Scovill companies introduced the commercial production of photographic roll film and created a new organization of the combined companies. On 23 De-

[6] Ibid., 14 August 1901. The essence of this statement was later confirmed by Frederick A. Anthony in his testimony, *Goodwin* v. *EKCo.*, p. 1296.

cember 1901, the Anthony and Scovill Company was chartered as a New York corporation.[7]

The motivations behind the merger were at least two-fold. The opportunity and need for joint cooperation on the promotion of the Goodwin patent was, of course, one. The other was the increasing pressure brought on by Eastman Kodak's sales policy; by creating exclusive dealerships, it stimulated both the old-line companies and Eastman Kodak to manufacture a full line of photographic materials and apparatus for amateurs and professionals. The Anthony and Scovill merger addressed itself to the dealer who, wanting to be independent of the Eastman sales network, needed a full line of photographic goods.

The Anthony and Scovill Company, a holding company, combined a variety of companies and contracts that assured the new company of nearly a full product line. The old Anthony company contributed manufacturing capacity in chemicals (Jersey City plant), collodion photographic paper (Monarch Paper Company of Cortland, New York, and Binghamton, New York), roll film cameras (Greenpoint, Long Island, camera works), and miscellaneous supplies such as albums and developing trays. The old Scovill Company contributed the capacity to produce plate cameras (American Optical Company, New Haven, Connecticut) and miscellaneous supplies. The Goodwin Film and Camera Company held the potential for production of photographic roll film. In addition, Anthony and Scovill purchased the Columbian Photographic Paper Company of Westfield, Massachusetts, a producer of DOP. Furthermore, the company held a jobbing contract with the Hammer Dry Plate Company of St. Louis. The weakness of the new combine, however, was its lack of capital, a problem with which the company grappled for the next two decades.[8]

The new combined company immediately faced the problem of wide dispersal of production facilities and offices. The principal offices of the company were on Fifth Avenue in New York, with production facilities located at Binghamton, New Haven, Westfield, and Jersey City. The executive officers of the company, Alexander C. Lamoutte and the Anthonys, remained at the New York offices, leaving managers in charge of each of the production plants. The Monarch Photo Paper plant, located in Binghamton, in time became the focal point for the centralization of the company's production and office facilities.

The Goodwin Film & Camera Company attracted the immediate attention of the leadership of the newly organized consolidation. Following his failure to interest George Eastman in the purchase of the company, Frederick Anthony sought someone with experience in the coating of celluloid film to supervise production, and he hired a former emulsion maker for the Blair Camera Company. After having surveyed other production sites, he and the company executives

[7] "The Anthony and Scovill Company," New York Corporations, Abstract of State Records, File No. 8/0–48, Bureau of Corporations, Dept. of Commerce and Labor, National Archives, Washington, D.C.

[8] Ibid.

Fig. 11.1
Ansco film cartridge.

decided to install the plant temporarily in the Monarch building in Binghamton and eventually build a new plant there.[9] The Anthonys and Lamoutte assumed the company could avoid production of the celluloid film base by purchasing it from the Celluloid Company in Newark. Frederick Anthony initiated negotiations with the Celluloid Company, seeking to acquire a supply of celluloid film base. Soon he learned that the Celluloid Company did not follow a process covered by the Goodwin patent. Since the principal reason for building the plant was to demonstrate the feasibility of the Goodwin process, Frederick Anthony insisted that the Celluloid Company produce the base under the patent and disclose the process. Celluloid, suspicious of the motives of the photographic firm, feared that disclosure of their formulae would either make them vulnerable to suit under the Goodwin patent or provide the fledgling company with important celluloid formulae secrets. Consequently, although the negotiations continued into the late spring, Celluloid finally abandoned them. Frederick Anthony then decided that the company should undertake production of the film base in Binghamton.[10]

The first Goodwin roll film was produced on a crude wooden frame machine in December of 1902.[11] A number of production problems with both the base and the emulsion plagued the new film. It was difficult to enter the market with an unknown product, but during the next three years sales, although small, grew steadily.[12] With the introduction of film production, Anthony and Scovill initiated a decade of considerable rivalry with Eastman Kodak. The contest was hardly evenly matched, as George Eastman well knew. As he did so often in his letters to Strong, Eastman responded with tongue-in-cheek to the Anthony and Scovill merger: "Whatever Kodak stock you desire to unload you should put it in the hands of Bert Fenn at once. . . ."[13] "I send you also a clipping from a Springfield paper giving details of the organization of this powerful competitor."[14] Yet, Anthony and Scovill proved to be more than a mere nuisance to the growing Rochester giant. In less than a week following the successful commercial production of film, Goodwin Film & Camera filed a suit against Eastman Kodak, charging infringement of the Goodwin patent. This action initiated more than a decade of prolonged court testimony and argument. Eastman Kodak responded quickly to the original charge.

Although George Eastman could state, "We do not trouble ourselves about the Anthony Co. nearly as much as they seem to think,"[15] Eastman Kodak did take specific actions against the new combine. When in 1903 Anthony and Scovill completed a camera factory at Binghamton and introduced roll film cameras (fig. 11.2),

[9] F. A. Anthony testimony, *Goodwin* v. *EKCo.*, pp. 367–68.

[10] Marshall C. Lefferts testimony, in ibid., pp. 2431 and 2447–78.

[11] F. A. Anthony testimony, in ibid., p. 368.

[12] Sales: in 1903, $58,226; in 1904, $98,229; in 1905, $115,374; in 1906, $102,129. From Sales Books of the Goodwin Film & Camera Co., in ibid., p. 1315.

[13] GE to Strong, 17 December 1901 (GECP).

[14] Ibid., 8 January 1902.

[15] GE to Philipp, 19 March 1904 (GEC).

Ansco Camera

No. 1

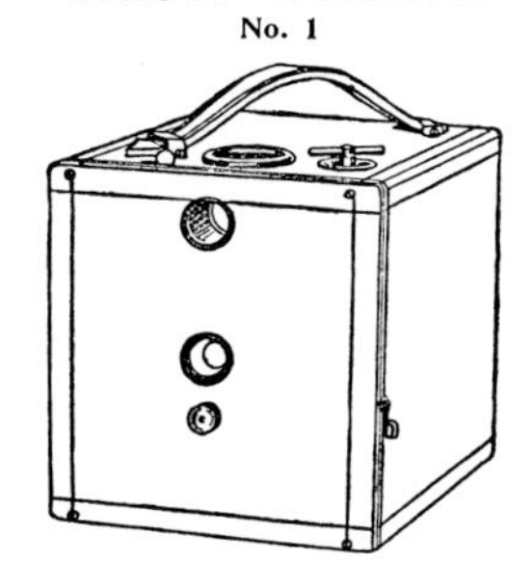

A thoroughly practical and exceptionaly desirable camera for snapshots or time exposures.

Shutter is always set and ready for any emergency.

So simple that a child can handle and produce perfect and satisfactory pictures.

All objects are in focus at any distance. Fitted with very best acromatic lense and brilliant view finder.

Uses 3½ x 3½ 6 or 12 exposure day-light Loading Films.

PRICE, - $5.00

The Anthony & Scovill Co.

Binghamton, N. Y. AND New York City

Fig. 11.2
Ansco camera no. 1.

Eastman Kodak filed suit against them for infringing the Turner patent on the daylight-loading feature. Moreover, in December 1904, George Eastman showed considerable displeasure with Anthony and Scovill, stating that they were flagrantly infringing the Kodak trademark. He insisted that M. B. Philipp, the company's principal patent attorney, write his strongest letter demanding an apology. Eastman went on, "We would like to publish a letter from such a concern that will be humiliating enough to show that they cannot do this sort of thing with impunity, or else put them to enough expense to impress upon them the same thing."[16]

Anthony and Scovill, however, possessed further retaliatory power. In 1899, Senator Thomas F. Donnelly had introduced and directed to passage in the New York State legislature the Donnelly Act, which declared as unlawful any agreement or combination creating a monopoly in the manufacture, production, or sale of any article in the state of New York. Twice, in 1903 and in 1904, the Anthony and Scovill interests promoted complaints to the attorney general's office in Albany. Although this action created considerable consternation and concern in Rochester, in both cases the charges were dismissed after lengthy hearings.[17]

The interfirm rivalry, especially through legal actions such as those pursued by the two New York photographic concerns, proved to be, of course, quite expensive. In such a competition, the larger and better-financed company had considerable advantage. Even losing the defense of a patent suit for a large company might, because of the financial burden to the complainant, prove advantageous in the long run. In 1906, for example, Eastman Kodak paid between three and four thousand dollars a month to defend itself in the Goodwin suit.[18] Expenses, especially for expert witnesses, were also quite high for the Anthony and Scovill Company. Indicative of the financial plight of the company were the periods of inactivity in the suit during which Anthony and Scovill sought to recover financially. The short-term financial advantage of these breaks was offset, of course, by the long-term disadvantage that each passing year prior to the decision represented the loss of about 1/17th of the value of the patent. All advantages, however, did not automatically fall to the Rochester giant. The Eastman Kodak suit against Anthony and Scovill for infringing the Turner patent failed both in the federal district court and in the court of appeals, for want of inventive novelty.[19]

Nevertheless, the financial burdens of litigations compounded the financial and production problems that Anthony and Scovill faced in the opening years of the new century. The new company found that the costs of outstanding bonds inherited from the old Scovill & Adams Company and the capital expenditures for new plant and

[16] Ibid., 9 December 1904.

[17] "Curbing the Trusts," *New York Times*, 26 September 1902, p. 3, and 7 August 1903, p. 3; GE to Strong, 8 February 1904; and GE to Sir DesVoeux, 23 July 1906 (GECP).

[18] GE to Strong, 30 January 1906 (GECP).

[19] 139 *Fed. Rep.* 36 (1905) and 145 *Fed. Rep.* 833 (1906).

machinery were not offset by adequate income. The film plant had clearly not been organized with production profits in mind, but merely to produce film quickly according to the Goodwin process so that suit could be filed against Eastman Kodak. None of the other products manufactured by the company proved to have any sales success in the first half of the decade. Furthermore, the jobbing function had all but disappeared for the former jobbing leaders of the industry. Only the Hammer plate agency proved to be of any real value to the company. Consequently, Anthony and Scovill failed to produce profits and dividends but instead, because of its speculative policy with regard to the Goodwin patent, found itself falling deeper into debt. The resulting internal crises drew a series of significant responses.

As the financial burdens became heavier, the Anthonys tendered a series of offers to George Eastman to sell both Anthony and Scovill and Goodwin Film & Camera. Two Anthony and Scovill offers in 1904 were spurned by Eastman. With all the publicity over trusts and the two actions against Eastman Kodak pending with the state, the wisdom of Eastman's attitude could hardly be questioned. A year later, in the summer of 1905, following the disposition of the Turner case in favor of Anthony and Scovill, a further offer was made, and this time Eastman made it very clear that he was not interested. He stated flatly to his legal counsel: "We do not want to buy out the Anthony business and never have since we investigated it, and I can see no probability of our wanting to. . . . [With regard to the patent] if it is not [declared good], the invention is open to everybody and we do not consider the Anthony outfit any more formidable than plenty of others who would rise up to succeed them if we should buy them out."[20] Clearly, Eastman Kodak did not covet its New York rivals, and certainly not at the price the Goodwin interests expected.

As the Goodwin patent litigation continued and Anthony and Scovill's financial woes mounted, the company's leadership had to develop new strategies. During the period from 1904 to 1906, major changes in the leadership of the company occurred, including the departure of the two Anthonys and an increase in the representation of the banking interests that held the company's notes.[21] With the changes in leadership and ownership in Anthony and Scovill and Goodwin, the general direction and policy of the companies changed, taking on a more distinct production orientation. Moreover, the widely separated operations of the company were centralized at Binghamton (fig. 11.3).[22] Although the new banking and financial leadership was undoubtedly speculating on the Goodwin patent, it took a long-range view of the operation, in contast to that taken by the Anthonys. It directed attention to production economies and efficiencies and sought to derive much-needed profits from the production operations of the company. Furthermore, the new leadership believed that before the primary problems of constricted

[20] GE to Philipp, 21 July 1905 (GEC).
[21] Richard A. Anthony testimony, *Goodwin* v. *EKCo.*, p. 442.
[22] Alexander C. Lamoutte testimony, in ibid., p. 1302.

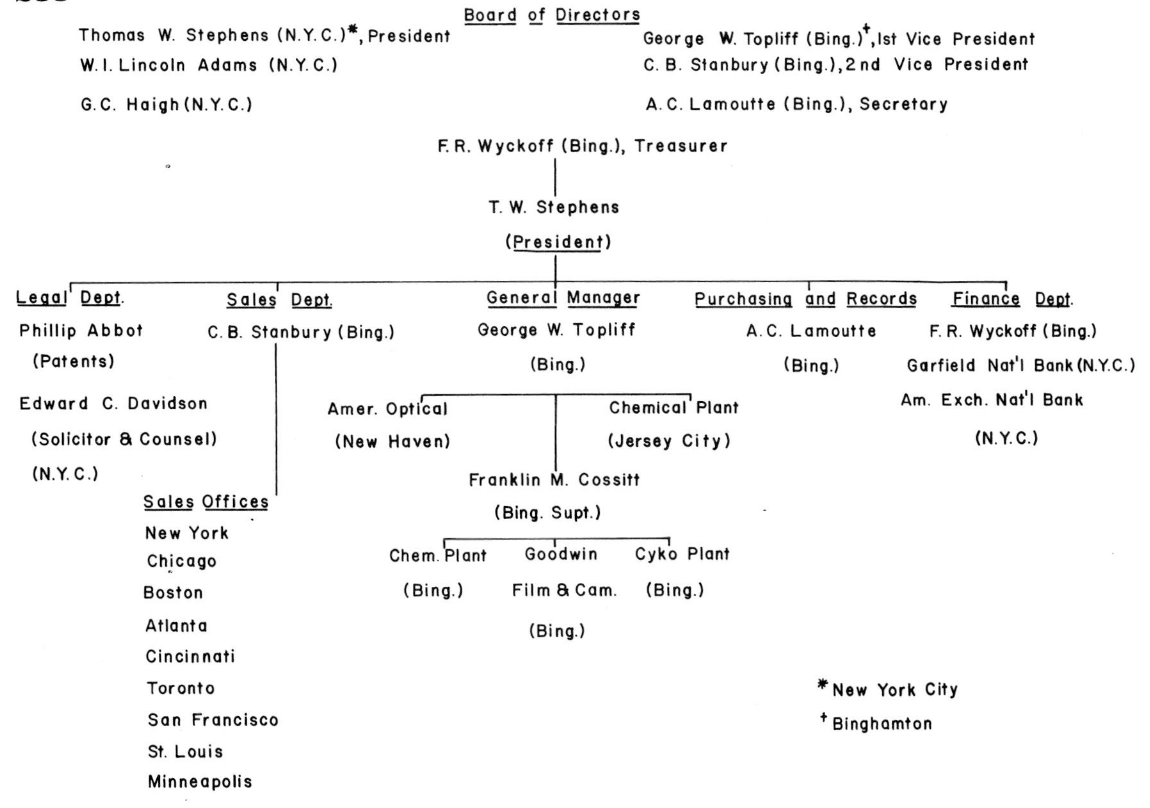

Fig. 11.3
Ansco Company Organization, 1909.

marketing opportunities could be attacked, profitable products of reasonable quality had to be produced. Upon the success of this policy depended the continued prosecution of the Goodwin patent, the company's only major asset at this time.

From 1906 through 1909, the Anthony and Scovill Company underwent reorganization but with considerable personnel stability (fig. 11.3). In 1907 Anthony and Scovill reorganized as the Ansco Company (ANthony, SCOvill), employing as the company name the trademark it had used for several years. The board and top executives and production personnel changed very little during the second half of the decade. Both Anthonys had left and even sought unsuccessfully to sell their stock to George Eastman.[23] The new leadership moved the company in the direction of efficient production of a full line of photographic goods. George Eastman observed late in 1907: "The present indications are that the Anthony Co. is being conducted on more business like lines than it has been before in many years and I would not be surprised if they were in a way to make some money. . . ."[24] The company also continued to defend and prosecute the patent cases then in court. The Federal District Court found in their favor in the Turner patent case, and in the spring of 1906 the appeals court dismissed the Eastman Kodak petition.[25] Buoyed by these favorable decisions, Ansco continued the

[23] GE to Philipp, 2 December 1907 (GEC).
[24] Ibid.
[25] 139 *Fed. Rep.* 36 (1905) and 145 *Fed. Rep.* 833 (1906).

Goodwin suit against Eastman Kodak, though not without considerable expense.[26]

The new leadership focused on improving the company's products and methods of production. The gelatin DOP, Cyko, began to ride the wave of popularity of developing-out papers, and production and sales increased markedly during the second half of the decade. Cyko became the principal product asset of the Ansco company. Attention also focused on the all-important Goodwin roll film. Quality improvements were introduced, including giving the film two coats of gelatin to produce noncurl film in 1905.[27] In the production area, two film-making machines of iron construction replaced the wooden ones, reflecting the company's increased commitment to production. In 1908, as the independent movie producers sought desperately for alternative supplies of cinematographic film, they turned to Ansco, which produced some film on the oilcloth-belt machines. Although the process worked, it was not fully satisfactory for cinematographic purposes. Soon, construction of a drum system of production such as the one employed by Eastman Kodak and Celluloid Company was begun and later significantly advanced the Goodwin company's method of film production.[28] Following the decision on the Turner patent, Ansco continued its production of inexpensive roll film cameras under the trade name Buster Brown.[29]

The sales department of Ansco faced the difficult problem of trying to cope with the effects of the Eastman Kodak sales policy and of Ansco's less-touted product line. Eastman Kodak's exclusive dealerships and the network of Eastman-owned stores in major cities greatly curtailed the avenues of retail distribution open to Ansco. Ansco's leadership developed a sales strategy that closely followed that of Eastman Kodak in its broad outline while deviating in its details. It established a network of its own offices and retail stores in major American cities and in Toronto. In addition, like Eastman Kodak, Ansco introduced a policy of exclusive dealerships for Ansco products. Other companies, such as Defender, complained of the policies of Ansco and Eastman Kodak but had to rely on the photographic dealers—approximately 3 percent—who remained independent.[30]

Although the new Ansco leadership brought substantial organizational and strategic changes to the struggling company, and although it did meet with growing success in creating a sales network and in selected production areas such as photographic paper, the company made no net profits during the first decade of the new century. Progress was made in curtailing the rate of losses, but the expensive litigation proceedings and production inexperience, aggravated by limited capital resources, still made profits difficult to

[26] Ansco Company, *Journal*, pp. 179 and 190. Located at the Photographic Dept., Smithsonian Institution, Washington, D.C.

[27] Lamoutte testimony, *Goodwin* v. *EKCo.*, p. 1340.

[28] Frank M. Cossitt testimony, in ibid., pp. 462–64 and 482–85.

[29] Carl Bornmann testimony, *U.S.* v. *EKCo.*, p. 514; and Ansco Company, *Journal*, p. 49.

[30] Ansco Company, *Journal*; and Frank W. Wilmot testimony, *U.S.* v. *EKCo.*, p. 191.

generate. The stockholders and bond holders continued to be speculators hoping for a favorable decision in the Goodwin trial. The principal difference between the old Anthony leadership and the new financially-oriented leadership was that the latter speculated on the basis of a broader horizon and, therefore, successfully battled—with eventual success—to preserve the corporation's financial integrity.

During these years when Anthony and Scovill were struggling to respond to and survive the economic and business consequences of the introduction of gelatin emulsions and the amateur roll film system, technical developments in a new direction, that of cinematography, were laying the foundations for a major redefinition of the industry. Technologically dependent upon both gelatin emulsions and celluloid film, these cinematographic developments were but the prelude to a new era in which both cinematography and amateur photography shared in stimulating explosive sales growth and in dominating both the technological and business aspects of the industry.

12 / The Emergence of the Cinematographic Industry

During the last years of the nineteenth century and first decade of the twentieth, a major new growth sector of the American photographic industry, cinematography, made its appearance and soon came to surpass in terms of sales and profits the traditional sectors of the industry.[1] Despite the failure of the established photographic firms to enter this sector except to supply the raw cinematographic film, they were profoundly affected by the rapid growth of sales of that film. Although the underlying technology of motion pictures depended upon cameras, projectors, and film, the emergence of commercial cinematography had to await the development of key changes in photographic materials technology and in the fundamental tech-

[1] The role of technology in the development of the American motion picture industry alone deserves a separate detailed study, but that is beyond the scope of this investigation. Therefore, this chapter will only outline the broad pattern of development in order to show the place of this important sector in the overall history of the American photographic industry. In keeping with the limited scope of the chapter, considerable reliance has been placed on certain secondary sources including: Josef M. Eder, *History of Photography*, trans. Edward Epstean (New York: Columbia University Press, 1945); Kenneth MacGowan, *Behind the Screen* (New York: Dell Publishing Co., 1965); Gordon Hendricks, *The Edison Motion Picture Myth* (Berkeley: University of California Press, 1961), *Beginnings of the Biograph* (New York: Beginnings of the American Film, 1964), and *The Kinetoscope* (New York: Beginnings of the American Film, 1966); Mae D. Huettig, *Economic Control of the Motion Picture Industry* (Philadelphia: University of Pennsylvania Press, 1944); and F. Paul Liesegang, *Zahlen und Quellen zur Geschichte der Projektionskunst und Kinematographie* (Berlin: Deutsches Druck-und-Verlagshaus, 1926).

nological mind-sets of motion picture experimenters. In a very real sense, commercial cinematography, like mass amateur photography, became possible only after the development of gelatin emulsions and continuous roll film.

Creation of commercial cinematography depended upon the convergence of at least three areas of technical understanding and development: projection, persistence of vision, and photography. Projection techniques for still transparencies were introduced during the middle of the seventeenth century, a period of considerable interest in optical instruments. Natural philosophers and instrument makers combined their interests and talents to make this a period of remarkable development in optical instrumentation. For the first time they employed lenses in combination—in the telescope and the microscope—and they studied binocular vision. Among the instruments created was the magic lantern, a box consisting of a lamp, internal reflecting mirrors, a condensing lens, a strip or disk of transparencies, and an objective (or projecting) lens (fig. 12.1). The instrument was so constructed that the light from the lamp was concentrated on the transparency, and the transparency and projecting screen were at conjugate foci of the objective lens. While considerable uncertainty shrouds the origin of this important optical instrument, there is clear evidence of its use as early as 1656 by Christian Huygens, a prominent Dutch natural philosopher. Soon thereafter, published references to the magic lantern became common, and it was produced commercially. From the seventeenth century to the middle of the nineteenth century, the design of the magic

Fig. 12.1
Kircher's magic lantern. From Athanasius Kircher, Ars magna lucis et umbrae, *2d ed. (Amsterdam: J. J. a Waesberg & H. E. Wayerstraet, 1671).*

lantern remained quite stable, with only slight improvements in lens combinations and light sources.[2]

Just as the first half of the seventeenth century was a period of considerable interest and activity in optical theory and instrumentation, the first half of the nineteenth century was another period of intense interest in optics and visual phenomena. One of the most notable new areas of interest, *persistence of vision*,[3] captured the attention in the 1820s of a number of leading European natural philosophers. The two most thorough and significant investigations of persistence of vision were those of Plateau and Stampfer. Joseph Antoine Ferdinand Plateau, a young Belgian physicist with a particular interest in physiological optics, developed in 1833 a stroboscopic device consisting of a circular disk with radial slots through which the observer peered at a mirror that reflected a series of painted figures on the reverse of the disk, drawn so as to produce, when the disk was rotating, the illusion of figures in motion. This Plateau disk was at first called a phantasmascope (literally, "instrument for viewing images"), later a phantascope (fig. 12.2). Simultaneous to and yet independent of Plateau's studies and creation of

[2] Eder, *History of Photography*, pp. 51–52; and Liesegang, *Projektionskunst und Kinematographie*, pp. 7–9.

[3] *Persistence of vision* is the physiological-psychological phenomenon wherein the visual sensation in the retina of the eye caused by a light stimulus does not immediately cease upon the removal of the stimulus but persists for some fraction of a second thereafter. Consequently, to a human observer a series of related images seen rapidly (16–30 or more images per second) appears as a continuous image rather than as a set of discrete images. The illusion created by this phenomenon is the foundation for motion pictures, television, stroboscopes, and the like.

the phantascope, Simon Ritter Stampfer, a Viennese professor of mathematics, developed a similar double-disked device that became known as a stroboscope (literally, "instrument for viewing the action of whirling"). From Plateau's and Stampfer's studies of human persistence of vision and the development of instruments based on this phenomenon, all subsequent methods of creating the illusion of motion were ultimately derived.[4]

Shortly after the publication of Plateau's and Stampfer's work, the English mathematician, William George Horner, published a description of a modified phantascope, which he called a Daedaleum (fig. 12.3). The device, consisting of a slotted rotating drum with figures on the interior surface, produced the illusion of moving figures without the necessity of a mirror (as with the phantascope) or of a double disk (as with the stroboscope). Horner's Daedaleum, which became popularly known as the zoetrope (literally, "life turning"), combined, as did the devices of Plateau and Stampfer, these elements: (1) a sequence of figures; (2) a rapid-acting shutter; and (3) utilization of persistence of vision.[5]

Fig. 12.2
Plateau's phantascope or disk. From Henry V. Hopwood, Living Pictures: Their History, Photo-Production, and Practical Working *(London: Optician & Photographic Trades Review, 1899).*

Within a generation, the curious devices employing persistence of vision were combined with the magic lantern by an Austrian artillery officer and teacher, Franz Uchatius. In his teaching he had utilized Stampfer's stroboscope as a means of explaining certain kinds of motions to students of physics, but the delays in passing the instrument from one student to another in the classroom prompted him to project the moving figures. In 1845 he constructed a projector that retained the basic elements of the magic lantern but replaced the transparency slot with a stroboscope that had one disk made of glass with transparent figures drawn on it. This device did not serve the function for which it had been designed because the weakness of the illumination restricted projection to a very short distance; however, by 1853, Uchatius had designed a totally new projector consisting of a fixed disk of twelve transparencies, twelve lenses, and a rotating light source. This device, which produced satisfactory projection and the illusion of movement, was sold commercially in Vienna, primarily as a curiosity. Although the second projector was more complex and cumbersome than Uchatius's first, it did make use of an important principle: the requirement of arrested motion. The transparency had to remain stationary for at least a short time. Other persons working with persistence of vision, such as Charles Wheatstone, also recognized at this time the importance of arrested motion to the illusion.[6]

Fig. 12.3
Horner's daedaleum or zoetrope. From Hopwood, Living Pictures.

Thus, by mid-century the key conceptual elements of projecting drawn figures to produce the illusion of motion were understood: (1) the projection of transparencies; and (2) the utilization of persistence of vision, including the sequence of figures, the intermittent shutter, and the arrested motion of the figures. Yet, the third

[4] Liesengang, *Projektionskunst und Kinematographie*, pp. 32–40; and Eder, *History of Photography*, pp. 496–97.
[5] Liesegang, *Projektionskunst und Kinematographie*, pp. 35–36.
[6] Ibid., pp. 51–52; Eder, *History of Photography*, pp. 497–99; and MacGowan, *Behind the Screen*, p. 38.

key factor in cinematography, photography, remained inapplicable. The long exposure times required by the daguerreotype and early collodion processes prevented the successful application of photography to the creation of motion pictures.

The decade from the early 1860s to the early 1870s witnessed the first efforts to use photography to create the illusion of motion, but these efforts, because of the insensitivity of the photochemical materials, depended upon posed shots. Among the subjects so photographed were a stationary steam engine, two children, a waltzing couple, and an acrobat. During the 1870s photography, which was proving a useful tool in astronomy, spectroscopy, and medicine, enjoyed popularity as a recording device. In that context, photographic studies of slow motions for scientific purposes arose. In December of 1874, the French astronomer, Pierre Jules C. Janssen, employed a clockwork device with a daguerreotype plate to make a sequence of forty-eight exposures on the periphery of four disks recording the transit of Venus. The disk was stationary during each exposure. He did not, however, intend this sequential record for eventual projection. The most important scientific study of motion of this period, a photographic analysis of the movements of animals, however, eventually led to projection.[7]

The man who put together all three of the key elements, photography, a persistence-of-vision device, and projection, embarked first upon an exclusively photographic project. This pioneer, Eadweard James Muybridge, emigrated to the United States from England and in the early 1860s entered upon a career as a professional photographer in California. When in 1872 a controversy arose between Leland Stanford and a friend over some drawings of the motions of race horses made from investigations conducted by the French physiologist E. J. Marey, Muybridge trained his camera, specially equipped with a rapid shutter, on one of Stanford's horses at a track in Sacramento. From the resulting photographs, Stanford, former governor of California and a major railroad investor, found that at certain points in a horse's gait all four feet simultaneously leave the ground. Muybridge executed this demonstration by means of a series of isolated photographs taken on wet collodion plates.[8]

During the middle 1870s Muybridge, inundated by domestic problems, turned his attention to other matters and even left the country for a time,[9] but he returned and at Stanford's request resumed in 1877 the investigation of the motion of the galloping race horse. Muybridge inaugurated a new approach by employing up to twenty-four cameras positioned at equal distances along a special track on Stanford's Palo Alto estate. Utilizing wet collodion plates of

[7] MacGowan, *Behind the Screen*, pp. 44–45 and 56; Helmut Gernsheim and Alison Gernsheim, *History of Photography: From the Earliest Use of the Camera Obscura in the Eleventh Century up to 1914* (London: Oxford University Press, 1955), p. 324; and Eder, *History of Photography*, p. 506.

[8] Eder, *History of Photography*, pp. 501–505; Liesegang, *Projektionskunst und Kinematographie*, pp. 71–72; MacGowan, *Behind the Screen*, pp. 46–52; and *National Cyclopaedia of American Biography*, s.v. "Muybridge."

[9] He was charged with murder and acquitted in the death of a suitor of his wife ("Muybridge," p. 152).

especially high sensitivity and electrically actuated shutters tripped by the horse as it sped by the battery of waiting cameras, he produced a sequential series of photographs of the horse in motion. He then published a set of prints, *The Horse in Motion*, which stirred considerable attention among artists and zoologists. By placing the prints in order, Muybridge noted that he could create the illusion of motion by riffling the photographs. Clearly, the success in capturing sequential photographs soon suggested the creation of the illusion of motion from them. As he continued his photographic studies of the motions of other animals, he applied the sequential photographs to the inner surface of the zoetrope, thereby improving upon the riffling technique.[10]

Soon Muybridge placed transparencies of his photographs on the periphery of a Plateau disk and created a projection-stroboscope for exhibition of the motion of the horse. After public demonstration of these pioneering pictures in Palo Alto and San Francisco, he went to Europe where, in the late summer of 1881, he exhibited motion pictures at the laboratory of E. J. Marey in Paris. In March of 1882, prior to his return to the United States, Muybridge gave several public exhibitions in London, attracting numerous prominent political, social, and scientific persons to the halls of the Royal Society and Royal Institution.[11] Even with the limitations of the wet collodion process, he had succeeded in uniting photography with stroboscopic projection. While his equipment was primitive compared to cinematographic projection equipment of even a generation later, it incorporated for the first time all the basic principles of cinematography.

Although Muybridge clearly pioneered in motion photography, his primary interest continued to be study of animal motions. While he soon adopted the new gelatin dry plate to considerable advantage in his work and developed automatic electrical devices for actuating the camera shutters, he did not concentrate upon the mechanical enhancement of his equipment for anything beyond his own immediate purposes. During the middle 1880s, he successfully undertook a major photographic study of animals, humans, and birds in motion under the auspices of the University of Pennsylvania. Employing gelatin dry plates,[12] he acquired tens of thousands of motion sequences, twenty thousand of which were reproduced in his eleven-volume study, *Animal Locomotion*, published in 1887. For the remainder of his life, Muybridge continued his photographic studies of animal motion. This work brought him considerable public attention because of its scientific importance and popular interest, but it was his pioneering work in motion photography from 1878 to 1881 that stimulated a large number of persons in the Atlantic community to modify and seek to improve Muybridge's equipment for commercial purposes.

The decade of the 1880s witnessed a broadening of interest in the

[10] Eder, *History of Photography*, pp. 501–5; and Liesegang, *Projektionskunst und Kinematographie*, pp. 71–72.

[11] Liesegang, *Projektionskunst und Kinematographie*, pp. 71–72.

[12] Eder, *History of Photography*, p. 504.

Fig. 12.4
Marey's revolver camera and disk. From Hopwood, Living Pictures.

application of photography to the study of motion—a major change in the fundamental technological conception of the method of projecting photographic images serially—but no effort to introduce commercial cinematography. Among those who were most directly influenced by Muybridge's work and lectures in Europe and America was Ottomar Anschütz, who adopted the multiple camera technique and continued to employ it through the 1880s.[13] However, the widespread adoption of the rapid, dry gelatin plates in photography at the beginning of the 1880s permitted and encouraged the abandonment of multiple cameras, lenses, and photographic plates in favor of a single camera and lens with a movable multiple image photosensitive plate. Yet the old method of arranging transparencies for projection, the Plateau disk, dominated the form of the plate for the photosensitive material, requiring in the photographic camera the coordinated motion of the photosensitive disk and intermittent shutter in addition to a mechanical method of arresting motion for the duration of exposure of the photosensitive material. Completing the circle of mutual influence, after Muybridge's visit to his Parisian laboratory, Marey, the French physiologist, developed a "cinematographic revolver" for photographing birds in flight. Adopting the single camera approach, he employed a photosensitive disk and clockwork mechanism for the coordinated motion of the disk, the shutter, and the aperture (fig. 12.4). Utilizing an arrested movement of the photosensitive surface, the "revolver" procured exposures of 1/720th of a second and twelve frames per second.[14]

Another important pioneer in photography of motion during this period, William Friese-Greene, likewise employed the single-camera technique and the Plateau disk. Friese-Greene, an Englishman, worked from the early 1880s with John Rudge, a producer of magic lanterns in Bath, in the application of photography to the projection of motion pictures. In the mid-1880s, Friese-Greene exhibited projection machines that employed glass disks.[15]

The single most important development in breaking the technological mind-set that had dominated the projection of motion figures, i.e., utilization of the Plateau disk, was the introduction of the Eastman-Walker roll film and roll film holder. Though Leon Warnerke had introduced the roll film idea into photography in England in the 1870s, its lack of commercial success and its identification with the slow collodion process deterred conceptualization of its use in cinematography; however, the Eastman-Walker system proved commercially feasible, was distributed in both America and Europe, and received considerable attention in the leading photographic journals. The roll film system of photography was introduced with

[13] Ibid., pp. 512–13; and Louis Walton Sipley, *Photography's Great Inventors* (Philadelphia: American Museum of Photography, 1965), pp. 92–93.

[14] Liesegang, *Projektionkunst und Kinematographie*, pp. 72–73; Eder, *History of Photography*, pp. 507–12; and Sipley, *Photography's Great Inventors*, pp. 90–91.

[15] *Dictionary of National Biography, Supplement*, 1927, s.v. "Friese-Greene"; and Eder, *History of Photography*, pp. 515–18.

Fig. 12.5
Thomas A. Edison. From Photog. T., *1891.*

paper film in 1885, and within two to three years several of the cinematographic pioneers had abandoned the disk and adopted paper film. In so doing, they took a vital technological step toward transforming motion photography projection from a curiosity to a technique with commercial potential.

Not unexpectedly, Marey was one of the first to experiment with and adopt paper film as a means of obtaining long, continuous sequences of motion. His camera, cranked by hand and employing the vital arrested motion, could take exposures in 1/1000th of a second and produce sixty frames per second. Augustin LePrince employed at about the same time a continuous film made of gelatin. Late in 1887 or early in 1888, William Friese-Greene employed a band of sensitized paper made transparent by having been treated with castor oil. The problem with the use of paper film in cinematography, as in roll film photography, was the poor quality of the positive prints caused by the grain of the paper.[16] Yet the paper film served adequately for nonphotographic motion pictures. The French physicist, Emile Réynaud, found his continuous perforated band of paper quite satisfactory for projecting animated drawings. He developed a projector system based on the principle of the zoetrope and succeeded in making a commercial success of the display of his animated motion pictures in Paris between 1892 and 1900.[17]

Those enthusiasts that employed the roll film approach to cinematography anxiously sought an alternative to paper film. At about the same time that Reichenbach developed a celluloid substitute for paper film, several workers in cinematography also began to experiment with celluloid, most notably Friese-Greene in the spring of 1889 and Marey in 1890.[18] In the long run, of course, the celluloid film proved to be one of the most important developments in cinematography.

Some of the most famous developments in cinematography were those supervised by Thomas Edison at his new West Orange, New Jersey, laboratory (fig. 12.5). Edison's inventive work in connection with telegraphy, the incandescent light, and the phonograph had brought him both fame and fortune. With some of the fortune, he organized and built in West Orange, New Jersey (near Newark), a large laboratory where he and a staff numbering from forty-five to sixty persons sought to develop new technical ideas for commercial exploitation. Edison, serving as the entrepreneur and director, placed the stamp of his ideas on the activities conducted by his large staff. Although he supervised the work on cinematography and directly influenced the general direction of development, William

[16] Eder, *History of Photography*, pp. 507–12; Sipley, *Photography's Great Inventors*, pp. 90–91; Liesegang, *Projektionskunst und Kinematographie*, pp. 76 and 80–85; and "Friese-Greene," p. 225.

[17] Sipley, *Photography's Great Inventors*, pp. 91–92; and MacGowan, *Behind the Screen*, pp. 60–63.

[18] "Friese-Greene," p. 225; Eder, *History of Photography*, p. 510; and Sipley, *Photography's Great Inventors*, p. 91.

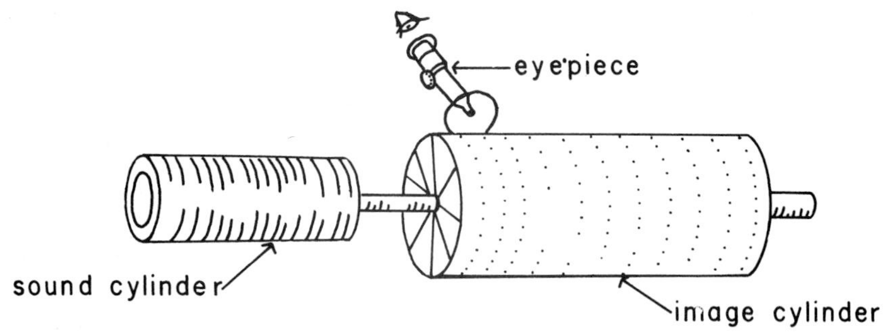

Fig. 12.6 Dickson-Edison synchronization of sight and sound.

Kennedy Laurie Dickson is credited with major responsibility for the specific technical innovations.[19]

Although Dickson knew photography well, Edison, unlike most of the other pioneers in cinematography, came to the project with little background in photography. In the late 1880s, as competing phonograph systems were commercially introduced, Edison sought to extend his phonograph to include motion pictures. His conception of motion pictures as an extension of the phonograph, not a new or separate invention, influenced his conception of how to create and observe motion pictures.[20] By starting with the phonograph, he at once faced the problem of coordination of sound with the motion picture. His initial conception of a dual cylinder apparatus on the same axle with one cylinder for sound and the second cylinder for vision dominated work on the project during the two years 1888–89 (fig. 12.6). Edison and Dickson thus sought to develop a means of carrying transparent images on the cylinder that was analogous to the sound grooves on the phonograph cylinder. When it became evident that projection of a sequence of microimages was virtually impossible, they turned to the use of a magnifying eyepiece to observe directly the tiny images spiraled onto the cylinder. The carrier for the images was, from the beginning, of course, a primary concern. At one point, Dickson procured materials for employment of a daguerreotype plate curved onto the cylinder. Later, he experimented with paper film and strips of celluloid film wrapped around the cylinder; but none of these efforts proved fruitful.[21]

While the need to use the cylinder dominated Edison and Dickson's ideas, certain ideas from other cinematographic pioneers also

[19] Matthew Josephson, *Edison* (New York: McGraw-Hill Book Co., 1959), pp. 313–17. In connection with Edison and Dickson's work I am following the interpretation of Hendricks, *Edison Motion Picture Myth.* Although this study does raise many questions about the traditional interpretations of Edison's cinematographic work, it does not detail the specific work and ideas of Dickson but merely indicates that Edison could not have done most of the work, and therefore, Dickson must have been the party responsible.

[20] Hendricks, *Edison Motion Picture Myth*, pp. 11–12 and 151; Josephson, *Edison*, p. 317; and interview with Edison, *Photog. T.* 24 (1894): 209.

[21] "Motion Picture Caveat I," reproduced in Hendricks, *Edison Motion Picture Myth*, pp. 28, 72–73, and 158–61; and sketch by Dickson claimed by MacGowan to be in the Earl Theisen Collection and reproduced in his *Behind the Screen*, p. 68.

influenced them. Gordon Hendricks, a recent historian of Edison and Dickson's cinematographic developments, suggests that Edison's interest in working with motion photography may have been stimulated directly by Muybridge, who, in February, 1888, gave a lecture in Orange, New Jersey, and a couple of days later visited the Edison laboratory, where he and Edison spoke with each other about combining the phonograph and motion pictures. Hendricks also suggests that Edison may have received from Friese-Greene a description of his projecting machine in the spring of 1889. Although there were numerous possibilities of influence from the published literature in the 1880s, the most important influence was the work of Marey in Paris, which changed the entire technological conception within which Edison and Dickson worked.[22]

Edison made a trip to Europe in the late summer and early fall of 1889, during which he visited Marey at his laboratory in Paris. Less than a month after his return to West Orange, Edison filed with the U.S. Patent Office a caveat in which he clearly indicated the shift from the cylinder to the continuous roll film carrier for transparencies, an approach that Marey had adopted prior to Edison's visit. The caveat contained a description of a system employing reel-to-reel perforated film, sprocket wheels, a continuous light source, and a revolving shutter. Although the caveat did not indicate that these elements had been successfully employed, it is clear that Edison and Dickson's approach had changed to the use of a continuous band.[23]

Although Edison and Dickson devoted most of their attention during the next two years to Edison's ore mill experiments, late in 1890 and early in 1891, with the new approach to cinematography, Dickson resumed serious work on the motion picture project. Although Eastman film had been used in place of Carbutt film plates, which had to be cut and fastened together, the Edison group abandoned the thin Eastman film in favor of the heavier and thicker Blair film. While thinness was an advantage for film for Kodak cameras, the thicker film proved advantageous for the perforated cinematographic film. By spring of 1891, the Edison group had developed a kinetoscope, a cinematographic film viewer for use by only one person at a time. The first public viewing of the kinetoscope occurred when the National Federation of Women's Clubs visited the West Orange laboratory in the spring of 1891. During the remainder of the spring and summer of 1891, further improvements were developed in the basic device, and the necessary application papers for patents were prepared. At that time, Edison decided to proceed to commercial production with the hope that he could have the device ready in time for display and promotion at the Chicago World's Fair in 1893. During the summer of 1891, Dickson also developed an analogous motion photographic camera for production of films for the kinetoscope. By December of 1892, construction was begun on a separate studio building for production of motion picture films at the West Orange compound. This building became

[22] Hendricks, *Edison Motion Picture Myth*, pp. 4–6 and 178–79.
[23] Ibid., pp. 52 and 82–83; and "Motion Picture Caveat IV," pp. 162–63.

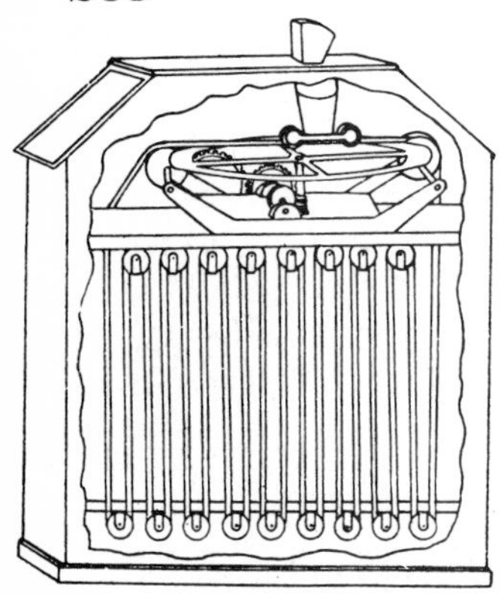

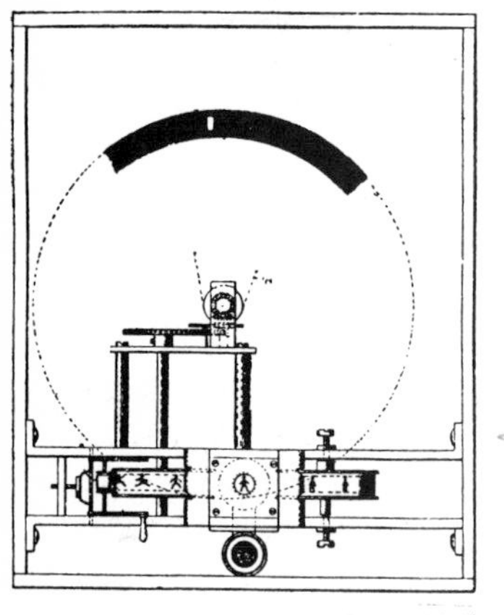

Fig. 12.7
Kinetoscope. From Hopwood, Living Pictures.
A. *Cutaway side view.*
B. *Top view showing shutter wheel and roll film passing through the aperture.*

popularly known as the Black Maria because of the interior color of the studio.[24]

The kinetoscope was a relatively simple device incorporating all the basic elements of cinematography except projection. It consisted of a large wooden box that sat upright on the floor and contained a continuous loop of perforated film about forty to fifty feet in length (fig. 12.7). The observer peered into an eyepiece on the top of the box and could see images as the film passed between the eyepiece and an electric light at a rate of about forty-six frames per second.[25] The motion of the film was continuous, but the rotary shutter provided an aperature ratio of nine to one. Electricity powered both the light and the motor that drove the film and the shutter.[26]

Delivery of the kinetoscope to commercial parlors began in the spring of 1894. Dickson soon devoted a large part of his time to producing films for the hungry devourers of the new popular entertainment (fig. 12.8). At first, customers paid their money to an attendant who turned the machine on for one period of viewing, but soon coin devices added to the machines introduced the era of the "nickel-in-the-slot."[27] Edison's commercial goals had propelled the infant device to the fore as a popular form of entertainment in the mid-1890s, but his business strategy deterred further technological development and soon placed the Edison enterprises on the defensive.

Edison sought for some time to attract financial interest in the kinetoscope, but his earlier financial supporters such as Henry Villard hesitated to invest in machines for peep shows, perhaps wary of arcades or faddish entertainment schemes. Early in 1894, however, Norman C. Raff and Frank Gammon, speculators in the entertainment business, organized the Kinetoscope Company and secured a contract from Edison for the supply of machines and films. From the spring of 1894, the Kinetoscope Company created a chain of kinetoscope parlors like the phonograph parlors in major cities across the continent. The public received the kinetoscope enthusiastically, with crowds waiting in long lines for an opportunity to view the latest prize fight, comedy routine, or documentary (fig. 12.8).[28] The popularity of the kinetoscope in the arcades and elsewhere, clearly demonstrating the commercial potential of motion pictures as a form of entertainment, attracted imitators and proved suggestive to innovators. Dickson himself was involved in the introduction of one of the most obvious, although technically different, imitations.

Questions regarding Dickson's relationship to another American group working on motion pictures arose in the spring of 1895. Eugene Lauste, the French experimenter from the Edison motion pic-

[24] Ibid., pp. 100, 103, and 111–41.

[25] The rate of forty-six frames per second may very well have been suggested by the work of the German natural philosopher Helmholtz in his *Physics.* See Hendricks, *Edison Motion Picture Myth,* p. 107.

[26] Hendricks, *Edison Motion Picture Myth,* pp. 106–7; and MacGowan, *Behind the Screen,* p. 71.

[27] MacGowan, *Behind the Screen,* pp. 72–73.

[28] Josephson, *Edison,* pp. 394–95.

Fig. 12.8
Kinetoscope parlor with inset view of frame from fight film. From Frank Leslie's Popular Monthly, *February 1895.*

ture research group at West Orange, had left Edison and later joined the Lathams, of New York City (discussed below). Upon suspicion that Dickson might be planning a similar move, Edison, at the urging of one of his financial advisers, reluctantly dismissed Dickson; Dickson then did join the Lathams. In the summer of 1895, Dickson, who disapproved of the personal conduct of the Latham brothers and their father, left New York City and joined another group of motion picture workers in Syracuse, New York. This group originated the mutoscope, a rival of the kinetoscope.[29]

Dickson, who claimed the idea of the mutoscope as his own, may have prior to his departure from West Orange suggested it to Herman Casler, a skilled mechanic and draftsman, who then constructed and filed a patent application for the motion picture viewing device in the fall of 1894.[30] The mutoscope resembled an elaborate riffling book, except that photographic cards were attached to a central cylinder that the viewer turned by a hand crank (fig. 12.9). The device was enclosed in a wooden box with apertures to admit external light for illumination of the photographs and for observation of the sequence of flipping photographs. Casler began late in 1894 to build a camera for the mutoscope, and by late spring of 1895 he had completed and tested the camera, which was called the mutograph (figs. 12.10 and 12.11). From the mutograph, Dickson began work on the biograph, which, in contrast to the mutoscope and the kinetoscope, *projected* images taken by the mutograph or biograph camera (fig. 12.12). By early fall of 1895, four men were associated with the production and sale of the mutoscope: Elias *K*oopman, Harry *M*arvin, Herman *C*asler, and William *D*ickson (KMCD).[31]

[29] From testimony of Dickson reported in Terry Ramsaye, *A Million and One Nights* (New York: Simon & Schuster, 1926), p. 184; and Josephson, *Edison*, p. 397.

[30] Quoted from Dickson in Hendricks, *The Beginnings of the Biograph*, p. 2.

[31] See illustration in *Scientific American*, 17 April 1897; and Hendricks, *Beginnings of the Biograph*, pp. 9–15 and 25.

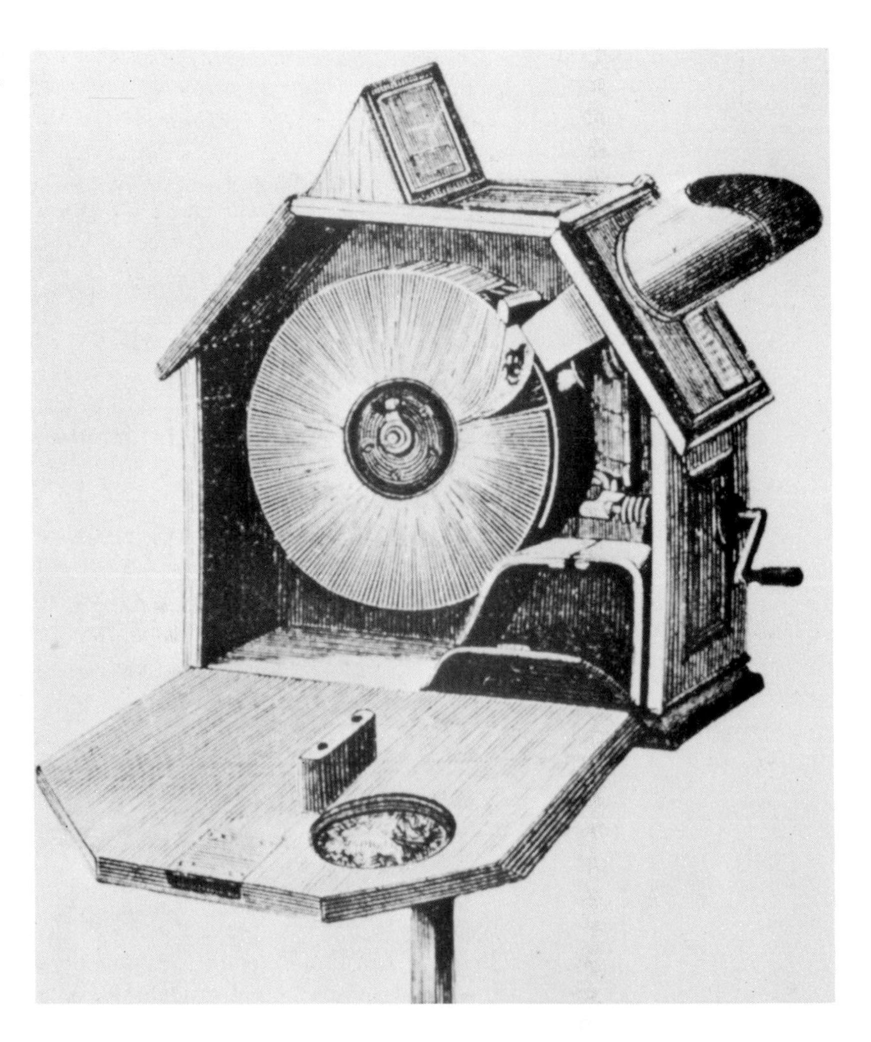

Fig. 12.9
Mutoscope. From Scientific American, *17 April 1897.*

Fig. 12.10
First mutograph or biograph camera, c. 1895. From Gordon Hendricks, Beginnings of the Biograph *(New York: The Beginnings of the American Film, 1964). Courtesy of Gordon Hendricks.*

Late in December of 1895, the American Mutoscope Company was organized in Jersey City, New Jersey, with Koopman, Marvin, Casler, and Dickson as the principal stockholders. Casler's patent on the mutoscope, his application for a hand-held mutoscope, and Mar-

Fig. 12.11
Biograph camera in use. From Hendricks, Beginnings of the Biograph. *Courtesy of Gordon Hendricks.*

Fig. 12.12
First biograph or projector. Original in Smithsonian Institution. From Hendricks, Beginnings of the Biograph. *Courtesy of Gordon Hendricks and Photographic Department, Smithsonian Institution, Washington, D.C.*

vin's application for improvements in the mutoscope constituted the basic asset upon which the New York Security and Trust Company floated the company a $200,000 loan. Beginning in January of 1896, the company established production facilities for the mutoscope on Broadway in New York City, and photographing of subjects began the following summer.[32]

The projector of the American Mutoscope Company, initially called the Mutopticon, was premiered in Pittsburgh in mid-September 1896 and was used at theaters in Philadelphia and Brooklyn prior to its New York debut in mid-October. The New York public gave the projector system an enthusiastic reception. During the next year the use of the biograph spread to theaters across the country, and it soon surpassed other projectors in popularity (fig. 12.13).

[32] Hendricks, *Beginnings of the Biograph*, pp. 30–33.

Fig. 12.13
Biograph in use in theater.
From Scientific American,
17 April 1897.

Thus Dickson's liaison with Casler, Marvin, and Koopman led to the creation of a motion picture viewer that came to compete seriously with Edison's kinetoscope and led to Dickson's development, in connection with Casler, of an immensely successful motion picture projector.[33]

Stimulated by the initial commercial success of the kinetoscope, other groups in the United States and abroad in the middle 1890s began to give serious attention to projection of motion pictures through the use of continuous bands of transparencies. One group of technical pioneers, the Latham family of New York City, became interested in the kinetoscope and began working in a small laboratory in New York to create a projection machine. To the native mechanical ingenuity of Major Woodville Latham, a professor of chemistry, and his two sons was added the knowledge and experience of other technical personnel including Eugene Lauste, a French technician who had worked with Dickson at Edison's laboratory; William K. L. Dickson, who, after leaving the Edison laboratory in early 1895, joined the Latham group for a few months; and Enoch J. Rector, an engineer. By the spring of 1895, the Latham group had developed a projecting kinetoscope that they called a Panoptikon. It featured a large reel, requiring the famous "Latham loop," which allowed a slack loop in the film so that the film could be pulled intermittently through the aperture frame without tearing. Although this projector possessed many advanced features and received an enthusiastic reception at its inauguration in New York in April 1895, its continued production did not prove commercially successful.[34]

[33] Ibid., pp. 40–51; and *National Cyclopaedia of American Biography*, s.v. "Casler."

[34] MacGowan, *Behind the Screen*, pp. 77–78; Josephson, *Edison*, pp. 395–97; Hendricks, *Edison Motion Picture Myth*, pp. 196–97; and Hendricks, *Beginnings of the Biograph*, p. 13.

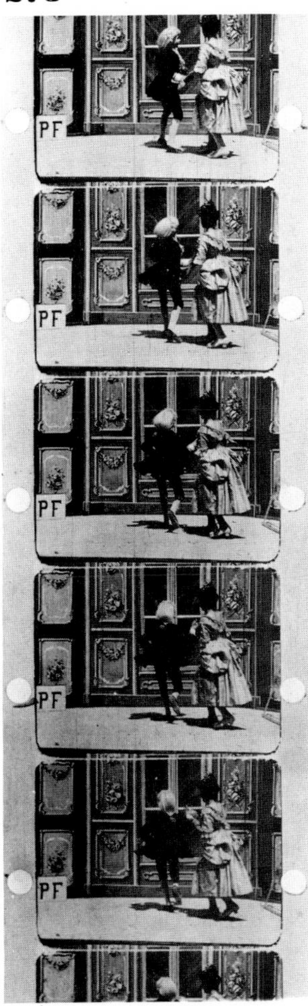

Fig. 12.14
Lumière 55-mm film stock with circular perforation. Courtesy of Kodak Limited.

At about the same time that the Latham and KMCD groups were working on projection machines in the United States, the Lumière brothers of Lyon, France, also developed a motion picture projector. Their dry plate company, established in the early 1880s in Lyon, rapidly grew into the leading and most progressive photographic materials company in France. Early in 1894 the senior Lumière, still active in the family business, learned about the kinetoscope. He invested six thousand francs in the purchase of a machine and, under his urging, his sons turned their attention to the American device and the problem of projecting motion pictures. By Christmas of 1894, they had succeeded in constructing a working model of a motion picture projector with some characteristics borrowed from the kinetoscope, some borrowed from the work of other pioneers, and some of their own devising.[35]

The Lumière projector employed an arc light or other lamp; a bulb of water as a condenser (to protect the film from the heat of the lamp); celluloid film of width, length, and frame size identical to that devised by Edison, and perforated with one hole per frame on each side of the film (Edison's film had four holes); and a claw-gripper mechanism devised by Louis Lumière to move the film intermittently (fig. 12.14). The film moved much more slowly than the Edison film: fifteen frames per second, being in motion one-third of the time and at rest two-thirds of the time. The machine was initially powered by a hand crank.[36] Following the application for a French patent early in 1895, the Lumières presented the projector at a meeting of the Société d'Encouragement à l'Industrie Nationale in March 1895. They held the commercial premiere on 28 December 1895 in Paris at the Salon Indien du Grand Café. Their projector soon enjoyed enormous commercial success. They also introduced to the market early in 1896 a motion picture viewer similar to the mutoscope for individuals and known as the kinora. A short time later they introduced their commercially successful motion picture projector, and as it won wide acceptance throughout Europe, they soon began to export it to the United States.[37]

Other motion picture projector pioneers during the middle 1890s were Charles F. Jenkins and Thomas Armat. Both men had technical backgrounds. Jenkins had experimented with motion picture machines since 1887, and in 1893 he developed a projector he called the Phantoscope. Armat, while attending the Bliss Electrical School in Washington, D.C., heard of Jenkins's work and joined him in improving the projector. In September of 1895 they exhibited their projector at the Cotton States Exposition in Atlanta, and Jenkins gave a presentation and demonstration at the Franklin Institute in Philadelphia. Eventually, owing to the decision of the Supreme Court in 1901, the patent on their improved projector, the vitascope,

more motion picture pioneers

Jenkins at institute

[35] Henri Kubnick, *Les Frères Lumière* (Paris: Librairie Plon, 1938), pp. 23–24 and 33.

[36] MacGowan, *Behind the Screen*, p. 80; Liesegang, *Projektionskunst und Kinematographie*, p. 92; and Kubnick, *Les Frères Lumière*, pp. 29–32.

[37] Kubnick, *Les Frères Lumière*, pp. 35–48; Eder, *History of Photography*, p. 520; and *Photog. T.* 29 (1897): 112.

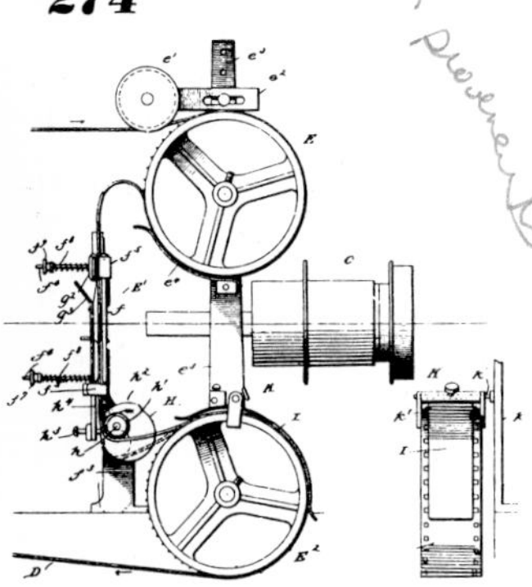

Fig. 12.15
Vitascope. From U.S. Patent # 673,992.

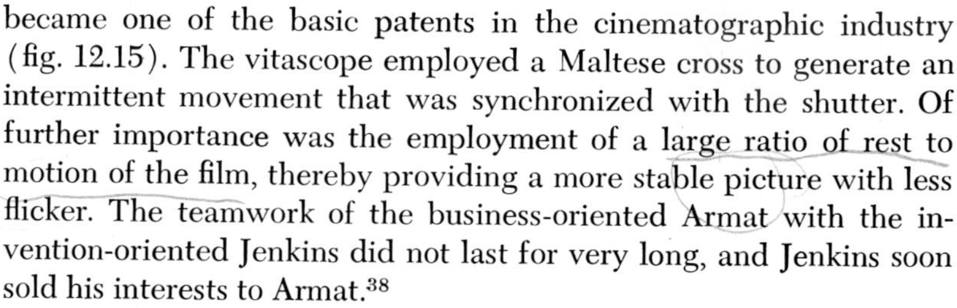

became one of the basic patents in the cinematographic industry (fig. 12.15). The vitascope employed a Maltese cross to generate an intermittent movement that was synchronized with the shutter. Of further importance was the employment of a large ratio of rest to motion of the film, thereby providing a more stable picture with less flicker. The teamwork of the business-oriented Armat with the invention-oriented Jenkins did not last for very long, and Jenkins soon sold his interests to Armat.[38]

Meanwhile, work on development of projecting apparatus received low priority at the West Orange laboratory.[39] Although Dickson did work on a projector and may have carried some of the ideas to the Lathams and to the KMCD, Edison, still committed to the "peep show" approach on the basis of business principles, argued that the market for films was so small that if they were projected and shown on a screen before a large audience, audiences would soon reach saturation and not attend any longer. Also, there would be substantially fewer orders for projectors than for kinetoscopes.

However, by 1896, the sales and popularity of the kinetoscope began to wane, while the mutoscope and screened films grew in public attention. In the face of the deteriorating sales, Edison and his associates reevaluated their position on projectors and turned their attention to Armat's vitascope. Recognizing its technical superiority to the projectors developed at the West Orange laboratory, Edison accepted an arrangement to market the vitascope under his name (Edison Vitascope) and to pay royalties to Armat. Armat agreed, despite the lack of recognition he was to receive, because of the commercial advantage of Edison's name. The Edison Vitascope was given its first showing in April 1896 in New York, and soon thereafter the screening of films began to grow in popularity and success.[40]

Thus, the development of commercially practical cinematography resulted from the creation of a *system* of individual but closely interrelated elements: camera, projector, and film. In the period from the late 1870s to the middle 1890s, during which commercial projection was developed, the technological frameworks out of which each pioneer technician or group of technicians worked shifted both with regard to the techniques of photographing motion and the methods of viewing or projecting the sequence of photographs of motion. First, the character of the photosensitive material required the employment of multiple cameras and projection with a rotating disk of transparencies. Second, with the introduction of gelatin emulsions, single cameras with multiple exposures became possible, but disk projection remained. Third, the introduction of roll film brought the employment of a continuous band of photosensitive material in the single camera and the "play-back" through a

[38] *National Cyclopaedia of American Biography*, s.v. "Jenkins" and "Armat"; and MacGowan, *Behind the Screen*, pp. 78–79.

[39] Josephson, *Edison*, pp. 395–96. Edison commented, "If we make this screen machine that you are asking for, it will spoil everything."

[40] Ibid., pp. 397–99; and MacGowan, *Behind the Screen*, pp. 82–83.

viewer such as the kinetoscope, mutoscope, or kinora. Finally, during the middle 1890s, with the creation of the necessary mechanical coordination of the film movements and the requisite illumination, commercial projectors were introduced. In this sequence, the two key photographic innovations that prompted technicians to depart from their older technological frameworks were the gelatin emulsion and the continuous roll film, neither of which was introduced into photography with cinematography in mind. The mechanical innovations, such as the film loop, sprockets and wheels for creating intermittent motion, synchronization of the motion of shutters with the motion of the film, and brighter illumination, followed in response to the two key technological developments. Of particular importance in the final stage of the emergence of cinematographic projection, however, was the commercial success of the kinetoscope, which demonstrated that motion pictures were not only of interest to scientists and artists as a tool and to a curious populace as a passing toy or fad but that they represented a commercial entertainment opportunity. This demonstration of the commercial potential gave the final impetus to the technological development of commercial cinematography.

The commercial introduction of the kinetoscope in the middle 1890s inaugurated the entry of motion pictures as a major force into the entertainment industry in the United States. Initially, kinetoscopes, mutoscopes, and kinoras were located in special parlors, penny arcades, shooting galleries, and other centers of incidental entertainment. While the initial enthusiasm expressed during the middle 1890s began to wane later in the decade, these peep shows continued in operation well into the twentieth century.

From the beginning, two of the major producers of apparatus and films, Edison and Lumière, sought to limit the production of both new apparatus and new films. At first, viewers and projectors were available only on a rental basis, and one could neither buy nor rent motion picture cameras. This situation, of course, encouraged technical entrepreneurs to seek new designs that would circumvent the early patents controlled by Edison; but of equal importance in the American market were American manufacturers who openly infringed existing patents, and European producers who were exempt from patent restrictions because of failures of Americans, most notably Edison, to file patent applications in European countries. Soon there were many competing "-scopes" and "-graphs" throughout the American entertainment market.[41]

After the initial enthusiastic reception of projected motion pictures, screen entertainment became closely attached to the vaudeville theaters. Soon the short films became repetitive and were boring to audiences; consequently, in the early years of the twentieth century, vaudeville theaters placed movies at the end of the program, as "chasers"—to bore the audience so that it would leave

[41] MacGowan, *Behind the Screen*, p. 132; Huettig, *Economic Control of the Motion Picture Industry*, pp. 11–12; and Josephson, *Edison*, p. 391.

and make room for the next program. In the early years of the century, motion pictures definitely declined in popularity.[42]

About the middle of the first decade of the new century, the screening of films in America underwent a major revival. The combination of the availability of projecting equipment, the introduction of longer films and films of improved variety, and the rapid growth of specialized store theaters with low admission prices stimulated an enormous wave of popularity. Although store theaters were not new to the period, the first theater specifically called a nickelodeon opened in Pittsburgh in 1905. Soon the entertainment centers with the catchy new name became very popular, with the nation boasting about 5,000 in 1907 and 10,000 by 1910. The store theaters, often equipped with 100 to 200 simple chairs on a flat floor, a simple screen, one motion picture projector, and one slide projector, featured shows from thirty minutes to over an hour in length, including several single reel (about sixteen minutes in length) comedies and perhaps one or two reels of a serial. Films such as *The Great Train Robbery* with its story, chase, dance sequence, and sentimental close-up at the end did much to generate the renewed interest in motion pictures.[43]

The rapid growth of motion picture houses in turn stimulated demand for new films. The single-reel film remained the mainstay of the movie houses, but the early methods of film distribution oriented to the peep shows were no longer appropriate to the newly emerging screen industry. With the peep shows that were individually viewed through machines, outright purchase of the film was appropriate; however, with the screening of a film, the potential audience for a given film was met with only a few showings, and then the film became worthless to the owner. Faced with this situation, exhibiters early in the century began informally to exchange films. The standardized cost and length of the single-reel films facilitated such exchanges. More formally, Harry J. Miles of San Francisco established in 1902 one of the first commercial exchanges, purchasing films from producers and then leasing them to exhibiters. With the rapid growth in the number of movie houses, the number of exchanges increased dramatically, reaching between 125 and 150 in 1907. This rapid expansion of both exhibition houses and exchanges provided opportunities for easy entry into the business and a crucial launching ground for later entrepreneurs such as Carl Laemmle, William Fox, Marcus Loew, and Adolph Zukor.[44]

By 1908 three distinct and highly competitive levels of the motion picture business had emerged: (1) exhibition (movie houses); (2) distribution (the exchanges); and (3) production of films and apparatus. During the period of rapid growth in the middle of the decade, the increase in demand attracted new businesses to all three levels but most particularly to the first two. In the production area, nearly a dozen relatively large producers had emerged by 1908,

[42] MacGowan, *Behind the Screen*, pp. 122–23.

[43] Ibid., pp. 124–29; and Huettig, *Economic Control of the Motion Picture Industry*, pp. 17–21.

[44] Huettig, *Economic Control of the Motion Picture Industry*, pp. 12–13.

largely concentrated in New York and Chicago. The largest producers included the Edison Manufacturing Company (West Orange), Biograph Company—formerly American Mutascope and Biograph (New York City)—Vitagraph Company of America (New York City), Essanay Film Manufacturing Company (Chicago), and Kalem Company (New York City and Chicago). Importers and smaller producers included Pathé Frères (New York City), George Kleine (Chicago), George Melies Manufacturing Company (Chicago), Selig Polyscope Company (Chicago), and Lubin Manufacturing Company (Philadelphia). Important projector patents held by Biograph, Edison, and Vitagraph prompted considerable litigation against alleged infringers during the period 1899 to 1907.[45] The patent conflict and the small number of companies had stimulated a decrease in the competitive conditions in the production level of the industry by the end of the decade.

The rapid growth in popularity of motion pictures from the middle of the decade not only generated nearly a whole new industry but also created an enormous demand on one sector of the older photographic industry: the sensitized celluloid film sector. In the years from 1906 to 1909, American cine film sales increased more than sevenfold.[46] Initially, the film for the experiments at Edison's laboratory came from John Carbutt of Philadelphia, but during the period 1893 to 1895, Blair Camera of Boston supplied the Edison firm. The celluloid base for cinematographic film needed to be about twice as thick as cartridge film in order to provide the strength and toughness for pulling it through the projector on sprockets. In 1896, the demand for film for cinematographic purposes was sufficient that Eastman Kodak began to produce a special, double-thick film base for cine film, quickly capturing from Blair this market that soon proved to be so important.[47]

Eastman Kodak's three types of photosensitive film required different emulsions and production facilities. Two types of cinematographic film, positive and negative, had to be produced. The negative film for motion cameras was of high sensitivity, but the positive film onto which the negative film was printed carried a much slower emulsion. Since cine film was kept rolled on reels, it did not tend to curl as did the third type, cartridge film. Therefore, the back of the cine film was not coated with gelatin.[48] Despite these differences, the general similarities—thin celluloid film and the sensitive photographic emulsions—proved of crucial importance, allowing the Eastman Kodak Company to transfer its superior production capacity for celluloid cartridge film to the new cine film market. Moreover, the introduction in 1900 of the continuous drum method of production of film base and continuous coating methods, for reasons

[45] Companies and important patents are listed in: Agreement between Motion Picture Patents Company, Edison Manufacturing Company, and Eastman Kodak Company, dated 1 January 1909, *U.S.* v. *EKCo.*, pp. 3126–46.

[46] Eastman Kodak, virtually the only film supplier, saw cine sales in the United States go from $284,718 in 1906 to $2,097,733 in 1909.

[47] GE to George Dickman, 2 July 1896 (GECP).

[48] GE testimony, *U.S.* v. *EKCo.*, p. 273.

of economy of production,[49] gave the company a further advantage over its potential competitors in the first decade of the twentieth century. At first, most companies, including the leaders of the German chemical industry, Bayer and Agfa, could not produce the film. Later, after they had learned how to make it, Eastman Kodak still maintained its dominant position because, as George Eastman stated: ". . . It is not that anybody cannot make the same kind of film, but it is making film exactly the same everyday, and the man that can do it must get the trade, because there is so much dependent upon it."[50]

Although other companies, particularly European ones, entered the American and European cine film markets, Eastman Kodak, because of its superior product, still held more than 90 percent of the world's market for cine film in 1910. Although Austin-Edwards of England and Gevaert of Belgium purchased Celluloid Company film and coated it, they were unable to meet the quality and price of the product made by the integrated operation at Kodak Park. The Lumière company proved to be a more active competitor, not only producing cine film at its factory in Lyon but also late in the decade opening a plate factory and outlet in the United States—North American Lumière. Although Lumière did not produce cine film at the Burlington, Vermont, plant, the company fought to acquire a substantial share of the American cine film market. The key difficulties the company faced, however, were the poor quality of its product and its method of producing film on tables, the method abandoned by Eastman Kodak in 1900.[51] This method was more costly and also provided only short lengths of film that then had to be spliced for cinematographic purposes. Quality, however, was the key. The cost of film compared to other studio costs was small and, therefore, *quality* and *reliability* took precedence in the mind of screen producers. The reputation of the Lumière film among film producers for streaks and spots and lack of uniform sensitivity gave the natural advantage to Eastman Kodak.[52] Thus, the lead that Eastman Kodak had in celluloid film production proved insurmountable to its competitors throughout the first decade of the twentieth century, and by the end of the decade, cinematographic film emerged as the largest sector of sales (and, no doubt, profits) for the company (table 12.1).

Eastman Kodak concentrated on the production of the cine film and avoided entry into the other rapidly growing sectors of the cinematographic industry. Eastman carefully charted the company's course away from production of either apparatus or screen plays. He calculated that Eastman Kodak's involvement in these spheres

[49] There is no evidence that considerations of production of long continuous strips of cinematographic film played any role in the decision to switch to the continuous flow methods of film making and coating. Reductions in the cost and speed of production were primary considerations.

[50] GE testimony, *U.S.* v. *EKCo.*, pp. 272–73.

[51] Brulatour testimony, in ibid., pp. 387 and 392–96; and Claudius Poulaillon testimony, in ibid., p. 194.

[52] Brulatour testimony, in ibid., pp. 389–90 and 397; and GE to Gifford, 23 January 1911 (GEC).

Table 12.1 Eastman Kodak Film Sales in the United States, 1897–1909

Year	*Cine Film Sales*	*Cartridge Film Sales*	*Cine Film Sales as % of Cartridge Film Sales*	*Cine Film Sales as % of Total Co. Sales*	*Total Film Sales as % of Total Co. Sales*
1897	$129,383	$514,699	25%	8%	38%
1898	72,546	660,237	11	4	37
1899	134,654	720,081	19	6	38
1900	104,425	732,174	14	4	33
1901	85,317	787,820	11	3	26
1902	89,153	900,348	10	2	25
1903[a]					25
1904[a]					25
1905[a]					25
1906	284,718	1,578,853	18	4	27
1907	628,907	1,880,950	33	7	30
1908	1,448,434	1,925,002	75	16	37
1909	2,097,733	2,098,673	100	22	43

SOURCE: *U.S.* v. *EKCo.*, pp. 2565–69.
[a] Disaggregated data unavailable.

would only antagonize customers with whom the company would be competing, eventually destroying good will, markets, and profits.[53] Eastman also considered the profits on film to be much greater than on apparatus. Some of the large American and European film producers, however, were anxious about their dependence upon Eastman Kodak for film. In particular, Pathé Frères sought from the middle of the first decade of the century to develop alternative supplies of film by purchasing film from Lumière of Lyon, by resensitizing exposed film at the old Blair Camera film factory in Pawtucket, and by sensitizing film base acquired from Celluloid Company of Newark. None of these strategies proved entirely successful for Pathé during the first decade, and the French company continued to depend at least in part upon Eastman Kodak.[54]

Thus, by the closing years of the first decade of the twentieth century, the cinematographic industry included four important sectors: (1) exhibition; (2) distribution; (3) apparatus and screen play production; and (4) sensitized film production. The first three could be characterized as highly competitive, with relative ease of entry; the fourth as completely dominated worldwide by one firm, with high technical barriers to entry. As the *rate* of growth of the movie industry slowed at the end of the decade and as patent fights and price competitiveness suggested to its leadership the rationalization of the industry, the structure and technical character of the industry indicated the direction the first rationalization efforts were to take.

[53] GE to Frank Seaman, 5 January 1900; GE to Thomas H. Blair, 28 December 1909; GE to Frank L. Dyer, 31 May 1911; and GE to Joseph T. Clarke, 5 September 1911 (GECP).

[54] GE to Henry A. Strong, 13 March 1906; and GE to William S. Gifford, 14 September 1909 (GECP); and GE to Baker, 22 June 1910 (GEC).

Period of Silent Cinematography 1909 to 1925

The reason for this [Cine film being the largest and most profitable item of output] is not on account of the existence of patents, or other restraints on competition, but is owing to the difficulties of manufacture which have been surmounted by us in a larger degree than by anyone else. We were the original introducers of film and have always succeeded in keeping a long distance in advance of our competitors. The whole industry was built up upon paying a certain price for the film and owing to increasing economies in the manufacture the profit has come to be very large notwithstanding reductions made from time to time in the selling price. . . .

George Eastman
1916

13 / The Cinematographic Industry: Integration and Innovation

The rapid growth of the cinematographic industry during the first decade of the twentieth century signaled the commercial potential of the new industry and stimulated efforts to organize and rationalize its structure. During the first fifteen years of commercial exploitation of cinematography, the industry, which comprised the producers of the apparatus and films, the distributors, and the exhibiters of motion pictures, grew increasingly competitive. Then came a period of effort within certain sectors of the industry to create an oligopolistic market structure through the pursuit of technological and marketing strategies.

From 1907 to 1915, the leadership of the industry, especially the Edison Manufacturing Company, organized and rationalized the operations of the production sector of the industry by pursuing a largely technological strategy somewhat reminiscent of that pursued by General Electric and Westinghouse in the electrical industry a decade earlier. By 1907 the cinematographic industry in the United States was facing not only unparalleled opportunity because of its very rapid growth but also increasingly intense competition. In the mid-1890s, the Edison Manufacturing Company possessed an initial lead in the industry, a lead it undergirded by obtaining several patents on projection equipment.[1] From the late 1890s until 1907, however, the company found itself challenged in the market by a number of producers of both projection equipment

[1] For example, U.S. Patents #578,185 (2 March 1897), the Armat Vitascope; #580,749 (13 April 1897), the Armat Vitascope; and #586,953 (20 July 1897), the Armat and Jenkins phantascope.

and motion pictures. Accordingly, Edison Manufacturing pressed a number of patent suits including significant ones against American Mutascope and Biograph of New York and against Selig Polyscope of Chicago. At the same time, other manufacturers introduced important improvements and modifications. Consequently, the technical leaders of the industry faced both the costs of lengthy litigation and the complexity of cross licensing on patents in order to produce high-quality equipment.

A further problem facing the leaders of the American industry was how to obtain what they regarded as sufficient remuneration for the patents on cinematographic apparatus. They could obtain only very limited remuneration through the sale of projectors, because the maximum initial market consisted of only about fifteen thousand theaters. The motion pictures were a mass product, but the projection equipment was not. While all the apparatus manufacturers also produced motion pictures, not all picture producers manufactured apparatus. Therefore, if the apparatus producers sought to compensate for lost apparatus royalties by increasing their prices on the motion pictures, other motion picture producers, both American and foreign, could afford to forgo the compensation and undersell them.

At the same time, the producer of raw photosensitive cine film, Eastman Kodak, while doing very well in this rapidly expanding market both at home and abroad, felt increasingly anxious because of the growing independence of its largest customer, Charles Pathé. Therefore, George Eastman and the leadership of Edison Manufacturing considered methods of reorganizing the industry domestically and internationally in order to assure themselves of their continued prominent positions in an increasingly competitive market.

As early as the spring and early summer of 1907, the strategy of pooling the important patents in the industry and using the licensing power of the patent pool to enforce restrictions on the price and use of equipment and films was considered. In June, Eastman, conferring in Rochester with his largest customer, Charles Pathé, found him rather indifferent to the patent pool proposition. Still, the Edison people persisted in discussions with Eastman and manufacturers in the movie and equipment industry.[2] The key to the success of any effective plan was Eastman Kodak. The patent pool could enforce conditions upon the production of the equipment, but the key bottleneck in the industry lay with the raw film, which was virtually controlled by Eastman Kodak. By cooperation in the control of raw film, both domestic and foreign producers of motion pictures might effectively enforce price and supply controls upon themselves.

A court decision in Chicago in the fall of 1907 stimulated the consummation of the first phase of this strategy to reorganize and restructure the industry. In a patent case between Edison Manufacturing and Selig Polyscope, the court upheld an Edison-held patent covering the loop and sprockets used in motion picture apparatus and set the stage for Edison Manufacturing to threaten court action against many other companies alleged to be infringing.[3] However, early in 1908 Edison Manufacturing signed patent pooling agreements with Kalem Company, Essanay Film Manufacturing, Lubin Manufacturing, George Méliès Manufacturing, Selig Polyscope, and Vitagraph. Edison Manufacturing also hoped later to bring into the agreement foreign producers of equipment and films who would, like the domestic producers, pay to this Motion Picture Patents Company a royalty of approximately one-half cent per foot on all raw cine film consumed by licensees. Under this plan, Eastman Kodak served as the accountant and the sole supplier of film for all licensees.[4]

George Eastman went to Europe in February 1908 to join with Charles Pathé in creating a new, international order in the industry. While everyone, including Pathé, agreed to an arrangement similar to the patent pool created in the United States with the royalty collection by Eastman Kodak, formal contracts were not concluded because Eastman's private counsel questioned the legality of this cartel arrangement under French law. The trip was not for naught, however, because Eastman did succeed in convincing Pathé that the Frenchman's plan to eliminate all his European competitors was not likely to succeed. Consequently, nearly all the European participants in the American market

[2] GE letters quoted in Carl W. Ackerman, *George Eastman* (London: Constable, 1930), pp. 210–11.

[3] Ibid.

[4] Reference in Agreement: MPP, Edison Mfg. Co., and EKCo., 1 January 1909, in *U.S.* v. *EKCo.*, p. 3127.

agreed to enter the American patent pool under the regulations of Eastman Kodak and the Motion Picture Patents Company.[5] Pathé Frères, the American agent of the French cinematography company, signed an agreement with Edison Manufacturing in May 1908, and by the end of the year the terms of Eastman Kodak's participation were settled. Thus, on 1 January 1909, formal agreements between Eastman Kodak, the Motion Picture Patents Company, George Kleine (Chicago), American Mutascope and Biograph, and the Edison Manufacturing Company were concluded.[6]

The terms of the agreement benefited all production participants in the industry. Fourteen patents on cameras and projection equipment were pooled. The largest part of the collected royalties were to be paid to Edison Manufacturing and American Mutascope and Biograph. Eastman Kodak was to collect about one-half cent per foot of cine film as royalty paid to the Motion Picture Patents Company. A fixed price on the raw film was established for all licensees, and the increased cost of the royalty was to be passed along by the licensees to the exhibiter through an increase in the price of finished films of one cent per foot.[7] Eastman Kodak insisted that all major cinematographic producers in the United States as well as those from abroad who participated in the American market had to be in the agreement in order to prevent price cutting on finished films and entry of new producers into the moviemaking sector.[8] Edison Manufacturing and other equipment manufacturers holding patents were to receive an assured royalty without court action. Eastman Kodak, with its major customers under contract, was assured of a stable or growing market for raw film.

Underlying the entire program of the Motion Picture Patents Company were a set of assumptions about the industry that must be understood in order to perceive some of the problems the plan encountered. The industry was seen as consisting of four parts: (1) producers of raw film; (2) producers of apparatus and movies; (3) distributors of movies; and (4) exhibiters of movies. The first two sectors were seen as the key sectors, control over which, because of the disunity of the other two sectors, meant control over the whole industry. The assumption was made that the technological complexity of raw film manufacture and the importance of the fourteen pooled patents could serve as a high enough barrier to entry to deter outside competition. A key assumption, also, was that the cinematographic industry was selling a standardized, undifferentiated product. Movies were one reel in length and could be sold interchangeably by the foot. The American public would attend theaters with little attention to the personality and talent of the actors, the quality of the play, or the ability of the director. Based on these assumptions—assumptions that nearly everyone in

[5] Ackerman, *George Eastman*, pp. 217–21.

[6] Reference in Agreement: MPP, Edison Mfg. Co., and EKCo., 1 January 1909, in *U.S.* v. *EKCo.*, pp. 3126–45.

[7] Ibid.

[8] GE to Great Northern Film Co., 30 December 1908 (GEC).

the industry made—the basic strategy developed by the leaders of the industry made sense as a method of maintaining their leadership position and assuring a steady flow of profits to participants in the agreement. Soon, however, some of these assumptions were proven invalid.

The Motion Picture Patents Company invited the operators of the exchanges and some exhibiters to New York early in January 1909 and simply informed them of the patent pool, telling them that thereafter their films would be limited to those who paid for licenses under the apparatus patents and that distributors and exhibiters would have to be licensed as well, at a license fee for each exhibiter of two dollars per week. The simple announcement without previous consultation caused indignation and resentment among the distributors and exhibiters. In a few cases this was manifested in action to destroy the new organization of the industry. Within a year after the establishment of the Motion Picture Patents Company and the agreement between its licensees and the Eastman Kodak Company, a group of new, independent motion-picture-producing companies appeared in New York and Chicago and came to pose a serious challenge to the established industry.[9]

One of the principal rebels against the new organization was the largest customer of the Motion Picture Patents Company, Carl Laemmle of Chicago, who operated a Midwestern exchange and a small chain of nickelodeons and theaters in the Chicago area. After obtaining an initial license, Laemmle quickly set plans to become independent. In the spring of 1909, he withdrew from the Motion Picture Patents Company and in June announced the establishment of his own production facilities for motion pictures under the name Yankee Films Company. After holding a contest for a better name, his company became the Independent Moving Pictures Company of America, shortened to IMP. Until the company could begin production, it depended upon foreign films; however, in late October 1909, it released its first picture: *Hiawatha*.[10]

The strategy of IMP consisted of more than the linking of distribution with production. Laemmle's partner, R. H. Cochrane, had experience in advertising and soon developed an advertising campaign against Motion Picture Patents Company in the trade magazines that attracted the attention of Motion Picture Patents Company's more than seven thousand exhibiters and distributors and soon began to influence the image of the leadership of the industry.[11] Every week the IMP assailed Motion Picture Patents with invective such as this:

Good morning! Have you paid two dollars for a license to smoke your own pipe this week? . . . Have you paid your two dollars for a license to breathe this week? . . . I will rot in Hades before I will

[9] John Drinkwater, *The Life and Adventures of Carl Laemmle* (New York: G. P. Putnam's Sons, 1931), pp. 70–71.

[10] Ibid., pp. 73–81.

[11] Ibid., p. 85.

join the Trust, or anything that looks like a Trust. How is your backbone?[12]

Although Motion Picture Patents decided to ignore this attack, the persistence of Laemmle and Cochrane gave moral courage to other independent producers, exhibiters, and distributors.

Perceiving the potential problem with the independent distributors or exchanges, which numbered between 115 and 120 in 1909, the Motion Picture Patents Company initiated forward vertical integration in 1910 by organizing and combining all the exchanges except one into the General Film Company. This move put additional pressure on the independent picture producers, who then found it nearly impossible to distribute through the traditional channels.[13]

Laemmle and the independents responded in two ways. First, Cochrane turned his advertising attack on the new distributing company, personifying it in General Flimco, a fierce, military cartoon character constantly grabbing and intimidating others.[14] When in 1912 the federal courts dismissed a suit against IMP on infringement of the Latham patent, Cochrane ran an advertisement featuring his despicable military character in a cartoon captioned "General Flimco's Last Stand." Second, the IMP interests spearheaded the creation in 1910 of an independent distribution company, Motion Picture Distributing and Sales Company, paralleling the General Film Company. Participants in the independent Sales Company included IMP, Rex Motion Picture Manufacturing, and Nestor Film. Two years later, Motion Picture Distributing and Sales became the core of a new consolidation of independent producers under the name Universal Film Manufacturing. Early presidents of the Sales Company and Universal Film Manufacturing included Carl Laemmle and Jules Brulatour. The association of Brulatour with the independents proved crucial not only to their initial success in surviving against the Motion Picture Patents but in the ultimate disintegration of that organization's strategy.[15]

The effective control of the supply of raw film loomed crucial to the Motion Picture Patent strategy. Without question, Eastman Kodak produced the toughest and most uniformly sensitive cinematographic film in the world. Although many firms throughout the world had sought to compete with the Rochester giant in cine film production, none could approach the standards of quality and uniformity. The effectiveness of the Motion Picture Patents strategy depended upon the difference in quality between Eastman film and that of other film makers such as Lumière, Austin Edwards, and Ensign. Yet, Laemmle and other independent American producers soon turned to the inferior Lumière film and were able, though with added cost and frustration, to market an accept-

[12] Ibid., pp. 76, 85, and 87.

[13] Ibid., pp. 123–24; and Mae D. Huettig, *Economic Control of the Motion Picture Industry* (Philadelphia: University of Pennsylvania Press, 1944), pp. 15–16.

[14] Drinkwater, *Carl Laemmle*, p. 86.

[15] Ibid., pp. 102, 105, and 131.

Table 13.1 Leading Independent Motion Picture Companies Ranked by Contractual Consumption of Film, 1911

Consumption	Company
100,000 ft. per week	Powers Company (N.Y.C.)
	IMP Company (Chicago)
	Carlton Motion Picture Company (N.Y.C.)
	Thanhouser Company (N.Y.C.)
	New York Motion Picture Company (N.Y.C.)
75,000 ft. per week	American Film Manufacturing Company (Chicago)
40,000 ft. per week	Gaumont Company (N.Y.C.)
30,000 ft. per week	Atlas Motion Picture Company (N.Y.C.)
	Rex Motion Picture Manufacturing Company (Chicago)
25,000 ft. per week	Champion Film Company (N.Y.C.)
Consumption unknown	Centeur Film Company (N.J.)
	Nestor Film Company (N.Y.C.)

SOURCE: Contracts, *U.S.* v. *EKCo.*, pp. 3212–56.

able product. The Lumière Company and its New York agent, Jules Brulatour, succeeded in developing a sizable business for the French film from among the American independents. Although the quality of the Lumière film was unreliable and the supply undependable, the independents had little alternative. By early 1911, Brulatour had obtained contracts from the independents for the sale of more than 35 million feet of positive cine film annually at a time that Eastman Kodak sales of cine film to the American market were at the rate of 91 million feet.[16]

The leadership of Eastman Kodak became nervous about the loss of sales to the newly emerging and rapidly growing independents. Obviously, the technical superiority of the Eastman Kodak film was not sufficient to serve as a barrier to entry of foreign raw film suppliers. Furthermore, Charles Pathé, well aware of the precarious position of his large business because of its complete dependence upon Eastman Kodak as the film supplier (no one else had the capacity to begin to supply Pathé), sought frantically from 1906 to 1912 to create alternative sources of film by regenerating previously used cine film and by producing virgin film from Celluloid Company base.[17] During this period Pathé continued to purchase large quantities of film from Eastman Kodak, but the latter eyed Pathé's moves to backward integration, together with the growing Lumière sales to the independents, as ominous signs of a deteriorating market position that the company would have to deal with.[18]

Yet, Eastman Kodak and Motion Picture Patents had entered into their agreement with one additional advantage: a new, non-

[16] The Brulatour figure is based on contracts with total sales of 700,000 feet per week. The contracts were published as Government Exhibits #278–79 and 311–19, in *U.S.* v. *EKCo.*, pp. 3212–56. Brulatour told Eastman in early 1911 that he needed forty million feet annually, although he was not selling that much at the time (Brulatour testimony, in ibid., p. 392). The Eastman figure is from George Eastman's statement, GE to William S. Gifford, 5 December 1910 (GEC).

[17] GE to Gifford, 14 September 1909 (GEC).

[18] Ibid., 26 October 1908 and 1 October 1909.

Fig. 13.1
David Reid. Courtesy of Eastman Kodak Company.

inflammable film. Early in the century a series of theater fires caused by ignition of the highly flammable nitrocellulose film had created considerable public attention and consternation, especially in Europe, because of the high loss of life. Initially fearing government restrictions because of these fires, the Eastman Kodak management had encouraged chemists and engineers at Kodak Park from the turn of the century to develop a suitable noninflammable cine film.[19] In 1906 David E. Reid was hired to work on the production of raw photographic paper in an experimental laboratory at Kodak Park (fig. 13.1). After the renewal of the contract with General Paper, news that Pathé was working on a noninflammable film stimulated Eastman to have Reid work along the same lines.[20] In a little over a year, he had made such progress with an acetate-cellulose film that production tests were begun. Eastman saw this new film as a means of staying ahead of European competitors—Lumière, Austin Edward, and Ensign—and potential competitors—Pathé and the German chemical companies Agfa and Bayer. When the new film came up in the discussions with the leaders of Motion Picture Patents in late 1908, it was seen as a means of raising even higher the entry barriers to independents.[21]

The introduction of the noninflammable film required dual efforts on the part of Eastman Kodak. The first effort involved achieving actual production capacity for the film. The road from Reid's laboratory to the large casting wheels at Kodak Park was long and tortuous. The Kodak Park engineers and chemists had to find the solvents that would maximize the toughness and transparent qualities of the acetate film. Then, when they had to make the best product, they had to modify the process in order to avoid certain American and European patents. Once they had finally decided upon the materials to be used, they confronted the problem of obtaining an adequate supply of acetic anhydride.[22] Finally, they had to make slight modifications in the conditions of production in order to maximize the desirable properties and minimize the unwanted ones. The second part of the company's introduction of the acetate film required a campaign to convince state and municipal authorities to pass legislation requiring exhibiters to use only noninflammable film. The Eastman Kodak organization proved much more adept in reaching the technical goal than the political one. Efforts to stimulate the required use of noninflammable film in the interest of safety failed, especially in those cities in America where the vaudeville interests predominated.[23]

[19] GE to H. Senier, 15 and 24 May 1899; and GE to Frank W. Lovejoy, 10 July 1899 (GEC).

[20] GE to W. H. Oglesby Paper Co., 3 August 1906; GE to H. M. Smith, 21 July 1906, 19 November 1906, and 12 February 1907; and GE to James H. Haste, 7 April 1908 (GEC).

[21] GE to Gifford, 1 and 6 April and 30 November 1908; GE to Haste, 27 April 1908; and GE to Andrew Pringle, 28 May 1909 (GEC).

[22] GE to Pringle, 28 May 1909 (GEC); GE to Moritz B. Philipp, 27 March 1907; GE to Joseph T. Clarke, 28 April, 3 July, and 2 September 1908; and GE to Senator Payne, 15 December 1908 (GEC).

[23] GE to Baker, 31 August 1910; and GE to Gifford, 5 December 1910 (GEC).

Fig. 13.2
Jules E. Brulatour. Courtesy of Eastman Kodak Company.

Nevertheless, late in the fall of 1908, just prior to the formal contractual agreement between Motion Picture Patents and Eastman Kodak, the Rochester firm introduced the acetate film to the American cine market.[24] Despite the extensive testing, there soon were complaints from exhibiters that the film was not as tough as the nitrate film, the sprockets of the projectors tending to tear it more readily. Later, complaints arose that it did not wear as long. At once, production engineers and laboratory chemists sought to increase the strength of the acetate film; yet, the combination of the failure to gain legislation requiring the use of the noninflammable film and the failure to overcome the prejudice of exhibiters and producers against making a change spelled the ultimate defeat for the new product. Although Eastman Kodak did succeed in improving the strength of the film, its higher cost of production, the indifferent attitude of the trade in the United States, and the company's failure to introduce it successfully in Europe prompted the company to withdraw the acetate film early in 1911.[25]

At the same time that Eastman Kodak was feeling the sting of the failure of its expensive gamble with acetate film, it was gradually losing its largest customer (Pathé) and seeing the growth of domestic sales of Lumière film to the independents. Jules Brulatour began to place some options in the hands of George Eastman and the leadership in Rochester. Although Brulatour was initially an agent for Lumière, his interests moved increasingly toward those of the American independent film producers, his source of livelihood, and he sought to protect their interests. Nearly every independent producer wanted Eastman film if he could get it. Some Eastman cine film was even surreptitiously brought from Europe, but such supplies were irregular. Brulatour made numerous personal, but unsuccessful, overtures to George Eastman during 1909 and 1910, seeking to acquire a supply of Eastman film for the independents.[26] He also sought, with considerable finesse and energy, to have Eastman Kodak purchase or acquire an interest in the Société Lumière in Lyon. Eastman had given consideration early in 1908 to overtures from Lyon to sell the company to the Rochester firm because of his interest in Lumière's work with noninflammable film and color processes, but intelligence reports indicated that neither product was commercially viable. Therefore, Eastman ignored the overtures.[27] Brulatour sought, however, to rekindle Eastman's interest in the French company, hoping that if Eastman Kodak acquired an interest in it, his independent customers could obtain Eastman film. The Lumière interests favored the sale of the family firm. Negotiations reached an advanced stage with Eastman visiting the plant in Lyon, but the sale was not consummated.[28]

Although Brulatour lost his commission on the sale of the French

[24] GE to Pringle, 28 September 1908 (GEC).

[25] GE to Mattison, 23 January 1911 (GEC).

[26] Brulatour testimony, *U.S.* v. *EKCo.*, p. 388.

[27] GE to Darragh de Lancey, 20 April 1908; and GE to Gifford, 18 July 1908 (GEC).

[28] Brulatour testimony, *U.S.* v. *EKCo.*, pp. 386–400; William A. Neale testimony, in ibid., pp. 502–3; and Henri Kubnick, *Les Frères Lumière* (Paris: Librairie Plon, 1938).

film company to Eastman Kodak, he continued to pursue his principal objective: Eastman film for the independents. Finally, after he had shown Eastman contracts with the independents for substantial quantities of cine film, Eastman indicated that he would seek some arrangement.[29] Early in 1911 he made a modification in the contract with Motion Picture Patents that allowed Eastman Kodak to sell to the independents. Eastman then asked Brulatour to act as the Eastman Kodak company representative to the independents on a straight commission basis.[30] This concession by Eastman to Brulatour and the independents was the admission that the original strategy of Motion Picture Patents and Eastman Kodak had failed. The flow of Eastman film to the independents spelled the first of a series of defeats for Motion Picture Patents and signaled the ultimate failure of the first attempt at the rationalization of the cinematographic industry.

Although the Motion Picture Patents had lost what it regarded as the key to control of the motion picture industry, it persisted with its strategy of technical and patent control. The most obvious target for attack among the independents was Laemmle's Universal Film Company with its continued antagonistic advertising. Motion Picture Patents launched a barrage of patent suits and injunctions against Universal that culminated in a three-year total of 289 suits and required of Universal defense costs of more than $300,000. Although the attack caused some major discomfort for the company, Universal survived to triumph over Motion Picture Patents. During a large part of the year 1911, the company's production facilities were located in Cuba, where they were free from the American court orders.[31] The combination of the warm climate, the variety of terrain and scenery, and the proximity to the Mexican border soon drew the attention of Universal and other independent producers to southern California. From 1911 to 1915, many of the independents moved their production facilities from Chicago and New York to the Hollywood area.[32] The Motion Picture Patents Company litigation strategy may have contributed to this localization of the industry.

One of the defenses of the Universal attorneys in one of the patent cases struck directly at the heart of the Motion Picture Patents Company patent problem. The company had sought to obtain remuneration for its projector patents by requiring the use of certain materials from which it could profit. The Circuit Court of Appeals ruled in 1916 that such a tying arrangement was in violation of the Clayton Act; however, the Supreme Court in 1917 ruled the arrangement illegal solely under the patent statutes.[33] These rulings indicate that even had Motion Picture Patents

[29] Brulatour testimony, *U.S.* v. *EKCo.*, p. 388.

[30] Ibid., p. 389; and Agreement, 14 February 1911, Modifying Agreement of 1 January 1909, Government Exhibit #242, in ibid., pp. 3185–89.

[31] Drinkwater, *Carl Laemmle*, pp. 97, 110, 153, and 159–60.

[32] Kenneth MacGowan, *Behind the Screen* (New York: Dell Publishing Co., 1965), pp. 138–39.

[33] *Motion Picture Patents Co.* v. *Universal Film Manufacturing Co.*, 235 *Fed. Rep.* 389 (1916); and *Motion Picture Patents Co.* v. *Universal Film Manufacturing Co.*, 243 *U.S.* 502 (1916).

been able to keep exclusive control of the unexposed cine film, its hold on the industry would not have prevailed. Moreover, with encouragement from Universal and other independent producers, the U.S. Attorney General's Office filed an antitrust petition against Motion Picture Patents Company in 1912. The issues included the pooled interest of the producers in the distribution company and many of the sales conditions imposed on the exhibiters, including the weekly two dollar license fee. The verdict of October 1915 recognized these practices as unlawful and required that they be halted. In addition, Motion Picture Patents was ordered to return all the license fees. However, despite government efforts to attach the company's bank accounts, the Motion Picture Patents Company ceased operations before any of the fees were recovered.[34] General Flimco's last stand was on the site he selected, before the bar of justice, and he was soundly defeated at his own game. Most of the constituent companies soon disappeared, and an important era in cinematographic history passed. Although the adverse antitrust and patent decisions climaxed the decline of this company and its once powerful position in the industry, of at least equal or greater importance in that decline was the development by independents of a new counterstrategy that revolutionized the industry.

The new marketing strategy involved a series of product innovations that broke substantially with many of the previous assumptions and attitudes. Initially, most of the industry could see little reason to pay high salaries to stars or to publicize these salaries. Nearly until its demise, Motion Picture Patents did not believe that the public would sit through more than a one- or two-reel picture. This adamant position may have reflected not only the company's conservatism but also, because of its concentration on production and distribution, its remoteness from the consumer.[35] Instead of conceiving of motion pictures as single-reel, standardized products with anonymous actors, actresses, and directors, the new leaders of the industry regarded them as highly differentiated products of feature length (at least five reels) with known plays and famous or well-known actors, actresses, and directors.

The leadership in the introduction of feature pictures and the star system came from a few independents and rapidly spread among them from 1911 to 1915. The importance of the personalities and salaries of the stars, the fame of the directors, and the quality and length of the plays converged to prove important elements of a new marketing strategy. Laemmle introduced in 1909 and 1910 the "Biograph Girl," who serendipitously became the first screen star, Florence Lawrence. Soon names such as King Baggot, Mary Pickford, and Tom Ince became well known to moviegoers. The first feature motion pictures came from Italy in 1911 and 1912, but soon the Famous Players Company, organized in 1912 by Adolph Zukor and Daniel Frohman, combined the star system with the feature film. Within the next two years, other companies such as

[34] Drinkwater, *Carl Laemmle*, pp. 120–29.

[35] Huettig, *Economic Control of the Motion Picture Industry*, pp. 17–18 and 24–26.

Kalem, Jesse L. Lasky Feature Play, and William Fox followed Zukor. As the films of these companies met with considerable box office success and as the power of Motion Picture Patents waned, many independents also imitated Zukor.[36]

At the same time, reflecting both the influence of General Film Company policies and the influence of the feature and star system, the character and structure of the distribution network changed. As the principal product of the industry became a high-cost and highly differentiated product, the distribution network shifted from a state and regional system to a national system, thus giving the high-cost film the maximum exposure and return. In addition, producers no longer sold, but rented, films, not on a per-foot basis as before, but according to a classification scheme that rated theaters according to their size, location, and potential audience and that rated film rentals on the basis of the quality and cost of the film and the time since it had first been released.[37] Successful product innovation broke the set of conceptions and assumptions concerning the industry and its products and, in turn, stimulated structural change.

An even more fundamental structural change occurred with the emergence of new leadership, as new approaches and strategies were introduced in an effort to secure a rationalization of the industry. The industry began to undergo the first stages of horizontal and vertical integration. In 1912 the Motion Picture Patents Company reorganized the General Sales Company and acquired the cine positive-printing facilities of all its constituent members. This venture, into which George Eastman personally invested $3 million, represented only a partial backward vertical integration.[38] Carl Laemmle led a more complete backward vertical integration with the creation in 1912 of Universal Film Company, a combination of his old production and distributing companies and those of several other independents.[39] Two years later William Fox, who owned a small chain of theaters and a very important film exchange, moved backward to include production facilities. At about the same time, in an effort to obtain acting and directing talent, Jesse L. Lasky Feature Play acquired Oliver Morosco Photoplay and Bosworth, Incorporated, in a horizontal integration.[40] This acquisition was, however, but a prelude to a much more significant movement toward horizontal and vertical integration among the new strategists in the industry.

The two most important new production companies consummated a merger in 1916. William W. Hodkinson, founder of Paramount Pictures, as the representative of five independent regional distributors, served as the principal distributor for both Zukor's Famous Players and Lasky Feature Play; however, when approached by Zukor, he would not consider merging his company

[36] MacGowan, *Behind the Screen*, pp. 140 and 155–59.
[37] Huettig, *Economic Control of the Motion Picture Industry*, pp. 24–26.
[38] GE to Gifford, 7 March 1912 (GEC).
[39] Drinkwater, *Carl Laemmle*, pp. 102 and 131.
[40] MacGowan, *Behind the Screen*, pp. 164–65.

with those of Zukor and Lasky. Consequently, Zukor obtained stock control of Paramount and then, in conjunction with Lasky, created Famous Players-Lasky, with himself as president, Lasky as the head of production, and Paramount as a distributing subsidiary.[41]

Zukor's strategy soon placed Famous Players-Lasky in the dominant position in the American motion picture industry. Zukor sought to control not the unexposed film or the key projection apparatus patents but what at this time became the key commodity: the stars. In 1916 Famous Players-Lasky held contracts with about 75 percent of the most popular motion picture stars in the industry. This position allowed the company to introduce block booking. This sales strategy presented to the potential exhibiter a group of films, some of which were of first quality with famous actors and actresses but some of which were of lower quality with little-known or unknown performers. The exhibiter had his choice of either taking the entire package or taking nothing at all. Accompanying block booking, Zukor raised the rental prices on films quite substantially. Then, in 1917, his company acquired an additional twelve independent producing companies, complete with their directors and stars. From 1915 to 1920 Famous Players-Lasky assumed the dominant position in the industry.[42]

As one producing and distributing company became dominant, the exhibiters became concerned for their autonomy. With the growing power of the producers and distributors from the beginning of the decade, individual theater owners had formed booking combines and even combined into small local and regional theater chains. This exhibiter combination movement achieved national scale in 1917, with the formation of the First National Exhibitor's Circuit. The First National focused particular attention on first-run theaters, starting as a purchasing agent for the twenty-six largest exhibiters. These first-run theaters, the most profitable sector, were the key to the success of any costly film. By the beginning of 1920, the First National controlled 224 first-run theaters and a total of 639 movie houses. It further strengthened its position by integrating backward to the production stage and obtaining the services of top stars such as Charlie Chaplin and Mary Pickford. First National, representing the first completely integrated firm of any large size, posed a serious threat to the dominant position of Zukor's Famous Players-Lasky Corporation.[43]

Not unexpectedly, Zukor responded by acquiring a large chain of theaters and also creating a completely integrated company. Zukor and the leadership of Famous Players-Lasky moved to theater ownership for two reasons: (1) the need to protect the ultimate market for its products; and (2) the desire to obtain a share of the handsome theater profits. In 1919 Famous Players-Lasky secured a report on the industry from the New York investment banking firm of Kuhn, Loeb & Company. It indicated that the

[41] Huettig, *Economic Control of the Motion Picture Industry*, pp. 30–31.
[42] Ibid., pp. 31–32.
[43] Ibid., pp. 32–34.

15,000 American theaters received a return of $800 million while the producers received only $90 million. The report made clear that first-run theaters represented the most profitable sector of the industry, some returning as much as 100 percent annually on the investment. Despite the success of Famous Players-Lasky, the company was not in a position to acquire a large chain of theaters without financial assistance. Kuhn, Loeb & Company floated a $10 million loan, and theater acquisition began; by August 1921, the company had obtained 303 theaters across the nation. Thus, Famous Players-Lasky achieved complete integration but from the reverse direction of First National Exhibitor's Circuit and at the sacrifice of the company's independence from Wall Street banking interests.[44]

The creation of completely integrated motion picture companies by Famous Players-Lasky, now known as Paramount, and First National brought pressure on Loew and Fox to follow suit by integration backward to production. By 1923 the structure of the industry reflected the achievement of the oligopolistic goal that Motion Picture Patents had sought fifteen years earlier. Four integrated companies, Fox, Loew, First National, and Paramount, held dominant positions in the industry. Paramount's strength lay in production; First National's in its large chain of theaters; and Loew combined limited strengths in production with a small chain of high-quality, well-located, first-run theaters.[45]

During the previous decade, Eastman Kodak had freed itself from the exclusive contracts with Motion Picture Patents and thereby retained its dominant position in the domestic cine film market, in spite of efforts by other members of the photographic industry to enter this lucrative field. In 1910–11, Gustav Cramer of Cramer Dry Plate in St. Louis and Henry Kuhn, formerly of Defender of Rochester, formed the Fireproof Film Company of Portland, Maine, and sought unsuccessfully to enter the noninflammable film market.[46] Domestic sales of unexposed cine film continued to increase during the decade while imported cine film sales remained steady (table 13.2). While Eastman Kodak maintained its domestic market position, its European markets became increasingly competitive as a consequence of the entry of Pathé and Agfa; however, the war curtailed the trade of Agfa and assisted Eastman Kodak in retaining its position of leadership in non-German cine markets.

Technically, the cine film innovations consisted of improvements in the quality of acetate film. When municipal authorities in Paris and Lyon threatened to require safety film, Eastman Kodak in 1914 resumed small-scale production of noninflammable film.[47] Moreover, during the war Eastman Kodak produced cellulose

[44] From quoted report in ibid., pp. 34–36.
[45] Ibid., p. 39.
[46] "Certificate of Organization" and "By-Laws," Fireproof Film Co., in Cramer Collection, Missouri Historical Society, St. Louis, Mo.; and GE to Henry A. Strong, 22 April 1911 (GECP).
[47] GE to H. D. Haight, 5 May 1914 (GEC).

Table 13.2 U.S. Exports and Imports of Cinematographic Film, 1913–20

Year	*U.S. Exports*		*U.S. Imports*		
	Unexposed Film	*Exposed Film*	*Unexposed Film*	*Exposed Film*	
				Negative	*Positive*
1913	$1,753,042	$2,276,460			$ 872,611
1914	4,264,722	2,282,924	$ 889,560	$402,704	1,009,469
1915	2,591,444	2,498,504	967,907	258,800	411,999
1916	2,220,118	6,757,658	750,023	225,690	256,332
1917	1,125,895	6,633,291	802,324	448,252	227,118
1918	1,385,291	5,132,448	739,135	166,033	177,148
1919	2,680,263	8,066,723	283,271	384,611	115,062
1920	1,698,248	7,900,198	1,697,976	728,899	204,217

SOURCE: U.S., Department of Commerce, *Commerce and Navigation of the United States*, 1894 to 1920 (annual).

acetate for coating the fabric wings of aircraft for the U.S. government. This led to extensive experimental work with cellulose materials and their production and proved valuable as the company slowly shifted to a cellulose acetate base for film. When it introduced cine film for home movies in 1923, it produced all such home-cine film only on the cellulose acetate "safety" base. Gradually, during the next quarter century, the nitrate base was withdrawn from all the company's film products, and other film companies did the same.[48] The supply of unexposed cine film never again played the kind of role in the motion picture industry that it had between 1909 and 1911.

During the mid-1920s the motion picture industry underwent two changes that brought a substantial alteration of its leadership. Paramount initiated the first change by acquiring an important part of First National's chain of theaters. Paramount's initiative against its principal domestic competitor eventually led to First National's disintegration. This left Paramount again the dominant leader of the industry with Marcus Loew and William Fox in secondary positions and Universal and Warner Brothers in distinctly tertiary positions. At the same time that Paramount strengthened its position in the mid-1920s, the popularity of the movies began to wane. Audiences became more discriminating, especially with the advent in the 1920s of a new outside competitor, radio.[49] Moreover, developments in radio technology influenced even more fundamentally the leadership of the motion picture industry.

The mid-1920s witnessed the introduction of a major technological change: sound movies. Though efforts to coordinate sound and motion pictures extend back to Edison and Dickson's early cylinder experiments in the late 1880s and Gaumont's commercial efforts in

[48] C. E. Kenneth Mees, *From Dry Plates to Ektachrome Film: A Story of Photographic Research* (New York: Ziff-Davis Publishing Co., 1961), pp. 142–46.

[49] Huettig, *Economic Control of the Motion Picture Industry*, pp. 40–41.

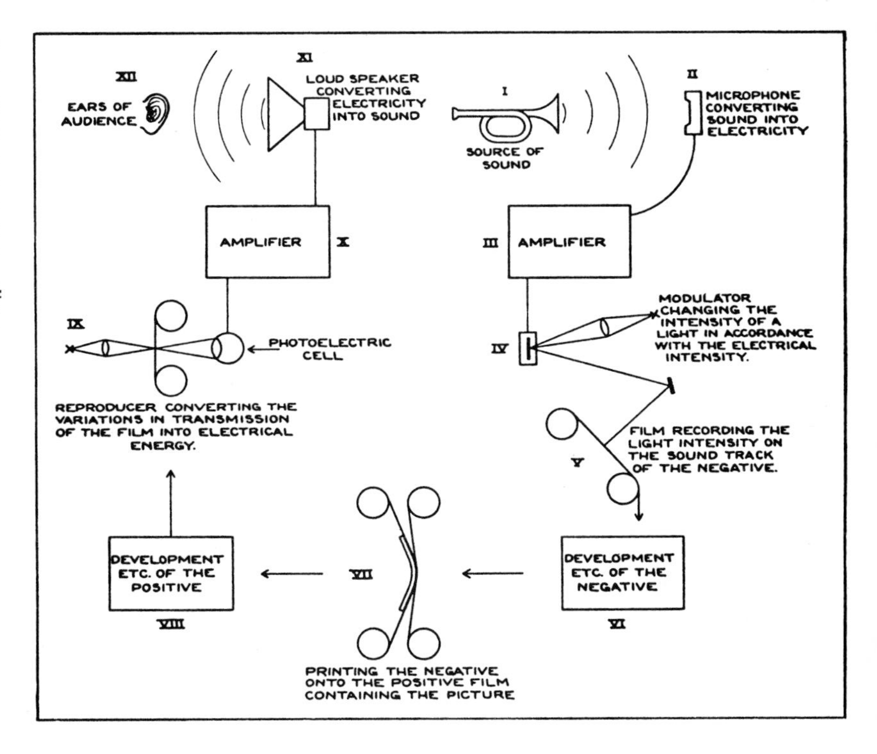

Fig. 13.3
Cycle of operations in sound recording. From C. E. Kenneth Mees, From Dry Plates to Ektachrome Film: A Story of Photographic Research *(New York: Ziff-Davis Publishing Co., 1961). Courtesy of Eastman Kodak Company.*

France at the turn of the century, the major problems remained the amplification and coordination of the sound with the picture. Western Electric and the Bell Laboratories developed in the early 1920s an optical method of sound reproduction through the utilization of amplifying circuits and Lee De Forest's recently developed gas discharge lamp method. These were employed in the translation of variations in sound into corresponding variations in the lamp brightness, which could then be photographed onto cine film as a sound track and upon replay could be reconverted to sound by reversing the process (see fig. 13.3). Although Western Electric solicited the major motion picture companies, the leadership of the industry chose to ignore the sound process. But two of the smaller companies, Warner Brothers, and to a lesser extent, Fox, paid attention. Warner Brothers, on the edge of financial disaster, gambled $800,000 on development of the sound system and introduced the first sound movie, *The Jazz Singer*, in 1927. The movie was a huge success, and Fox quickly followed Warner's lead. Within a year or two, the rest of the industry had adopted sound motion pictures. Nevertheless, Warner Brothers' initial lead gave it an advantage of about a year during which it gained substantial profit and leadership in the new technology. The company retained much of its profits for investment and issued considerable new stock and bonds in order to integrate forward into theaters. Between 1928 and 1930, Warner Brothers acquired 250 theaters by combining with the Stanley circuit, and it also acquired a large share of the remaining First National theaters. In the same period, the company's assets grew from $16 million to $230 million. Warner

Brothers combined its pioneer effort in sound movies with the unusually receptive Wall Street market of the late twenties to rocket itself into a major leadership position.[50]

Radio Corporation of America developed an alternative sound system for motion pictures and introduced its system by entering the industry with a fully integrated company. The new company, Radio-Keith-Orpheum (RKO), bought FBO Productions, a producing organization, and acquired controlling interest in Keith-Albee-Orpheum, a large theater chain with considerable strength in New York City. Though RKO possessed an independent system for sound movies and created an integrated corporation, much of its management, which came from RCA, had little feeling for the nuances of the motion picture industry. Consequently, RKO did not fare nearly as well as Warner Brothers.[51]

The impact of the new sound technology on the leadership of the motion picture industry is indicated by the statement of net income for the five leading firms in the industry in 1929:

Warner	$17 million
Paramount	15½ million
Fox	13½ million
Loew	12 million
RKO	2 million

In general, Paramount maintained its position of leadership, but Warner and Fox moved into prominent positions in the industry through their leadership in the introduction of sound motion pictures.[52]

Although technological innovation and the exclusive possession of certain technological knowledge proved significant for the pattern of development of the motion picture industry, innovations in marketing and corporate structure were at least equally important in shaping the pattern. Some of the most significant technical influences, such as roll film and the sound system, originated outside the industry; likewise, product innovations such as the feature play and the star system were patterned after the legitimate theater. Certainly, the vertically integrated motion picture company and the oligopolistic market structure were but imitations of a pattern common in many American industries at the time. The American motion picture industry's innovators, like the other members of the industry, initially operated out of a common set of assumptions and conceptions of the industry, but under the pressure of financial difficulty, legal restrictions, or threat to established positions or interests, they were willing to alter at least some small part of their assumptions and to risk changing their *modus operandi*. Established leadership preferred the security of the known to the risk of the unknown, but utilization of developments or models outside the industry lessened the gamble. For a few men such as

[50] Ibid., pp. 42–45; and Mees, *From Dry Plates to Ektachrome Film*, chap. 16.
[51] Huettig, *Economic Control of the Motion Picture Industry*, pp. 46–53.
[52] Ibid., p. 53.

Zukor and Warner, the rewards were great, but there were also many innovators who did not succeed. Innovation was no universal panacea for the business problems of the industry, but successful innovation often destroyed old conceptions and assumptions, shook the established industrial structure, crumbled the barriers to entry, encouraged new organizational structures, and ushered in new leadership to the industry.

While organizational and industrial restructuring and major innovations swept the infant cinematographic industry, the leadership of that more mature sector, with continued substantial commitment to still photography, faced a number of problems. These included government antitrust action, extended patent litigation, potential challenges from technological innovators, and imminent changes in personal leadership. Confronted with these challenges the leadership employed modifications of its traditional strategies to maintain its corporate integrity and to preserve the structure of the industry.

14 / Preservation of the Corporate and Industrial Structure

Eastman Kodak continued to dominate the still photograph sector of the industry during the period 1909 to 1925. Much of the internal activity at Eastman Kodak and the state of interfirm relations between the Rochester company and its foreign and domestic rivals reflected either the efforts of rivals to attack the Eastman position or efforts of Eastman Kodak to preserve and protect its position in the industrial structure. The responses of the Rochester firm included further development of its research and development capacity, defense of its previous patent and horizontal integration strategies, and major new commitments to vertical integration. At the same time, many of the functions of leadership previously conducted by George Eastman were gradually institutionalized in a developing corporate organizational structure.

Eastman Kodak

Research and Development

Since the founding and success of the Eastman companies rested on product and production innovations, emphasis continued to be placed upon technical innovation; however, the period when George Eastman and a few other mechanically or chemically oriented persons personally addressed technical problems out of their broad experience with the market and industry gradually passed. From the time of Eastman's personal withdrawal from inventive activity in the early 1890s, inventive and development work was gradually institutionalized. Eastman's strategy of institutionalizing technological innovation differed in the apparatus and materials sectors of production. During the 1890s and the early part of the twentieth century, Brownell served as the principal camera designer, although he was assisted by several men working under him. Gradually, as Eastman Kodak acquired plate camera companies during the period of integration, men such as John A. Robertson and William E. Folmer came to play increasingly important roles in camera and production design and development. Since the founding of Kodak Park, the center for materials production, the Experimental and Testing Laboratory, had served as a center for research and development work. Henry Reichenbach, S. Carl Passavant, Harriet Gallup, and Frank Lovejoy had worked as chemists at the laboratory in the 1890s. Although the prime functions of this laboratory were to test the quality and purity of incoming raw materials, especially silver nitrate, gelatin, "cotton," and halide salts, and to perform quality-control tests on finished films, plates, and papers, the college-trained chemists and chemical engineers who performed these routine tasks were also called upon to solve specific practical problems involving product quality and production engineering.[1]

During the early twentieth century, as production capacity at Kodak Park increased, the Experimental and Testing Laboratory

[1] Wyatt Brummitt, "Historical Notes," and "Kodak and Research," Brummitt MS, Office of Corporate Information, Eastman Kodak Co., Rochester, N.Y.

also grew rapidly. In 1906 David E. Reid, a pharmaceutical chemist originally hired to investigate the possibilities of making raw paper stock, was assigned to head the laboratory with a staff of eight persons. During the next decade, the responsibilities assigned to Reid and the laboratory mushroomed, and the staff grew to forty-five. Most of the day-to-day production problems and special work on product development that George Eastman had previously addressed were now assigned to Reid and his laboratory in coordination with the director of Kodak Park.[2] However, despite this institutionalization of development work, many of the individual plants and departments also exerted many unheralded and uncoordinated developmental efforts. Charles Hutchison described the situation: "Experimental work of a sort was carried on in the production departments that would eventually apply it, but much of its value was lost through lack of coordination. Duplication of effort resulted from a reluctance on the part of individual departments to share the fruits of their discoveries. The caliber of the research personnel was only average, and the long-range slant on research . . . had no place in the picture. 'Get out the goods' was the byword; research was an adjunct to that aim."[3] Nonetheless, major development projects were undertaken at Kodak Park, two of the most important of which were the noninflammable film and color photography.

The development of noninflammable film was begun prior to David Reid's employment at Kodak Park, but soon after Reid's arrival in the summer of 1906 he assumed responsibility for its development.[4] George Eastman, concerned that European chemical or film companies might develop a noninflammable film, monitored Reid's experiments closely.[5] Eastman was wary of changing the product too quickly. With the increase in size and in reputation of Eastman Kodak, the company became more conservative. Eastman noted, "We make it a rule not to change any of our products without testing in every possible way."[6] One error could seriously damage the company's reputation and position. By the spring of 1908, Reid, who had devoted a considerable amount of his own personal energies to the development work on noninflammable film, created a successful acetate support for cine film, and soon production on a commercial scale was begun.[7] Despite all of the advanced testing, the cine film proved weaker and less tough than the nitrate film and this difference, which was not successfully overcome during the two years of marketing, led to the failure of the innovation at this time.[8] Eastman had not been overly conservative.

[2] Ibid.; David E. Reid obituary, *Rochester Democrat and Chronicle*, 16 April 1934; and GE to W. H. Oglesby Paper Co., 3 August 1906 (GEC).

[3] Interview with Charles Force Hutchison by Jean Ennis, 17 October 1950, Office of Corporate Information, Eastman Kodak Co., Rochester, N.Y.

[4] GE to H. M. Smith, 21 July 1906 (GEC).

[5] Ibid., 19 November 1906.

[6] GE to Moritz B. Philipp, 15 May 1907 (GEC).

[7] GE to James H. Haste, 7 April 1908; and GE to William S. Gifford, 27 February 1909 (GEC); and GE to Henry A. Strong, 7 July 1908 (GECP).

[8] GE to Mattison, 23 January 1911 (GEC).

Color photography was the second major development project conducted during the early twentieth century at Eastman Kodak. George Eastman felt threatened by the Lumière Brothers' creation in 1904 of their Autochrome process. Immediately he began learning the technical details of color photography, a subject to which he had devoted little attention previously. In the spring of 1904, he alerted Joseph T. Clarke to seek new processes and to investigate old ones. Clarke began experimenting at Harrow on his own color process (fig. 14.1); at the same time, he also reported to Eastman on the work being conducted by other color experimenters in Europe. Meanwhile, early in the spring, J. H. Powrie, who, like the Lumières, sought to produce color photography by what was known as the screen method, approached George Eastman.[9] Although Powrie worked for several months with the resources of the chemical laboratory at his disposal, he failed to demonstrate that the method was commercially feasible; and the Eastman Kodak Company did not, therefore, exercise its option to acquire the process.[10] From the summer of 1904 until early in 1910, little systematic work was done on color photography at Eastman Kodak either in Britain or in the United States. Clarke provided frequent reports on new ideas to Eastman, and chemists in the Experimental and Testing Laboratory at Kodak Park made occasional investigations; but when Lumière's Autochrome process, marketed in 1907, did not receive wide commercial success due to low sensitivity, the pressure Eastman had felt in 1904 was gone.

During a European business trip early in 1910, Eastman once again became anxious about the possible threat of successful development of a color photographic process by someone outside the Eastman Kodak organization.[11] Two German companies, Neue Photographische Gesellschaft and Vereinigte Kunstseide Fabriken, had obtained patents on screen processes that Eastman thought might interfere with and nullify an option Eastman Kodak had acquired on a dichromated gelatin screen process developed and patented by Carl Späth.[12] At this time, Eastman's desire to develop a practical color process led him to establish at Kodak Park a color laboratory to work specifically on the development of the Späth process and to work generally on color photography. Late in the spring of 1910, the laboratory was established and a small staff placed under the direction of Emerson Packard, a recent graduate of M.I.T. in chemical engineering.[13] At the same time, Späth, working at Harrow, continued his experiments, probably working with Joseph T. Clarke. Although within a year Eastman sought to make a contract with Späth to bring him to Rochester to continue his work, there is no evidence that he ever came or that the color laboratory existed for very long.[14]

[9] GE to Joseph T. Clarke, 7 April 1904; and GE to Philipp, 26 April 1904 (GEC).

[10] GE to George Davison, 5 July 1904 (GEC).

[11] GE to Darragh de Lancey, 12 April 1910 (GECP).

[12] GE to Baker, 7 June 1909; GE to Clarke, 28 April 1910; and GE to Messrs. Church & Rich, 11 May 1910 (GECP).

[13] GE to Clarke, 18 May 1910; and GE to Baker, 31 August 1910 (GEC).

[14] GE to Clarke, 15 June 1911 (GEC).

Fig. 14.1 Physical and chemical laboratories at Harrow, England. Courtesy of Eastman Kodak Company.

Thus, during the period from the early 1890s to the early part of the second decade of the twentieth century, Eastman Kodak depended largely, though not exclusively, upon the creativity of its own personnel to develop new mechanical improvements in cameras and apparatus; however, in the chemical-materials sector of the business, new ideas were sought from outside the company, most notably from Europe, and then efforts were made to employ either the Experimental and Testing Laboratory or some specially designated person or group to implement the ideas. The difference in approach between mechanical and chemical creativity may have reflected the background and personality of Eastman who, no doubt, stood in much less awe of mechanical developments than of chemical ones and, therefore, could more readily conceive of the institutionalization of creative endeavor in the mechanical arts than in the chemical arts. The failure of the color laboratory and of the Späth experiments, combined with Eastman's increasing anxiety regarding the development of color photography, provided the setting for the introduction of a major new strategy and direction in the institutionalization of research and development—the creation of the Eastman Kodak Research Laboratory.

Scientific and technological research had become increasingly institutionalized during the last two-thirds of the nineteenth century. Building upon the models of the French Ecole Polytechnique, a research-oriented technical school founded during the Revolutionary Era, and the literary-philosophical traditions in the German universities, theoretical and experimental research became an integral part of French and German academic activity. During the last third of the nineteenth century, in association with the science departments of the German universities and technical schools, certain German manufacturing firms, especially those in the chemical and electrical industries, established their own laboratory research facilities. The most notable of these were those established by the chemical dye companies: Bayer, Badische Aniline- und Soda-Fabrik, Farbwerke Höchst, and Agfa. In their research laboratories, these companies sought to create new anilin based dyes, the development of which was accessible to theoretical methods because of the recently adopted molecular theory of organic structure. The German organic chemical manufacturers established research laboratories in order to insure a flow of new dyes, which were crucial in the highly competitive dye market.

The establishment of industrial research facilities spread to the United States late in the nineteenth and early in the twentieth centuries, especially among companies in the chemical and electrical industries. Independent research establishments such as those of Thomas Edison and Arthur D. Little received wide attention in the last quarter of the century, but the first decade of the twentieth century witnessed the founding of research laboratories associated with manufacturing firms such as General Electric, American Telephone and Telegraph (Bell), and Du Pont.

Eastman, who knew only vaguely of industrial research efforts in the United States, became increasingly aware of laboratory research in Europe, especially in Germany. In the winter of 1911 and 1912, with Pathé independently producing film and Germany's Agfa initiating film production, Eastman went to Europe to assist in solving the many problems facing his company in the cine film market. It is reported that during his continental journey, the executives of the Bayer chemical company of Elberfeld invited Eastman to tour their large facilities and entertained him at a formal luncheon. In conversation at the luncheon, one of the hosts told Eastman about the Bayer research facilities and then asked Eastman, ". . . And how many people do *you* have in your research work?" Apparently Eastman was embarrassed to admit the limited

Fig. 14.2
C. E. Kenneth Mees, first director of the Eastman Kodak Research Laboratory. Courtesy of Eastman Kodak Company.

nature of the company's commitment to research. Upon his return to England, he consulted with Joseph T. Clarke about the possibilities of establishing a research laboratory and finding a director for such a bold new venture. Clarke recommended a young English scientist and manufacturer, C. E. Kenneth Mees (fig. 14.2).[15]

Mees combined an excellent research education with considerable practical experience in manufacturing. Students of the renowned English physical chemist, William Ramsay, Mees and a close friend, Samuel Sheppard, early in the twentieth century had conducted research into the theory of the photographic process —research that gained doctoral degrees for both men. Mees had continued his basic and practical research, including fundamental work in color photography, while serving as joint managing director of the small but highly regarded English dry plate company, Wratten & Wainwright. As a consequence of this work and his earlier work in association with Sheppard, Mees quickly gained an international reputation in photographic circles.[16]

Early in 1912 George Eastman visited the Wratten & Wainwright plant at Croyden, ostensibly reciprocating a visit made by Mees to Rochester in 1909. Eastman indicated that he wanted to establish a research laboratory at Rochester and wanted Mees to serve as director. Mees indicated that he could accept only if Eastman were willing to purchase Wratten & Wainwright and to employ Wratten at Kodak Limited. To obtain one of the leading photochemists and "the highest authority in color photography in the world,"[17] Eastman willingly made the investment in the English company.[18]

Each of the two principals in the creation of the Eastman Kodak Research Laboratory pursued his own goals, but the two sets of

[15] This story of the immediate stimulus to Eastman's decision to establish a research laboratory is given in Brummitt, "Kodak and Research," Brummitt MS. Although I do not know the source of the information, elements of the description corroborate other evidence. (1) While it is not clear that Eastman was on the continent in late 1911, as reported by Brummitt, Eastman did make an extensive continental tour during January 1912 and specifically mentioned stops in Paris, Turin, Milan, Venice, Munich, Nuremberg, Jena, Berlin, and Nancy (GE to Strong, 16 February 1912 [GECP]). A stop at Elberfeld on the journey from Berlin to Nancy would have been entirely feasible. (2) Eastman's visit to Wratten & Wainwright was in January 1912, corroborating the report of Eastman's return from the continent to England, consultation with Clarke, and decision to try to hire Mees in late January (C. E. Kenneth Mees, *From Dry Plates To Ektachrome Film: A Story of Photographic Research* (New York: Ziff-Davis Publishing Company, 1961), p. 42. (3) A draft copy of a paper, "Kodak Research Laboratories," prepared for the Royal Society of London, 10 July 1947, by C. E. Kenneth Mees, reports that a visit to the Bayer company at Elberfeldt influenced Eastman's decision. As this book goes to press, I discovered corroboration of this story in an interview with Mees by RSJ and LA, 24 April 1952, in Mees file at the George Eastman house.

[16] C. E. Kenneth Mees, *An Address to the Senior Staff of the Kodak Research Laboratories*, 9 November 1955 (Rochester: Kodak Research Laboratories, 1956); Mees, *From Dry Plates*, pp. 19–34; Mees, "Fifty Years of Photographic Research," (acceptance speech for the Franklin Medal Award, Franklin Institute, 20 October 1954), draft in files of Office of Corporate Information, EKCo., Rochester, N.Y.

[17] GE to Strong, 16 February 1912 (GECP).

[18] Mees, *An Address*; Mees, "Fifty Years," and Mees, *From Dry Plates to Ektachrome Films*, pp. 19–34.

goals intersected in areas that allowed the two men to work together to mutual advantage. With Eastman, the establishment of the laboratory represented the climax of the institutionalization of that technical function he had conducted from 1879, which he and Walker shared in 1884, and which Eastman had begun to transfer with the employment of Henry Reichenbach in 1886; but the research laboratory also represented a major new direction: a long-range commitment to fundamental scientific research, with the hope that ultimately new fundamental understandings might provide the company with new products and processes; and a new kind of insurance against losing its prominent position in the industry. While the laboratory may have been founded as a consequence of Eastman's hurt pride in Elberfeld, only casual mention of its founding in the official biography of Eastman, which he carefully edited prior to publication in the late 1920s, indicates that it was not merely a public relations gesture.[19] Perhaps, too, in establishing a center for the scientific study of the photographic process, Eastman sought to make a contribution to the field that had brought him so many benefits. It may only be coincidental, but within a month and a half of his decision to establish the laboratory, he made his first major philanthropic contribution to science education: $2½ million toward the building of the new M.I.T. campus in Cambridge. He made this first of several major anonymous contributions to M.I.T. in acknowledgement of Eastman Kodak's debt for the large number of Kodak executive and engineering personnel who had been educated at the Massachusetts Institute of Technology and in support of the goals and philosophy of M.I.T. President Richard C. Maclaurin.[20] Both the science of photography and science education began to reap the benefits of Eastman's generous mood early in 1912; yet, both the M.I.T. gift and the commitment to the research laboratory reflected the philosophy and values out of which Eastman had operated for more than thirty years.

The opportunity to establish and direct a research laboratory for Eastman Kodak permitted Mees to address several of his goals. Although his business-technical career at Wratten & Wainwright had been most successful by both commercial and scientific standards, he had been sorry to have so little opportunity to pursue his real love: research into the nature of the photographic process. Moreover, unlike his friend and colleague, Samuel E. Sheppard, he had been encouraged to pursue not an academic career but rather an industrial career, a much less prestigious choice. For a former Fabian socialist and son of a Wesleyan minister, the shift from the academic and research life to the business world was not easy, but two elements eased the transition: (1) he continued his research work in conjunction with his business work; and (2) his manifestation of concern for the plight of England's poor was redirected from the socialists' redistribution plans to a philosophy that saw

[19] Carl W. Ackerman, *George Eastman* (London: Constable, 1930), pp. 228 and 240–41.

[20] Ibid., pp. 328–35; and GE to Richard C. Maclaurin, 6 March 1912 (GECP).

science and technology harnessed to industry in providing an ever greater amount of material goods.[21]

Influenced by the ideas of Francis Bacon, Mees began to consider:

> *. . . As we did more science and learned more about it, we could improve methods of production. If you could improve the methods of production, you could increase wealth. . . . Those . . . ideas of Bacon's struck me very strongly indeed. I thought that the important thing was the acquisition of knowledge but that perhaps the next most important thing was the application of the knowledge to industry, so that we could relieve the poverty that was strangling my country. . . . And so it very soon seemed to me that it was much better to increase the total amount of wealth than it was to redistribute it.*[22]

His position at Eastman Kodak allowed Mees to address directly the problem of harnessing science and industry; to direct, like a professorial head of a German academic laboratory or institute, a staff of professional researchers investigating the nature of the photographic process; and to continue the production of those products he had introduced at Wratten & Wainwright.

Eastman and Mees agreed initially on certain fundamental principles according to which the laboratory was to be run. Although Mees and the laboratory were "responsible for the future of photography,"[23] the two men agreed that nothing of major commercial significance was to be expected of the laboratory for ten years. Hence, the laboratory represented truly a long-range commitment of Eastman. Moreover, the operation of the laboratory was to be independent of the rest of the organization, with Mees reporting directly to Eastman and his protégé, Frank Lovejoy. All financial allocations were to be directed through the chief executive. Eastman also agreed that the scientific work of the laboratory could be published except when it might directly interfere with the commercial aspects of the business. Mees secured Eastman's agreement that only Eastman would review articles prepared for publication. Having obtained these important agreements, Mees quickly secured Eastman's confidence. In less than ten months from the time Eastman hired him, Eastman had not only made a substantial ten-year commitment but had authorized the transmission to Mees of the most highly confidential and valuable assets of the company: all of the company's emulsion formulae![24]

In creating a major new industrial research facility, Mees drew upon his own conception of the nature of science and of scientific discovery. Influenced by the empiricism in the writings of Francis Bacon, he adopted a positivistic view of science, holding the high-

[21] Mees, *An Address*, pp. 6 and 22.

[22] Ibid., pp. 8–9.

[23] C. E. Kenneth Mees and John A. Leermakers, *The Organization of Industrial Scientific Research*, 2d ed. (New York: McGraw-Hill Book Co., 1950), p. 35.

[24] Ibid., pp. 147 and 342; Mees, *From Dry Plates to Ektachrome Film*, p. 50; Mees, "Kodak Research Laboratories," p. 3; Mees, *An Address*, p. 29–31; and GE to W. G. Stuber, 7 October 1912 (GEC).

est respect for descriptions of observation and experiment. He saw scientific research as the "investigation of the relationship between cause and effect in natural phenomena,"[25] and the knowledge resulting from scientific research as growing progressively with the accumulation of new data. For Mees, one of the principal purposes in life was to increase this body of knowledge: "What one had to do was to add to the total of scientific knowledge, and when you went, as you would in the end, you could feel that you had done something in adding to that total."[26]

Regarding an industrial research laboratory as similar in many respects to a research institute, Mees saw the prototype in Bacon's "House of Solomon," which he described as follows:

A great number of observed facts would be collected, and from them the fundamental processes of nature could be understood. In this way, he believed, it was possible to attain to "the knowledge of Causes and secret motions of things, and the enlarging bounds of Human Empire, to the effecting of all things possible." This was a great vision, a new vision on earth, and a vision that has in some measure been realized. The method that Bacon suggested for carrying out this idea was the organization of a research institute, which he entitled the "House of Solomon" and described in his "New Atlantis." This institute contained a series of laboratories for experimental research. . . . This research institute was to be manned by a great company of Fellows to whom Bacon . . . allotted specific functions. . . . A noble dream, much before its time and greatly overorganized, it led to the idea of cooperation in the pursuit of knowledge. . . .[27]

With this cumulative-empiricist-experimentalist view of science and with the acceptance of the cooperative-institutionalized organization of research, Mees's conceptions reflected the waning of the heroic view of scientific and technological discovery that had so dominated the last two-thirds of the nineteenth century. Eschewing the dependence of research laboratories upon scientists of exceptional caliber, he held

that much scientific research depends upon the accumulation of facts and measurements, an accumulation requiring many years of patient labor by numbers of investigators, but not demanding any special originality on the part of the individual worker. . . . He can make valuable contributions to scientific research even though he be entirely untouched by anything that might be considered as the fire of genius.[28]

Furthermore, Mees even questioned the validity of the heroic view:

It is, indeed, a matter of doubt how many of the men commonly considered to be of great genius by virtue of some important discovery they have made really possessed any distinguishing ability

[25] C. E. Kenneth Mees, *The Organization of Industrial Scientific Research* (New York: McGraw-Hill Book Co., 1920), p. 1.

[26] Mees, *An Address*, p. 4.

[27] Mees and Leermakers, *Organization of Industrial Scientific Research*, pp. 24–25.

[28] Mees, *Organization of Industrial Scientific Research*, pp. 90–91.

compared with their fellows who did not have the fortune to make a similarly important discovery.[29]

In organizing and structuring the Research Laboratory, Mees not only drew upon his own conceptions of the scientific enterprise but also upon the limited but important experience of other pioneers in industrial research. The two most important influences upon him were the ideas of William Rintoul and William R. Whitney. Rintou, chief chemist at the Royal Gunpowder Works at Waltham, England, during the first decade of the century, left Waltham and joined Nobel Explosives Limited, where he established an industrial research laboratory that became a model in English-speaking countries. Mees, who had worked for Rintoul for a short time, had considerable respect for his ideas. Rintoul's highly organized and structured laboratory became a model for Mees's laboratory with respect to dissemination of information and reports and the transfer of results from the laboratory to pilot plants to manufacturing divisions. Mees felt, however, that the Rintoul laboratory was too tightly structured. In contrast to it was the laboratory of William Whitney, director of research since 1900 at General Electric in Schenectedy, New York, which Mees had visited in 1912. In the conduct of his laboratory Whitney sought to maximize the freedom of the scientists, encouraging them to pursue areas of their own interest. Although Mees adopted Rintoul's and Whitney's attitudes on the free exchange of information and prompt publication of scientific work, he regarded Whitney's organizational policies as too laissez-faire. Mees sought to create a laboratory with a style of organization intermediate between that of Rintoul and Whitney.[30]

The initial primary goal of the Eastman Kodak Research Laboratory was the scientific understanding of photographic processes, with the belief that eventually such understanding would lead to new and improved products and processes of commercial benefit to the company. Photography, unlike most other areas of industrial research effort, had not been subjected to extensive academic research. There were only two or three academic institutions in the world where photography was studied as a science and, therefore, Mees saw the Kodak Laboratory as serving in place of an academic center in the investigation of photochemical and physical processes.[31] Within this concentration on photographic processes, which gave it its reputation as a "convergent" laboratory,[32] Mees sought to approach an academic atmosphere with self-directed research and a free flow of information. The spectrum of departments in the laboratory, from physics and chemistry to pilot manu-

[29] Ibid., p. 91.

[30] Mees, *An Address*, pp. 27–28; Mees, *From Dry Plates*, pp. 50–51; and Mees and Leermakers, *Organization of Industrial Scientific Research*, pp. 92 and 134.

[31] Mees, "Fifty Years of Photographic Research."

[32] Mees classified laboratories according to the problems they addressed. Those investigating problems associated with one common subject, he called "convergent"; those investigating problems with no interconnecting subject, he called "divergent."

Fig. 14.3
Samuel Sheppard. Courtesy of Eastman Kodak Company.

facturing, brought both theoretically and practically oriented personnel in intimate contact with one another.

Drawing upon the experience of both Rintoul and Whitney, Mees initiated what he called the conference system, wherein each morning of the week a different subject would come under discussion in a conference of part of the staff, insuring that each week all the personnel of the laboratory would have an opportunity to discuss their work with others in related fields. Mees envisioned this interaction and the ultimate direction it would give to research as heightening the creativity of the laboratory workers as a whole and at the same time diminishing the possibility of director-dictated research.[33]

Recognizing that all the best plans for operating a laboratory were of little consequence unless the appropriate leadership personnel could be obtained, Mees gave careful consideration to whom he should employ. The initial personnel for the laboratory were largely employees of the National Bureau of Standards and technical people Mees knew in England. Dr. Perley G. Nutting, a physicist from the National Bureau of Standards who later became director of the Westinghouse Laboratory in Pittsburgh was placed in charge of the physics department. Nutting brought with him from the bureau several assistants, including Loyd A. Jones, a graduate of the University of Nebraska. The inorganic chemistry section was placed under Dr. A. S. McDaniel, also from the National Bureau of Standards. Mees called on his friend and former colleague, Samuel E. Sheppard, to head the physical chemistry section (fig. 14.3). The Wratten photographic department, which had responsibility for the manufacture of color filters, was headed by an Englishman, John George Capstaff, a science student and photographer who had learned Wratten & Wainwright techniques of color filter production in order to transfer the technology to Rochester. John I. Crabtree from Victoria University in England headed the photographic chemistry section. Also associated with the laboratory were personnel employed in the photographic studio and pilot plate-making plant. With the gradual employment of the staff during 1912 and the completion of the building at Kodak Park late in the year, Mees and his staff of about twenty began formal operations early in 1913 in the laboratory building equipped with $30,000 worth of scientific and production apparatus.[34]

During the decade following the founding of the Research Laboratory, Mees enlarged the staff and their activities and deepened the company's commitment to systematized research and development. Within a year of the establishment of the laboratory, Mees, organic chemist turned physical chemist, realized that he had originally created no section devoted to organic chemistry.

[33] Mees, *An Address*, p. 28.

[34] Mees, *From Dry Plates to Ektachrome Film*, p. 49; "C. E. Kenneth Mees Private Laboratory Notebooks," January 1912–January 1913 (photocopy); Mees, "John George Capstaff," (MS); and "Citizen of the Day," 11 October 1956 (RVF 2 MS), from the files of the Office of Corporate Information, EKCo., Rochester, N.Y.

Eastman at first refused to rectify this situation, but by the summer of 1914 Mees had prevailed upon Eastman, and an organic section was initiated. During the war new sections devoted to microscopy, photomicroscopy, and technical services were added. In less than a decade the staff of the laboratory more than quadrupled and expenditures increased more than sixfold.

Year	*Staff*	*Expenditures*
1913	20	$ 53,757
1915	40	126,745
1920	88	338,680

Such a growth in staff and expenditure indicated the deep commitment of the Eastman Kodak Company to the research and development activities of the Research Laboratory.[35]

In the conduct of the Research Laboratory, Mees followed the guidelines he had initially envisioned. While the staff directed about one-sixth of their energies to manufacturing and administrative matters, they devoted half of their time to the development of new materials and processes and nearly a third of their time to work on fundamental understanding of photographic processes.[36] Soon the scientific work of the laboratory became well known and respected throughout the world. Eschewing the publication of its own journal, the laboratory encouraged its staff to publish in leading professional journals—the *Journal of the American Chemical Society*, the *Journal of the American Optical Society*, the *Journal of the Franklin Institute*, and the *Photographic Journal* (Royal Photographic Society). Each of the scientific papers published by the staff received a special number. Beginning with Perley G. Nutting's "On the Absorption of Light in Heterogeneous Media," published in the *Philosophical Magazine* in 1913, the number of communications by 1925 exceeded 240. In addition, each year the laboratory published a volume of abridgements of the publications by the laboratory staff, *Abridged Scientific Publications*. Moreover, in 1915 the laboratory initiated publication of an abstract of all photographic literature: *Kodak Abstract Bulletin*. Mees's efforts to establish the laboratory as the institute for the scientific study of photography and his belief in the central importance of communications influenced the laboratory internally, with its conference system and collaborative work, and externally, with its extensive publication and bibliographical activity.

Likewise, the character of the scientific work at the laboratory reflected the interests and philosophy of Mees, especially during its early years. Much of the work of Mees and Sheppard when they were graduate students was devoted to extending the investigations of Hurter and Driffield in sensitometry, i.e., in the measurement of the relationship between (1) photographic density and (2) the duration and intensity of illumination and development.

[35] Mees, *From Dry Plates to Ektachrome Film*, p. 50.
[36] Ibid., p. 54.

Sensitometry represented the central focus of the laboratory until World War I. The work of the most prolific authors, Nutting, Jones, and Wilsey of the physics section, indicated concern with establishing standards of measure, creating new instrumentation, and making sensitometric determinations.[37] However, Nutting in a couple of papers directed attention to some of the "subjective" elements in photography, examining the relationship between physical and physiological phenomena.[38]

With the advent of World War I the Research Laboratory specifically and the company generally provided enthusiastic support for the Allied effort. Eastman, an Anglophile, openly criticized Wilson's initial reluctance to join the European war. Once the United States had entered the war, Eastman personally led the highly successful Rochester liberty bond drives and turned the company's production resources to the war effort. The moribund cellulose acetate plant was revived for production of coatings for airplane wings, and the apparatus division began production of airplane gun sights; by the end of the war, 38 percent of the company's products were directly or indirectly directed to the government.[39] The leaders at the Research Laboratory, many of whom were native Englishmen, fully supported the war effort. Among the most significant contributions was the production of fine chemicals. During the war, American chemists, dependent upon the German dye companies for synthetic chemicals, suddenly faced a crisis. The Research Laboratory felt the severance of supply and finally initiated production not only for the laboratory but for the rest of the country, beginning with a list of 265 chemicals. By 1921, the list had grown to 1,144. This production, undertaken at considerable financial loss, paid national dividends in maintaining American research.[40]

Following the war, Mees, having thought for some time that there was need for a more systematic approach to product innovation in the apparatus sector, asked Dr. Albert Chapman, a physicist who had graduated from Princeton, to create and head a new Mechanical Development Laboratory. Until this time, the managers of the three principal apparatus manufacturing divisions had sought in addition to their executive responsibilities the opportunity to serve also as camera designers or to supervise others in the development of new products. Chapman accepted the position and gradually gathered about him during the next four years men of

[37] For example: L. A. Jones, P. G. Nutting, and C. E. K. Mees, "The Sensitometry of Photographic Papers," *Photographic Journal* 38 (1914): 342; L. A. Jones, "A New Standard Light Source," *Transactions of the Illuminating Engineering Society* 9 (1914): 716; and L. A. Jones and R. B. Wilsey, "The Spectral Selectivity of Photographic Deposits. Pt. I. Theory, Nomenclature, and Methods," *Journal of the Franklin Institute* 185 (1918): 231.

[38] P. G. Nutting, "Effects of Brightness and Contrast in Vision," *Transactions of the Illuminating Engineering Society* 11 (1916): 939; and "The Retinal Sensibilities Related to Illuminating Engineering." in ibid., 11 (1916): 1.

[39] GE to John G. Milburn, 1 May 1918; GE to Alexander M. Lindsay, 8 February 1918; and GE to Mr. Carrol, Secy., Sub-Committee on Capital Issues, 8 July 1918 (GEC).

[40] Mees, *From Dry Plates to Ektachrome Film*, pp. 285–87; and Mees, *An Address*, p. 34.

high technical capacity, including Newton Green from Oberlin College, who had worked under Capstaff in the Research Laboratory; Dr. Fred Bishop, a physicist from Yale; and Dr. Edmund Fitts, who had a Ph.D. degree from the University of Illinois. The Development Laboratory worked on both amateur and commercial apparatus and met with some success and some failure in its initial efforts. While Mees carefully supervised the work at the Development Laboratory, Chapman reported directly to Lovejoy, the chief executive of the company under Eastman. In 1923, Newton Green assumed the leadership of the Development Laboratory, and Chapman moved to the inner sanctum, the top floor of the Kodak Office Tower (where George Eastman's office was located), where he became an assistant to Lovejoy. Gradually, Mees transferred the development and research sectors of the company into institutional organizations with highly educated personnel in leadership positions.[41]

After the war the research work of the Research Laboratory broadened, with organic and physical chemistry playing more important roles and with Sheppard emerging as the prolific contributor. Sheppard gathered about him a small group of workers, including A. P. M. Trivelli, a physical-organic chemist, and focused primarily on the causes of sensitivity in the creation of the latent image. His investigations took two general directions. One was the study of silver halide grains in the latent image, in particular the determination of their size and distribution. Microscopy and X-ray analysis of crystals became increasingly important in this work.[42] The second direction was the study of gelatin and its role in photographic processes.[43] No doubt Sheppard's earlier work at the Sorbonne and at Cambridge University had provided him with the necessary background to proceed in this area, and the difficulties the company was having in manufacturing its own high-quality gelatin may also have stimulated this effort. Ultimately, these researches led to the collaborative discovery by Sheppard, Hudson, and R. Punnett that certain sulfur compounds in gelatin react with the silver in minute quantities and serve as special sensitizing agents. These investigations continued throughout the 1920s and led to one of the most important fundamental and practical contributions of the laboratory in its first dozen years.[44]

During the early and middle 1920s work in sensitometry continued, though often without the spectacular consequences of

[41] Interview with Dr. Albert Chapman, 27 July 1970; Mees, *An Address*, pp. 32–33; and GE to Baker, 7 February 1919 (GEC).

[42] For example: S. E. Sheppard and A. P. H. Trivelli: "Note on the Relation Between Sensitiveness and Size of Grain in Photographic Emulsions," *Photographic Journal* 61 (1921): 400; S. E. Sheppard, E. P. Wightman, and A. P. H. Trivelli, "Studies in Photographic Sensitivity," *Journal of the Franklin Institute*, 198 (1924): 629; and S. E. Sheppard and R. H. Lambert, "Grain Growth in Silver Halide Precipitates," *Colloid Symposium Monograph* 6 (1928): 265.

[43] For example: S. E. Sheppard, "The Nature of the Emulsoid Colloid State," *Nature* 107 (1921): 73; S. E. Sheppard, "Gelatin in the Photographic Process," *Journal of Industrial and Engineering Chemistry* 14 (1922): 1025; and S. E. Sheppard, "Plasticity in Relation to Gelatin," *Journal of Physical Chemistry* 29 (1925): 1224.

[44] Mees, *From Dry Plates to Ektachrome Film*, p. 139.

Fig. 14.4 Spectroscopic laboratory room in the Research Laboratory at Kodak Park, c. 1913. Courtesy of Eastman Kodak Company.

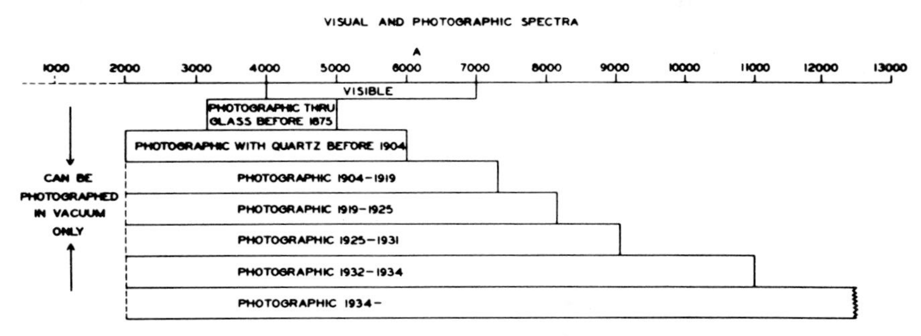

Fig. 14.5 Progress in the photography of the spectrum. From C. E. Kenneth Mees, From Dry Plates to Ektachrome Film: A Story of Photographic Research *(New York: Ziff-Davis Publishing Co., 1961). Courtesy of Eastman Kodak Company.*

Sheppard's work. With Nutting's departure in 1917, Jones assumed the leadership of the section. Development of additional instrumentation, including a nonintermittent sensitometer, became important in determining density, graininess, and resolving power. In connection with these studies, Jones published significant papers on tone reproduction, a continuation of the work initiated by Nutting.[45] During the twenties both the chemists and physicists became increasingly interested in spectral sensitivity and experimented extensively with sensitizing dyes as means of broadening the range of sensitivity of photographic emulsions (figs. 14.4 and 14.5).

In the fundamental work of the laboratory, Mees continued to play a role, although he had progressively less time for his own research work as the laboratory expanded and his responsibilities broadened. After the early years, the number of Mees's research papers had diminished rapidly, but as the chairman of the daily conferences, he not only kept abreast of the work of others in the

[45] L. A. Jones, "On the Theory of Tone Reproduction . . . ," *Journal of the Franklin Institute* 190 (1920): 39; L. A. Jones, "Photographic Reproduction of Tone," *Journal of the Optical Society of America* 5 (1921): 232; and L. A. Jones, "The Contrast of Photographic Printing Paper," *Journal of the Franklin Institute* 202 (1926): 177.

laboratory but also made his contributions through the informal structure of the conference, giving direction to much of the work and creating an environment that stimulated the researches of others.[46] Throughout this period, however, the emphasis was upon measurement and instrumentation. Only very occasionally did one of the section heads address the implications of some of the more fundamental theoretical ideas such as those in quantum physics. In a very real sense, the early years of the Eastman Kodak Research Laboratory under Mees's direction represented an updated version of "Solomon's House."

Like Bacon's research institute, the laboratory maintained a pragmatic posture and during its first dozen years made several major contributions to the Eastman Kodak product line. Sheppard's work with sulfur sensitization led to the development of a number of additives for emulsions. This work, which contributed to increased reliability as well as increased sensitivity of emulsions, was patented. The production of sensitizing dyes and the use of these dyes in spectral sensitivity studies led to several improvements in panchromatic films that were of particular importance in cinematography and in the scientific uses of photography.[47] The most significant new product, first announced in 1923, was the Cine-Kodak, a home movie system that included a coordinated camera, projector, and safety film. John Capstaff of the laboratory developed one of the key elements in the system, the controlled reversal process for the film. This process alleviated the need for making both a negative and a positive for home movies.

The research activities of Eastman Kodak during the first decade of the Research Laboratory's existence continued to reflect the personal stamp of George Eastman. Although the laboratory, like so many other staff departments, represented the institutionalization of the functions originally performed by Eastman, Eastman continued to act in certain areas as though the company were still a single proprietorship. For example, he continued to examine and make personal judgments with regard to new inventions and patents that came to his attention; consequently, he was flooded with correspondence relating to new inventions. One of the most notable of his responses was his 1914 decision to acquire rights on and ultimately to introduce the Autographic camera. H. J. Gaisman of New York conceived of the Autographic camera and film, which enabled the photographer to open a small slot in the rear of the camera and write with a stylus on the back of the film (fig. 14.6). The Autographic film consisted of a strip of film with a thin red backing paper and a strip of carbon paper in between. The combination of the carbon paper and the red paper provided protection from light exposure from the back of the film. When words were impressed upon the backing paper with a stylus inserted through the opening in the special Autographic camera, the carbon was transferred from the carbon paper to the red backing paper, removing the light protection where the word impressions were

[46] Mees, *An Address*, p. 28.
[47] Mees, *From Dry Plates to Ektachrome Film*, chap. 9.

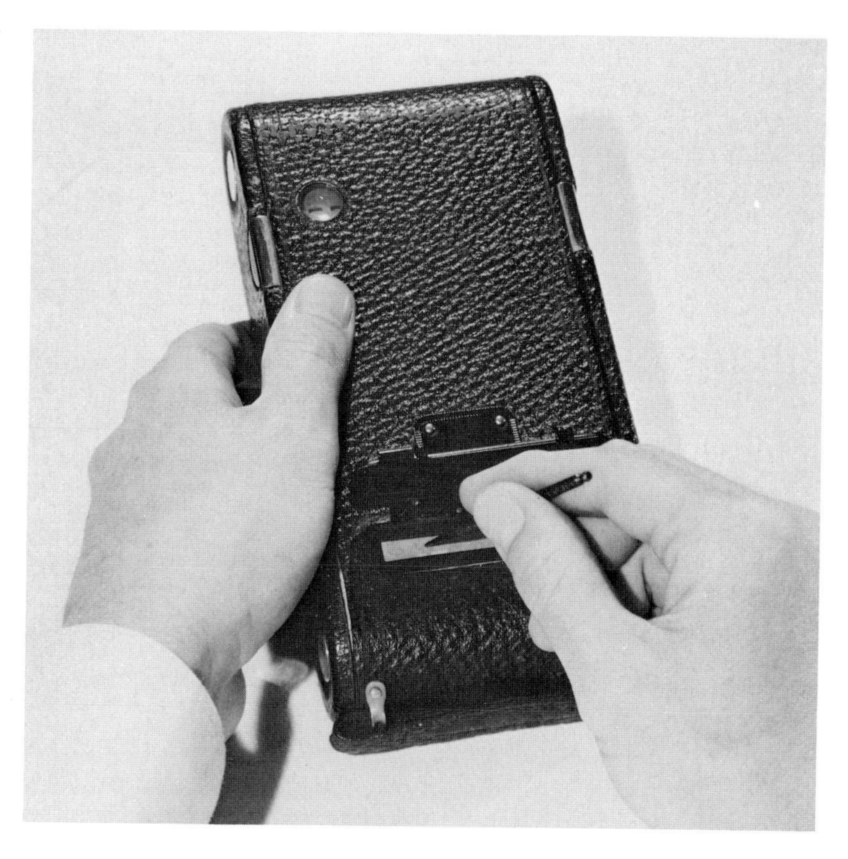

Fig. 14.6
Using an Autographic camera. Courtesy of Don Ryon, Eastman Kodak Company.

made. When the Autographic film was developed, the red backing paper was removed and the film was given a short exposure to light from the back, providing a photographic image of the written impression on the film (fig. 14.7). Eastman became quite enthusiastic about this invention because he saw in it a new system of photography that was covered by patents. If the public could be encouraged to adopt the system, the Eastman Kodak firm would have exclusive control of amateur photography for the following seventeen years. Eastman Kodak acquired the rights on the patents and developed and introduced the Autographic system. Although it had some popular acceptance during the late teens and early 1920s, the system never realized the expectations Eastman had initially envisioned.[48]

Fig. 14.7
An Autographic camera negative. Courtesy of Eastman Kodak Company.

Eastman also influenced the direction and intensity of activity within the Research Laboratory. One of his principal goals for the laboratory was the development of a commercial color process. From the outset the laboratory spent some time on color development work. Although this had not led to spectacular results by the middle 1920s, the company did acquire a color transparency process that employed lens filters: the Berthon process.[49] At Eastman's urging, considerable development effort was expended upon the

[48] GE to Clarke, 16 February 1914 (GEC); and GE to Strong, 14 March 1914; and GE to H. J. Gaisman, 19 July 1918 (GECP). I am indebted to Donald C. Ryon for the description of the Autographic film.

[49] Mees, *From Dry Plates to Ektachrome Film*, chap. 14.

system in the middle 1920s. Past seventy years of age, he was anxious that he should see his company introduce a commercial color process. A modified form of the Berthon process under the name Kodacolor was introduced in 1928 for home movies, but it did not prove a popular success because the high degree of light absorption by the color filters made very bright illumination manditory.

Nonetheless, when Eastman saw his laboratory serving as the world's leading photographic research institution and possessing some of the same innovative goals with which he had begun his photographic career forty-five years before, he rested satisfied with the gamble he had taken a dozen years before. Indicative of his perception of the central role of Mees and the Research Laboratory to the strategy of the company, Eastman appointed Mees in 1923 to the highly select Board of Directors of the Eastman Kodak Company. Hence, Eastman prepared a legacy that guaranteed the continuity of those business strategies that saw research and development as absolutely central to their successful execution.[50]

Integration: Defense and New Strategies

During the first two decades of the twentieth century, Eastman Kodak, like a number of other large corporations in America, came under serious attack from both state and federal governments, which charged it with violating the antitrust acts. The dominant position of the Eastman company in the photographic industry automatically drew attention to its fair trade and exclusive dealership policies and to its activities in horizontal integration of the photographic paper, plate camera, and dry plate sectors. These policies and activities prompted continued governmental investigation and legal action.

Anthony and Scovill instigated at the state level the first serious antitrust charges brought against Eastman Kodak. New York State's Donnelly Act prohibited agreement or combination creating a monopoly in the manufacture, production, or sale of any article in the state. Furthermore, it required witnesses to testify about known violations and required alleged violators to produce all corporate records and papers.[51] In August of 1903, a supply house in New York filed with the attorney general of New York charges against Eastman Kodak. Aided by legal counsel from Anthony and Scovill, the company also presented corroborating affidavits from several other supply dealers.[52] After a year and a half of hearings and legal conferences, the attorney general concluded that Eastman Kodak must prove that the companies it had acquired had not been competitors before the acquisition.[53] After another year of continued legal proceedings, Eastman Kodak was

[50] GE to J. L. Gorham, 26 June 1922 (GEC).
[51] "Curbing the Trusts," *New York Times*, 26 September 1902, p. 3.
[52] Ibid., 7 August 1903, p. 3.
[53] Ibid., 16 December 1904, p. 7.

exonerated from the allegations.[54] While these charges were being pressed, Anthony and Scovill filed, in February 1904, a petition with the New York State Attorney General's Office asking for a restraining order against Eastman Kodak.[55] Ultimately, the rival company asked that the state corporate charter of the Rochester firm be annulled. In this case, as in the earlier one, the attorney general decided in favor of Eastman Kodak.[56] The outcome of these two important tests renewed George Eastman's confidence that the integration activities of the company could withstand the critical scrutiny of the antitrust investigators. Nevertheless, he and the company's legal counsel were wary of the opportunities presented to the Rochester firm in the middle of the first decade to acquire its two principal competitors: Ansco and Defender.[57]

Meanwhile, the Eastman company also came under scrutiny from federal investigators. In March of 1901, the testimony of Walter Hubbell and Charles S. Abbott before the Industrial Commission in Washington left certain commission members highly critical of some policies of the company but conveyed a generally favorable impression to the press.[58] As the decade progressed, Eastman himself became increasingly nervous about the popular antitrust feeling: ". . . The power of this popular uprising against trusts . . . it is a thing that has to be now taken into calculations by anyone whose business comprises any large part of the total output in any given line."[59] This worry continued when the Standard Oil and American Tobacco companies were ordered in 1911 by the Supreme Court to dissolve.[60]

Under the Taft administration, the Department of Justice and the Attorney General's Office initiated an active investigation of the Rochester photographic firm. Attorney General George W. Wickersham sent investigators to Rochester in May of 1912, and George Eastman, following the advice of his legal counsel, opened virtually all the company's records and historical files to the investigators, confident that the company had nothing to hide.[61] While Eastman knew that the company's sales policy had come under attack because of the limitations requiring Eastman Kodak dealers to market the company's products exclusively, he thought that modifications in that sales policy implemented since 1907 should absolve the company of further criticism. The investigating team felt differently. The Justice Department wanted further modifications in

[54] GE to Philipp, 29 March 1905 (GEC); and GE to Des Voeux, 23 July 1906 (GECP).

[55] GE to Strong, 8 February 1904 (GECP).

[56] GE to Des Voeux, 23 July 1906 (GECP).

[57] GE to L. Goffard, 16 April and 23 August 1907; and GE to Philipp, 2 December 1907 (GEC).

[58] GE to C. S. Abbott, 27 February 1901 (GECP); and Abbott to GE, 12 March 1901 (GEC).

[59] GE to Goffard, 16 April 1907 (GEC).

[60] GE to Paul Fahle, 6 July 1911; and GE to Frank L. Babbott, 7 June 1911 (GEC).

[61] GE to Strong, 10 May 1912 (GECP); and GE to L. M. Antisdale, 14 January 1914 (GEC).

the sales policy. In order that it might receive public credit for the modifications it wished to instigate, the Justice Department reached an informal understanding with the Eastman Kodak attorneys that the company would delay certain modifications in its sales policy; the Justice Department would file a complaint against the company, and then the modifications would be implemented under a consent decree. This plan was not consummated, however, because of legal delays that extended into the term of the new Wilson administration.[62]

Although Eastman regarded Wickersham and his staff as "anything but friendly" and as holding "a grave misconception of the spirit and conduct of our company,"[63] Wilson's attorney general, James Clark McReynolds, and his team of investigators assumed a much stronger position and attitude against the company. Attempts on the part of the Eastman Kodak attorneys to renegotiate the understanding reached with the previous administration were unproductive,[64] and within three months of taking office the new administration had filed a petition in federal court charging Eastman Kodak with violation of the antitrust statutes. The position of the Justice Department was that the company had, through its acquisition of competing companies, its control of raw paper from Europe, and its sales policy, sought to gain a monopoly of the photographic business. Moreover, the Justice Department held that the company gained its position in the industry as a consequence of these alleged illegal practices, and therefore, modifications of the company's sales policies would not sufficiently redress the grievances.[65] Thus, the government sought continually to dissolve the company into component parts.

During the nearly twenty-seven months of the court trial against Eastman Kodak, the Justice Department presented an extensive set of historical records of the major companies in the industry and testimony from their officers. Meanwhile, the Eastman Kodak attorneys continued unsuccessfully to seek a settlement with the attorney general.[66] Nevertheless, George Eastman, the perpetual optimist in nearly every legal case the company entered, remained confident that the courts would provide exoneration.[67] Late in August of 1915, however, Judge John R. Hazel of the Western District Court of New York handed down a decision charging the company with intent to monopolize as indicated in its raw paper contracts, its acquisition of paper, dry plate, and plate camera companies and supply houses, and its employment of terms of sale.[68] Within six months, the company appealed the decision to the U.S. Supreme Court, hoping for a reversal of the decision and realizing, in any case, that the appeal would take time and give

[62] GE to Baker, 7 April 1913; and GE to Hon. James W. Wadsworth, 4 November 1921 (GEC).

[63] GE to Antisdale, 14 January 1914 (GEC).

[64] GE to Strong, 3 February 1914 (GECP).

[65] Petition filed 9 June 1913, *U.S.* v. *EKCo.*

[66] GE to Strong, 15 January and 19 February 1914 (GECP).

[67] Ibid., 17 February 1915.

[68] "Opinion," *U.S.* v. *EKCo.*

the company a respite from dissolution action.[69] During the five years the appeal remained on the Supreme Court docket, company attorneys made numerous efforts to reach a settlement with the Justice Department, but the changing personnel in the department during the war emergency complicated these efforts.[70]

During the period from 1916 until 1921, the attitudes of the public, the Justice Department, and the Eastman Kodak Company mellowed. With the coming of the war and the considerable contribution of large corporations to its successful execution, public attitudes toward large-scale enterprise softened, and this change in attitude was reflected in the personnel in the Wilson administration. In 1920 the Supreme Court's acquittal of United States Steel of violation of the Sherman Act under the "rule of reason" reflected a modification of the judicial position from that of the era 1911 to 1915. During the war, Eastman Kodak not only turned substantial attention to the military effort, like other large American corporations, but, in the face of rapidly rising prices on raw materials and labor, either maintained prewar prices or increased them only very modestly, effectively cutting them on its product line relative to the prewar prices.[71] Even George Eastman's attitude toward the old rival, Ansco, mellowed considerably. When in 1920 Ansco continued to have financial difficulty, Eastman wrote to a representative of the Ansco interests:

If the trouble is in the sales, or in any department of the business other than the manufacture of the goods, we are not only willing but would be very glad to furnish any assistance that we can; but in such a case it must be understood that we are not to receive any compensation or have any control of the business. We realize that we cannot be without competition and that any disaster coming to our largest competitor would be a disadvantage to us. To prevent this is the sole motive actuating us to offer assistance along the lines indicated. . . . If the difficulty is in the manufacture of the product, particularly in the manufacture of sensitized goods, the management of this Company could not in justice to its stockholders give to a competitor the results of more than twenty years' experience and what has cost this Company millions of dollars in research and experimental work.[72]

The attitudes and positions of the public, government, and the company had indeed mellowed.

In this new setting, Attorney General A. Mitchell Palmer pushed for a negotiated settlement. The Justice Department assigned different men to the case; in the meantime Eastman Kodak replaced its counsel working on the case, and during the deputy attorney general's visit to Rochester even George Eastman himself was conveniently vacationing at his retreat in North Carolina. In Janu-

[69] GE to Gifford, 2 April 1914 (GECP).
[70] Frank W. Lovejoy to GE, 8 May 1917; A. K. Whitney to GE, 25 July 1917; and GE to Kennedy, 20 September 1917 (GEC).
[71] GE to B. C. Forbes, 8 June 1920 (GEC).
[72] GE to D. E. Pomeroy, 21 December 1920 (GEC).

ary of 1921, just before the Supreme Court was to hear the case, Eastman Kodak withdrew its appeal and during the next several years sought to follow the terms of the consent decree for dissolution.[73] According to the decree, the company was to divest itself of many of the acquisitions made during the period of horizontal integration: Standard, Stanley, Seed, Artura, Century-Folmer, and Premo. Because the bulk of the company sales and profits came from its cine film and roll film system products, the business derived from the so-called violating companies represented only a small fraction of the total Eastman Kodak business.[74] The dry plate and plate camera companies addressed dying sectors of the industry. Consequently, Eastman Kodak had considerable difficulty in selling these companies. Century-Folmer was sold and soon changed its corporate name to Graflex, Incorporated; it became an important independent specialized camera company in Rochester. Defender purchased Artura Paper Company and the Standard, Stanley, and Seed brands, but Eastman Kodak continued to produce the dry plates for Defender at Kodak Park. This arrangement, necessary because no bids were submitted when the companies were placed on the auction block, allowed Eastman Kodak to retain exclusive possession of its highly valuable negative emulsion formula secrets.[75]

Because the Eastman company was under attack from the government for its horizontal integration activities during the late 1890s and the early 1900s, the company's strategy shifted during the period 1911 to 1920 to vertical integration, moving backward into the production of many of its requisite materials. This strategy gave the company increasing independence from the European cartels controlling many of the requisite supplies and also created major new sources of profit and operating economy. Because most of these supplies were not produced in the United States, the company did not acquire producing companies but sought instead to develop its own capacity for manufacture.

About 1911–12 efforts were directed toward the production of raw paper, gelatin, and lenses. Consideration had been given, of course, to the production of raw paper in 1906 before the General Paper contract was renewed. In 1911, after the General Paper contract had been terminated, David E. Reid once again investigated raw paper production and in 1912 visited European paper mills.[76] Upon his return, his experiments were sufficiently successful that early in 1913 the company constructed an experimental paper mill at Kodak Park. Within two years, this mill provided about 40 percent of Kodak Park's paper requirements, and plans were advanced to construct, at a cost in excess of $1 million, a

[73] GE to J. J. Kennedy, 24 May 1922 (GEC).

[74] Total sales for 1920 were approximately $70 million, with sales of products of companies included in the consent decree amounting to $5 million. The profits from the sales of those companies amounted to $900,000, while total profits of the company amounted to about $18.6 million (GE to F. C. Mattison, 14 February 1921 [GEC]; and *Annual Report, EKCo.*).

[75] Blake McKelvey, *Rochester: The Quest for Quality, 1890–1925* (Cambridge: Harvard University Press, 1956), pp. 344–45.

[76] GE to Lovejoy, 5 October 1912 (GEC).

paper mill sufficient for the company's entire needs.[77] The development of papermaking capacity came at an auspicious time because the experimental plant at Kodak Park provided a most important supplement to the irregular European supply of raw paper stock during the war.[78]

For some time Eastman and other company executives had considered initiating the production of photographic gelatin, but the complexities and difficulties of gelatin production, an art like papermaking, discouraged them from making the commitment. The variables in producing high-quality photographic gelatin were not well understood. Purchase of gelatin depended upon prior photographic testing of sample batches. Yet the specialized consumer of gelatin faced the problem of disposing of inferior grades of gelatin if he were to engage in commercial production. Despairing of the cartelization of the principal German gelatin producers, Eastman hired in 1911 a photographic gelatin maker who conducted experiments for more than a year. He succeeded in developing methods that allowed production even in hot, humid weather, conditions under which most gelatin plants could not operate. In 1912 Eastman Kodak constructed at Kodak Park a small-scale experimental gelatin plant with an initial output of a half-ton per week. By the end of the year the plant produced more than two tons per week.[79] During the next decade the company confronted difficulties in obtaining supplies of high-quality raw stock and in consistently maintaining the quality of output. Although by the end of the war the Kodak Park plant had a capacity of more than six tons per week, the company continued to depend substantially upon its large stockpile acquired from Germany before the war. With its stockpile depleted Eastman Kodak constructed in 1919 a $1½ million gelatin plant and also renewed its purchase of gelatin from abroad. Meanwhile, the quality of its own production remained inconsistent, a problem that no doubt stimulated to some extent the gelatin researches of Sheppard at the Research Laboratory in the early and middle 1920s. These researches, of course, led to the discovery of the crucially important trace quantities of sulfur compounds in gelatin that served as catalysts of photosensitivity. These gelatin researches ultimately served to rationalize gelatin production and permit the American firm to become independent of German suppliers.[80]

Prior to 1910, the company depended entirely upon outside sources for its optical supply. Since the late 1880s, when the company first marketed the detective and Kodak cameras, Bausch & Lomb produced most of the lenses for Kodak cameras, while German optical companies supplied the more limited quantities of high-quality optics for the most expensive Kodak cameras. While cordial relations prevailed between the two Rochester firms, the

[77] GE to Strong, 16 April 1915 (GECP).

[78] Ibid., 14 December 1917 and 12 March 1919.

[79] GE to S. V. Haus, 31 January and 1 June 1911; and GE to Baker, 6 July 1911 and 13 November 1912 (GEC).

[80] GE to Strong, 12 March 1919 (GECP); and GE to Haus, 10 June and 6 July 1920, and GE to J. J. Rouse, 23 July 1920 (GEC).

high profit margins enjoyed by Bausch & Lomb finally enticed the Eastman company into optical production. By 1911 the company employed a Mr. Graf, a practical optician who held a patent on a simple anastigmatic lens and placed him in an experimental department at the Camera Works. His success in producing satisfactory lenses encouraged the company to establish a small lens department on the top floor of the Blair Camera Building, where he initiated production of an f/8 anastigmatic lens to replace the rapid rectilinear lens then used in many of the better lines of cameras. In establishing the lens department, the company made it a policy to hire no one previously employed by Bausch & Lomb, hoping to minimize the friction that might arise as Bausch & Lomb began to lose one of its major markets.[81]

As the company established a pattern of backward vertical integration in the second decade of the twentieth century, Eastman became increasingly self-conscious of the strategy. Late in 1917 the company, which consumed 100,000 tons of coal per year, acquired its own coal mine at a cost in excess of $1 million. At that time Eastman jokingly suggested that the company should acquire a silver mine and an optical glass factory. A short time later the consumer of a quarter of a million ounces of silver annually actually gave serious consideration to purchasing a silver mine.[82] In 1920 the large consumer of wood distillation chemicals purchased a small alcohol plant in Kingsport, Tennessee, and established the Tennessee Eastman Company. It became an innovative chemical subdivision of the company, later leading the company into diversification of its product line.[83] Tennessee Eastman represented the climax of the decade of vertical integration. In that decade, partly in an effort to develop production economies and gain freedom from European cartels and partly in response to the severance of European supplies during the war, Eastman Kodak drew upon its large surplus account to complete the building of a truly integrated photographic materials and apparatus manufacturing corporation.

The Emergence of New Internal Leadership

The years from 1909 to the end of World War I saw George Eastman continue to dominate the general policy and direction of the company, but he also began to pass responsibilities to others within the organization, gradually diffusing the decision-making responsibility. The tempo of this change quickened in the years from the end of the war until 1925, when, after many years of preparation and experience, a new generation assumed command of the company. The transition was made gradually and smoothly.

1909–18. The problems of staffing a large corporation were ones that Eastman recognized early in his career, and increasingly

[81] GE to Clarke, 6 July 1911; GE to F. W. Barnes, 27 June 1912; and GE to Mattison, 28 October 1912 (GEC).

[82] GE to Strong, 14 December 1917 (GECP); and GE to E. E. Powell, 29 June 1918 (GEC).

[83] GE to J. F. Johnson, 10 June 1920; and GE to G. O. May, 4 August 1920 (GEC).

Fig. 14.8
William S. Gifford. Courtesy of Eastman Kodak Company.

he came to see the staff of the company as its major asset; yet from 1909 on Eastman became greatly concerned with finding qualified *young* men who could begin to fill the positions being vacated as the first generation of company leadership began to retire. From the beginning of the second decade of the century, Frank Lovejoy began to emerge clearly as George Eastman's successor. This tall and smiling New Englander, like Eastman, was somewhat shy and retiring; yet, he was a man of decisive action (see fig. 8.3). Unlike Eastman, who sometimes appeared cold and distant, Lovejoy conveyed a sense of warmth and concern that served him well in working with the increasingly complex staff at Eastman Kodak.[84] Lovejoy's keen technical ability and training, his capacity for work, and his respect for the production function no doubt endeared him to Eastman. In less than a decade, Lovejoy moved from Kodak Park (where he was replaced by James Haste, another graduate of the Massachusetts Institute of Technology) to the Kodak Office, where he served as assistant general manager directly under Eastman. At first his primary responsibilities were in coordinating production among the many manufacturing centers of the company, but increasingly broader responsibilities fell to him as the demands of the business increased and as Eastman sought more time for leisure. The influence of others who had previously played important counseling roles to Eastman gradually diminished: Strong, Hubbell, and Fenn. The foreign operations, which had presented so many problems for the company during the first decade of the century, troubled Eastman less during the second decade because of the leadership of William S. Gifford (fig. 14.8). The former employee of American Aristotype and former manager of the Camera Works assumed the burden of European responsibilities, and Eastman came to rely fully upon his judgment.[85]

The Board of Directors of the company continued to reflect the principal concentrations of ownership, retaining part of the group carefully selected earlier by Eastman and Strong. Despite Eastman's extensive philanthropic contributions, in 1913 he still held nearly a quarter of the shares in the company. Moreover, major blocks of stock remained in the hands of the families of original stockholders and of major officers in the company. The six Rochester board members were among the twenty largest shareholders. The combination of Eastman's major ownership in the company and his vast knowledge and experience in building the business helped to reinforce his nearly single-handed direction of the company. As a consequence, the administration of the firm was highly centralized. With the emergence of Lovejoy as a major officer of the company, decision making continued to be focused in the hands of either Eastman or Lovejoy.

1919–25. The year 1919 signaled a quickening of the pace of leadership change at Eastman Kodak. Eastman turned sixty-five years of age; Henry Strong died; and Gifford retired as Manager of

[84] *F. W. Lovejoy: The Story of a Practical Idealist* (Rochester, N.Y.: Eastman Kodak Co., 1947), pp. 10–11.
[85] GE to Gifford, 18 July 1919 (GEC).

Kodak Limited. Consequently, attention was focused both on requisite leadership and organizational changes. Early in 1919 Price, Waterhouse & Company conducted a study of Eastman Kodak's executive staff and organization and raised some serious questions about the high degree of centralization of management. The study questioned whether the responsibilities of direction of the company should not be shared by the president with the staff and whether some of the functions assumed by the assistant general manager (Lovejoy) should properly be his. In the production sector of the company operations, the study questioned whether one person should direct all of the company's production and suggested the possibility of organizing the factories as independent units. Clearly, the study questioned the wisdom of the highly centralized control and administration of the company.[86]

Eastman responded willingly to the questioning process, and some of the minor suggestions were immediately implemented, while some of the major suggestions were postponed or not acted upon after further study of their implications. Suggestions regarding the organization of a separate service organization were adopted at once and certain limited reorganization of the sales department introduced. More gradually, the decision-making process was diffused from the hands of Eastman to certain selected members of the staff.[87]

The staff functions at Eastman Kodak developed late in the corporation's history. Originally, Eastman and Miss Whitney exercised considerable power with little staff assistance. As early as 1914, however, Eastman sought some assistance for Frank Lovejoy, hiring Marion B. Folsom directly from the Harvard Business School. Folsom worked closely with Lovejoy for a few years as an assistant and then, upon his return to the company from military service, became personal assistant to George Eastman. Meanwhile, Albert Chapman, who founded the Development Laboratory immediately after the war, became, in the early 1920s, assistant to Lovejoy. These assistants to the two principal managers of the company soon developed their own staffs. Folsom's interest in statistics and the preparation of elaborate statistical accounts of the company's production and sales for Eastman's perusal soon led to the development of a statistical department, which became a major adjunct to the executive staff.[88] Another important staff function that had been largely externalized throughout the early history of the company was legal counsel. With the appointment of James S. Havens, a former student of law at Yale, a local Rochester politician, and an active attorney, Eastman created the company's first legal department. In 1919 Havens also became a vice-president of the company.[89] Stimulated by the Price, Waterhouse study, the leadership of the company accelerated the development of the staff departments.

[86] GE to Price, Waterhouse & Co., 19 May 1919 (GEC).
[87] Ibid.
[88] Interview with Mr. Marion B. Folsom, 28 July 1970.
[89] *Brit. J. Photog.* 74 (March 1927): 138.

Fig. 14.9
Eastman Kodak Company's Rochester facilities, c. mid-1920s. Courtesy of Eastman Kodak Company.
A. *Camera Works and Kodak Tower*
B. *Hawk-Eye Works (Optical)*
C. *Kodak Park*

The major suggestion of the study was also one of the most controversial. Late in the teens and early twenties, one of the more innovative organizational ideas for large corporate enterprises was that of breaking administratively centralized corporations into decentralized multidivisional structures. Most noteworthy of such decentralizations were those stimulated by Pierre S. Du Pont at Du Pont and, later, at General Motors.[90] Consideration was given to decentralized structure for Eastman Kodak in 1919, but it was not acted upon. No doubt the pending antitrust appeal before the Supreme Court discouraged the adoption of the multidivisional structure. For more than a half decade the Justice Department had sought to get agreement from Eastman Kodak officials to dissolve the corporation: to break it up into smaller units. The company leadership had argued with considerable vigor that because of the nature of production of photosensitive materials, this was not feasible. If for organizational reasons, however, the company decentralized its manufacturing operations into separate divisions, each with its own staff, that would simply facilitate the government's effort to break up the company into a group of autonomous companies. Moreover, the critical nature of photosensitive emulsions did, in fact, require coordination of research and production. The closely coordinated activities of Stuber and the small group of Eastman Kodak emulsion technicians scattered throughout the world was crucial in maintaining the uniform quality of the company's photosensitive products. Consequently, the plan for decentralization into a divisional structure was not adopted.

The reorganization of 1919 did create an organizational structure with five vice-presidents in charge of four principal functions: manufacturing, photographic quality, sales promotion, and legal matters. The organization chart aside, Lovejoy remained at the heart of the operation.[91] From 1919 to 1925, Eastman gradually withdrew from the active management of the company, and increasingly, Lovejoy assumed the responsibilities of management (fig. 14.11). The new generation of leadership that was emerging was also reflected in the Board of Directors of 1925. It included three honorific appointees, Eastman, Hubbell, and Stuber; only one Englishman, Watson; one local outside stockholder, Clark; and the five new managing executives, Lovejoy, Haste, Mees, Havens, and Jones. Eastman personally appointed Stuber president, an honorific position only; but in the appointment, Eastman recognized the central role of emulsion quality to the success of the company and also Stuber's years of loyal service.[92]

In 1925 Eastman's share of ownership in the company had decreased substantially as a consequence of his public philanthropy, especially to educational institutions such as M.I.T., the University of Rochester, and Tuskegee Institute, to medical and

[90] Alfred D. Chandler, Jr., *Strategy and Structure: Chapters in the History of American Industrial Enterprise* (Cambridge: M.I.T. Press, 1962). Alfred D. Chandler, Jr., with Stephen Salsbury, *Pierre S. DuPont and the Making of the Modern Corporation* (New York: Harper & Row, 1971).

[91] GE to Thomas Baker, 14 August 1919 (GEC).

[92] Interview with Dr. Albert K. Chapman, 27 July 1970.

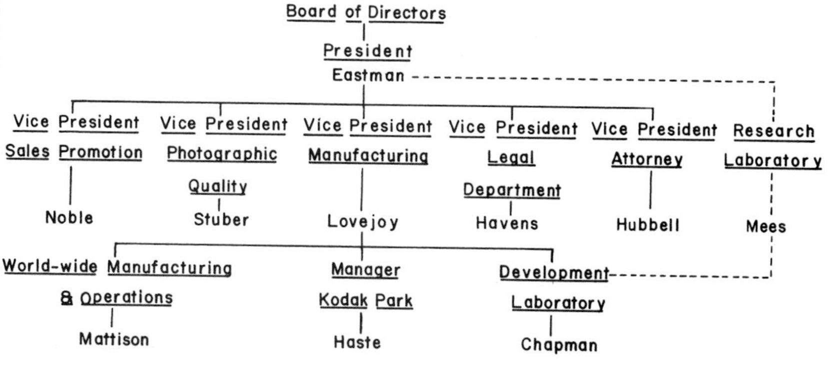

Fig. 14.10
Eastman Kodak Company Organization, 1919.

Fig. 14.11
Eastman Kodak Company management team, 1925. Seated, left to right, *Stuber, Eastman, Lovejoy;* standing, left to right, *Jones, Havens, Folsom, Mees, and Haste. Courtesy of Eastman Kodak Company.*

dental institutions, and to Eastman Kodak Company employees. As Eastman stepped down from the presidency and the general managership of the company, he assumed, like Strong before him, an honorific position where he was kept informed and was consulted about *his* company's affairs. Albert K. Chapman, whose office was adjacent to Eastman's on the top of the Kodak Office Tower, related: "Then there came the time when he was in his office less and less frequently. It came to be news when he was there. The word would go round—Mr. Eastman is in today. Somehow or other that made it a better day for all of us. Finally he came no more."[93]

The key figure in the growth of the American photographic industry passed from the scene. This shy, terse, and diminutive figure who had overshadowed his much more gregarious peers—Strong, Walker, Blair, and Abbott—uniquely combined talent, knowledge, and imagination in understanding and interrelating technology, finance, and business-industrial organization. Committed foremost to goals of sales and profit growth and market dominance, George Eastman had effectively pursued these goals initially by commitment to patented technological innovation, to

[93] Dr. Albert K. Chapman, "Address," Chamber of Commerce Luncheon, Rochester, N.Y., 8 July 1954.

Fig. 14.12
George Eastman. Courtesy of Eastman Kodak Company.

the financial integrity and independence of the corporation, to an international business perspective, and to the attraction of highly capable men to complement his areas of personal need in terms of capital, legal counsel, and technological knowledge. Later, he had modified these strategies, placing less reliance on patents and more on continuous innovation, less reliance on his own personal technical participation and more on academically trained chemists and engineers and ultimately on institutionalized research and development. Moreover, he had derived from his strategy of patent acquisition his first efforts in horizontal integration and from the horizontal integration his first acquisitions in forward vertical integration.

In the early twentieth century, as the judicially redefined legal boundaries threatened his strategies of horizontal and forward vertical integration, Eastman had turned to backward vertical integration as an alternate path for continued growth and investment opportunity. While he made investments in plant and equipment largely in terms of his intuitive anticipation of sales growth and cost savings in the context of technological changes and competitive conditions, in the acquisition of supply houses and competing manufacturers Eastman followed a carefully considered formula based on the acquired company's profits and sales. Yet, in the execution of these goals and strategies, George Eastman did not act single-handedly but depended upon the loyalty and finances of Henry A. Strong; the technological contributions of many unnamed pioneers in the art of photography and also the vital contributions of Walker, Reichenbach, de Lancey, Stuber, Mees, and many others associated with the Rochester firm; the legal counsel of Selden, Hubbell, and Philipps; and the executive capacities of his successor, Frank Lovejoy. Eastman's role was not that of sole creator of the modern photographic industry; but he was a shrouded symbol of power, able to awe and attract people and draw resources requisite for the unleashing of the new technological and business ideas and forces of his time.

Other Multiproduct Firms

While Eastman Kodak continued to maintain its dominance in photographic materials and apparatus production and marketing, two much smaller companies, Ansco and Defender, served as diversified materials and apparatus producers in the American industry. Ansco continued its policy of open antagonism with the Rochester giant, while the even smaller Defender firm, located within the shadow of Eastman Kodak, pursued a cooperative relationship.

Ansco

The Ansco Company of Binghamton, New York, continued during the period 1909–22 under the same basic leadership that had acquired control of the company in the middle of the first decade of the century. The company had established itself as the only competitor of Eastman Kodak in the United States with a full

Table 14.1 Ansco Company Net Sales, 1910–12

Year	*Cameras*	*Paper*	*Film*
1910	$142,657	$585,478	$232,859
1911	179,000	773,210	289,031[a]
1912	259,884	887,018	292,915[a]

SOURCES: Ansco Company, *Journal*, p. 483, Smithsonian Institution, Washington, D.C.; and *U.S.* v. *EKCo.*, p. 403.
[a] U.S. sales only.

product line; however, Ansco's financial position depended almost entirely upon the outcome of the Goodwin patent case. Nevertheless, the continued attention to production economies and efficiencies remained manifest.

During the period 1910–13, Ansco committed itself to new and improved areas of production. Stimulated by the new demand for cinematographic film, it installed the drum method of film base production, and soon the Goodwin Film & Camera Company, as opposed to the Ansco Company, began to return profits. Furthermore, the banking interests in the firm extended further loans, increasing the bonded indebtedness for construction of a new camera factory.[94] The company closed its New Haven plant, transferred many of the personnel to Binghamton, and in the new camera works focused attention on production of amateur roll film cameras rather than the traditional plate cameras, the mainstay of the old New Haven shop. Carl Bornmann, who had worked in the Scovill camera production shop since 1887, assumed the position of superintendent of the new camera works and, like Robertson and Folmer in Rochester, spent a good deal of time on design and development of new models of cameras. From 1902 to 1914 he obtained at least twenty-two patents on new cameras and their improvements. Although Ansco had sought since 1902 to create and market a full line of amateur cameras, it was not until the second decade of the century that the company became independent of other camera manufacturers like Century and Seneca and placed a premium upon camera design and extensive advertising as an important element of camera-marketing strategy.[95]

Ansco expanded its network of exclusive dealerships and company-operated offices in major cities. It opened a major office in Chicago, which served the sales network west of the Mississippi. Also, in 1911 an office and store were opened in London. This branch, operating as an independent company under the name Ansco Limited, acted as the foreign sales branch of the company in Europe.[96] During the early years of the second decade, the Ansco Company experienced a gradual growth in sales in general and a more marked increase in sales of the popular Cyko paper. Never-

[94] Alexander C. Lamoutte testimony, *Goodwin* v. *EKCo.*, pp. 1364 and 1379; and *Moody's Manual: Industrials* (1910), p. 2497, and (1911), p. 2713.

[95] Carl Bornmann testimony, *U.S.* v. *EKCo.*, pp. 511–16; and Lamoutte testimony, in ibid., p. 546.

[96] William Foote Seward, ed., *Binghamton and Broom County, New York: A History* (New York and Chicago: Lewis Historical Publishing Company, 1924).

theless, the company did not produce a profit or pay dividends. Its investors continued to rely upon a successful conclusion to the Goodwin suit as the best hope for a return on their initial investment.

The period from 1914 to 1917 encompassed the most financially successful years for the company during the first quarter of the century. The key to this financial success was the set of decisions favorable to the Goodwin Film & Camera Company handed down by the Federal District Court and the Federal Circuit Court of Appeals in the Goodwin patent case. These decisions brought a financial settlement with Eastman Kodak and other film manufacturers that substantially altered the Ansco balance sheets, at least for a short period of time.

The key issues in the entire proceedings were whether Goodwin's modified claims introduced in 1895 were equivalent to those of the original application and whether Eastman Kodak's use of camphor, even at a level of 14 percent, absolved the Rochester firm from the infringement charge. The Goodwin patent did not claim celluloid, interpreting that term to apply to nitrocellulose dissolved in 40–60 percent of camphor. Judge John R. Hazel decided that the utilization of less than 40 percent of camphor made a process fall under the Goodwin patent and that the terminology in the original application and the final patent issue were equivalent.[97]

Eastman Kodak appealed the decision to the Circuit Court of Appeals, where a three-man judicial board heard the appeal. Judge Alfred Conkling Coxe's opinion reflected much more emotion than that of the lower court. It began by picturing Goodwin as a poor, aged clergyman facing insurmountable obstacles such as "the five examiners who improperly deprived him of his rights during these eleven years. . . . Truly an extraordinary and deplorable condition of affairs. . . . The long delay and the contradictory rulings of the Patent Office would have discouraged an inventor who had not a supreme faith in the justice of his cause." Although Judge Coxe made no direct comparisons between Goodwin and the large corporate "trust," it seems apparent that his sympathies lay with the part-time, attic-inventor who was "hampered by his inadequate surroundings."[98] While the ultimate legal judgment may well have been a fair one, there is little doubt that in an age of considerable popular antitrust feeling, Ansco benefited from its attorney's portrayal of the poor, hard-working inventor-clergyman in his lonely garret laboratory fighting against the wealthy, powerful Rochester "trust."

Judge Coxe's argument basically followed that of Judge Hazel. He emphasized four points: his failure to find in the record any evidence of anticipation of Goodwin's claims; his feeling that the description of the process was adequate for production of the desired product; his belief in the equivalency of the solvents in the original application and the final patent; and also his concurrence

[97] 207 *Fed. Rep.* 351 (1913); and *Goodwin* v. *EKCo.*
[98] 213 *Fed. Rep.* 231 (1914).

that when Eastman Kodak reduced the camphor content of its film to 14 percent, it began to infringe the broad claims of the Goodwin patent.[99]

From a legal point of view, the case was closed; and the Goodwin patent stood as one of the key patents in the early development of photography. From the point of view of technological development and innovation, Hannibal Goodwin, too, occupies a significant position in history because he described, before anyone else on record, a method of employing celluloidal material as a film base for roll film photography. Despite their considerable interest and high desire for patents on new developments, George Eastman and his technical staff did not begin serious investigation of celluloid as a base until more than a year following Goodwin's application.

However, from the point of view of practical innovation, the work of Reichenbach and other technical staff members at Rochester must be regarded as the crucial technological development. There is no evidence that any of the Eastman staff were aware of the Goodwin work until after the Reichenbach-Eastman patent applications had been submitted. Therefore, the Reichenbach-Eastman work represented independent development. Most crucially, the Eastman staff actually created a celluloid film base and placed it in commercial production in 1889 and, despite technical difficulties in the early 1890s, produced the film continuously from that date. Despite Judge Coxe's view that with Goodwin's description of the process "any intelligent chemist would understand and . . . [could] produce the desired film," clearly the stream of improvements in the product and the process of its production proved to be the major barrier to followers of Eastman Kodak. The world-renowned German chemical industry, with its famous research and development laboratories, was unable to produce a competitive film base until just prior to World War I, and the quality of the Goodwin film introduced in 1903, both base and emulsion, did not approach that of Eastman Kodak at any time in the first quarter of the century. Recognition must be given to Hannibal Goodwin, but the legal opinion in the Goodwin case should not be interpreted as a valid judgment on the technological significance of Goodwin's work relative to that of Reichenbach, Eastman, de Lancey, Lovejoy, Haste, and a multitude of other unnamed contributors who introduced celluloid film to the marketplace and made it a quality consumer product.

The financial speculations of the New York attorneys and bankers who had invested in the Goodwin and Ansco companies were rewarded. After Judge Coxe handed down his opinion on 10 March 1914, Eastman Kodak applied for a stay of the injunction giving Eastman Kodak sufficient time to market about $150,000 worth of infringing film. At the conclusion of negotiations with Ansco, Eastman agreed on a $5 million settlement for all previous Eastman Kodak film sales and for licensing rights for the year-and-

[99] Ibid.

Table 14.2 Payments to Goodwin Film & Camera Company for Production Rights under Goodwin Patent

Single Payments Made for Previous and Subsequent Production under Goodwin Patent	
Eastman Kodak Company (26 March 1914)	$5,000,000
Raw Film & Supply Company (15 May 1914)	20,000
Celluloid Company (1 July 1914)	100,000
Jules Brulatour (Lumière) (24 December 1914)	17,500
Subtotal	$5,137,500
Periodic Payments Made for Current Production under Goodwin Patent	
Eclectic Film Company (1914)	$ 11,203
Pathé Frères	
(22 June 1914)	50,000
(1914)	94,011
(1915)	17,810
Pathé Exchange, Inc. (1915)	89
Subtotal	$ 173,113
Total	$5,310,613

SOURCE: Goodwin Film & Camera Company, *Cash & Ledger* (26 March 1914–31 December 1915), pp. 75–78, 87, 286, and 292.

a-half remaining life of the Goodwin patent.[100] Ansco not only received a settlement from Eastman Kodak but from several other companies who marketed or produced celluloid film in the United States (see table 14.2). These payments greatly enhanced the financial position of the Goodwin and Ansco companies.[101]

The problem the Ansco company faced was that speculators wanted their remuneration immediately; once settlements had been made on the patent and its term expired, only the fledgling production capacity of the company represented any potential for future profits. On the basis of performance to 1914, that potential appeared to be very low. From its share in the patent settlement, Ansco retired its debts, paid a 100 percent dividend on common stock, and paid quarterly dividends of 2½ percent from July 1914 to April 1917.[102] Although there is no evidence of the exact distribution of the settlement moneys, Ansco did find sufficient capital, reputedly from the Goodwin settlement, to initiate a building program for a first stage of backward vertical integration. This move consisted of the construction of a cellulose nitration plant at Afton, New York, a few miles east of Binghamton.[103]

After 1917, Ansco gradually returned to the condition of its financial plight prior to the Goodwin settlement. The settlement money went largely to impatient speculative stockholders rather than to capital investment that might have given the company the production technology and capacity to make a serious entry into the American photographic materials and apparatus markets. The Afton nitrocellulose plant proved a disaster, unable to produce the

[100] GE to Strong, 30 March 1914 (GECP).

[101] See Balance Sheets of Ansco reproduced in *Moody's Manual: Industrials*, (1914), p. 80, and (1915), p. 2122.

[102] *Moody's Manual: Industrials*, (1915), p. 2122; (1916), p. 3928; (1917), p. 2140; and (1918), p. 1992.

[103] Fritz Wentzel, *Memoirs of a Photochemist* (Philadelphia: American Museum of Photography, 1960), p. 104.

raw material at a price competitive with other American suppliers. Consequently, the plant remained idle for many years.[104] Although markets in chemicals and photographic products in general expanded rapidly during the war years, Ansco could not capitalize on the opportunity. The company's failure to produce a continuing flow of product innovations in cameras and photographic paper played a prime role in its difficulties. Even successful new products were soon imitated or superseded by new products from other manufacturers.

Following the spirit and style of the time, Ansco did boast of its "research laboratory." The appellation was a misnomer that derived from the inflated conception of laboratory employees and from the public relations office. We can catch two quick glimpses of the "research laboratory," one during the mid-teens and the other during the mid-twenties. Early in 1915 Professor Charles Chandler recommended to Ansco one of his former chemistry students at Columbia University, Frederick B. Gilbert, for a position in its research laboratory.[105] Gilbert accepted the position and then, shortly after beginning employment, provided Chandler with a description of the laboratory and his activities in it. He portrayed it as "splendidly equipped for a large amount and variety of chemical and physical research" and as having a considerable library. Yet, the staff consisted only of the supervisor, a man who mixed and controlled the emulsion, and Gilbert, whose duties consisted of "chiefly the routine daily testing of the film which is being made." He pointed out that he employed the Hurter-Driffield method, determining the density of the exposed film with a polarization photometer.[106]

Fritz Wentzel, knowledgeable of laboratories from his work in the photographic industry in Germany, briefly described the Ansco laboratory as it existed a decade later (fig. 14.13): "I began at Ansco as the assistant to the chief chemist . . . the many problems of production and research that were handled by my three or four American colleagues in the only existing laboratory which was filled with obsolete equipment and lacked any facilities for exact experimental work. The main task of this laboratory was testing raw materials. I made analyses of all the chemicals that were used and became acquainted with all the physical tests for gelatins. In addition, I made emulsions in small amounts in a somewhat crude way."[107] He also told of the staff's less routine work, including efforts to solve problems such as (1) postdevelopment streaks on coated papers; (2) unusual retrogression of the latent image on film; and (3) "the offsetting of the signs and figures of duplex paper on the emulsion of roll films." Furthermore, he reported that following publication of Sheppard's work on increasing the light sensitivity of emulsions by using organic latent sensitizers,

[104] Ibid.

[105] A. C. Lamoutte to Charles F. Chandler, 24 February 1915, Chandler Collection, Butler Library, Columbia University, New York City.

[106] Fred B. Gilbert to C. F. Chandler, 2 May 1915, in ibid.

[107] Wentzel, *Memoirs of a Photochemist*, p. 100.

Fig. 14.13 Ansco Research Laboratory in 1929. From Wentzel, Memoirs of a Photochemist, (Philadelphia: American Museum of Photography, 1960). *Courtesy of Minnesota Mining and Manufacturing Company.*

"we made endless trials in this direction with no remarkable results except that we ran into fog."[108]

Clearly, through the first quarter of the twentieth century, Ansco's laboratory served primarily as a routine analytical and testing laboratory where some developmental problems were addressed, but it had neither the facilities nor the staff to pursue basic photochemical research or even major product development. The management of Ansco did not adopt a strategy of promoting substantial research and development even though the company sought to compete in a market where the leader consciously sought to be a *moving* target. Ansco was not even equipped to imitate or follow that moving target.

By the early 1920s Ansco had returned to the position of financial losses, and the debts mounted once again; however, unlike the period of the first decade of the twentieth century, the company did not have a major asset such as the Goodwin patent.[109] Under financial pressure, leadership changes were initiated in 1922, but even though the new management group and ownership brought financial resources and promotional skills to the company, they lacked the crucial production and technical insights requisite for success in the technically complex photographic industry.

At the end of the first quarter of the century, Ansco made a good physical appearance (fig. 14.14). Wentzel described his first impressions of Ansco:

When I was first taken around the plant, everything looked wonderful and very active to me. I was especially impressed by the two big wheels on which the film-base was cast and by the film coating alley in which the film was dried by floating it horizontally on free air. I had never seen anything like this before. The paper plant was particularly interesting to me. It had five coating machines with straightaway drying equipment—something unknown in Europe—and operated day and night. Ignorant as I was of the financial status

[108] Ibid., p. 101.

[109] Ansco Balance Sheets in *Moody's Manual: Industrials*, (1917), p. 2140; (1918), p. 1992; (1919), p. 2167; (1920), p. 2243; (1921), p. 1067; (1922), p. 871; (1923), p. 1345; and (1924), p. 856.

Fig. 14.14
Ansco plant in 1926. From Wentzel, Memoirs of a Photochemist. *Courtesy of Minnesota Mining and Manufacturing Company.*

of the company and the conditions behind this glittering facade, I was confident of a brilliant future for Ansco and myself.[110]

He then pointed out that the company, with about five hundred employees, possessed fatal weaknesses not only in finances but in its narrow and weak line of products, in aging production facilities, and in great dependence upon European suppliers for chemicals, gelatins, and baryta papers.[111]

Four years after the reorganization that created Ansco Photoproducts, a change came that indicated that the new leadership had begun to recognize one of the company's major problems: its technical and scientific weakness. In 1928 the major German chemical cartel, I. G. Farbenindustrie, through its photographic division, Actiengesellschaft für Aniline Fabrikation, assumed control of the American firm, renaming it Agfa Ansco. The German company thus acquired desired entry to the American market and brought to Ansco engineering and production talent, capital, and access to its all-important scientific and technical research facilities in Germany. At once major changes were introduced into both the scale and methods of production and into the breadth and quality of the Ansco product line.[112] In the late 1920s the entry of the German chemical interests finally brought to this descendent of American pioneer companies the scientific and technological capabilities and leadership that had been necessary for success in the American photographic industry since the 1890s.

Defender Photo Supply Company

The diminutive Defender company of Rochester, New York, was in an even more vulnerable economic position than Ansco because of its small operation and even more limited product line. Conse-

[110] Wentzel, *Memoirs of a Photochemist*, p. 100.
[111] Ibid.
[112] Ibid., pp. 104–6; *National Cyclopedia of American Biography*, s.v. "Davis"; and *Moody's Manual: Industrials*, (1928), pp. 2663–65.

Fig. 14.15
Argo trademark, Defender Company of Rochester, N.Y.

quently, it pursued a much more conciliatory policy toward Eastman Kodak than did Ansco. In response to the forces of international relations, antitrust considerations, and internal policy changes, Defender passed through three distinct phases of development from 1909 to 1925.

From 1909 to 1913 the company sought to expand its product line and production facilities, addressing both the mass consumer market and the professional market for photographic materials: paper, plates, and chemicals (fig. 14.15). Early in the century, in response to the exclusive dealerships established by Eastman Kodak and Ansco, Defender created its own supply houses in Chicago, New York, and Boston and sought to attract both professional and amateur photographers.[113] The company broadened its product line from its strength in photographic paper to dry plates through the acquisition of Carbutt's Keystone Dry Plate Company of Philadelphia and the hiring of Rowland S. Potter from England as emulsion maker and manager of the dry plate plant, which was moved to Rochester in 1911. The company provided major facilities for the production of an outstanding X-ray plate.[114] Defender also marketed roll film under the brand name Vulcan. At first it imported Ensign film from England and relabeled it, but from 1909 to 1913, unbeknown to the trade, Defender film was produced at Kodak Park.[115]

The production relationship with Eastman Kodak grew out of a more intimate financial relationship. In 1909 a stockholder decided to sell his substantial holdings in the company, but other shareholders were in no financial position to purchase the stock. When no other buyer appeared, Frank Wilmot, the president of Defender, asked George Eastman if he would purchase the stock. Through his local attorneys, Eastman agreed to buy the stock if the other half-dozen stockholders would sell sufficient stock to give Eastman controlling interest in the company. Eastman's agreement to this purchase, when earlier he and the Eastman Kodak Company had eschewed purchase overtures from Defender, indicated the confidence Eastman had in 1909 that the antitrust issue with Eastman Kodak was dead.

Eastman personally acquired 60 percent of the shares in the small Rochester firm, and soon Eastman Kodak was shipping Vulcan film to Defender by means of an elaborately disguised supply route.[116]

From 1913 to 1919 Defender experienced a number of severe external and internal difficulties. With the outbreak of war in Europe, the flow of vital supplies, most particularly glass, raw paper, gelatin, and chemicals, slowed and finally ceased. Unprepared for the cessation of supplies and without an independent source of materials such as Eastman Kodak possessed, the com-

[113] *Defender Spotlight* 3 (May 1945): 5. [Available at Rochester Public Library.]
[114] Ibid., p. 3; and Wentzel, *Memoirs of a Photochemist*, pp. 17–18.
[115] Frank W. Wilmot testimony, *U.S.* v. *EKCo.*, p. 190.
[116] Ibid., pp. 189–91; and GE to Wilmot, 4 October 1919 (GEC).

pany faced diminished production capacity. The combination of loss of Belgian glass and the introduction of Eastman Kodak's double coated X-ray plates idled the company's X-ray plate production plant and virtually destroyed its small but vital area of strength.[117] Internally, Defender faced a series of crises as George Eastman, threatened in 1913 with the government antitrust suit against his company, disposed of his stock in the company. Although the company incorporated in 1914, its continued severe financial stringency brought about the resignation and withdrawal of Frank Wilmot.[118] The company faced these enormous problems and survived the war period by increasing its reliance on the jobbing of photographic materials and apparatus, made possible by the changes in the terms of sale and exclusive dealership policies of Eastman Kodak and Ansco.[119]

After the war, the Defender company examined its potentialities and redirected its strategies. Under the leadership of a new president, the company recognized its weakness in the mass amateur market, abandoned that market, and concentrated upon production and marketing for the professional. Renewing its active production of photographic papers, the company added a new professional enlarging paper, Velour Black, in 1920. During the middle 1920s, while Eastman Kodak was abiding by the terms of the government consent decree, Defender acquired the production facilities and formulae for Artura paper and also acquired three of the leading dry plate brands: Seed, Stanley, and Standard. The plates themselves continued to be produced by Eastman Kodak with the Defender label, and the small Rochester firm served as the marketing agent for them.[120] These acquisitions from Eastman Kodak did not, however, provide the firm with major new growth products. Dry plates were rapidly declining in favor, even with professional photographers, and the Artura paper no longer remained the popular seller it had been a decade and a half earlier. In an effort to follow the new direction of professional photographers, the company introduced its own brand of sheet film in 1927. The film base was produced by the Du Pont Corporation, initiating a relationship that culminated nearly a generation later in the acquisition of the Defender company by Du Pont.[121]

Despite the product-line expansion in the early and middle 1920s, Defender, like Ansco, was struggling to survive under the shadow of Eastman Kodak. Although the company narrowed its focus and although it was actively encouraged by Eastman Kodak and although it was in a better financial position than Ansco, its diminutive size and lack of technical and financial resources left it with only a very small share of the photographic supply market.

[117] "Defender Joins Du Pont," *The Du Pont Magazine* 39 (August 1945): 18–19; Wenzel, *Memoirs of a Photochemist*, p. 18; *Defender Spotlight* 3 (May 1945): 3.

[118] GE to Wilmot, 4 October 1919 (GEC); and Wilmot testimony, *U.S. v. EKCo.*, pp. 189–91.

[119] *Defender Spotlight* 3 (May 1945): 3.

[120] Ibid., pp. 3–5; and McKelvey, *Rochester: The Quest for Quality*, p. 344.

[121] "Defender Joins Du Pont," p. 19.

Conclusion

From the middle of the nineteenth century to the first quarter of the twentieth century, the American photographic industry grew from a small and diffuse industry to one that was large and geographically centralized. Initially, individual proprietors employed handicraft techniques to produce photographs and photographic supplies for a small domestic market, but by 1925 a few large-scale, integrated corporations were employing mass production techniques to dominate the supply of apparatus and materials for an international, predominately mass-amateur market. During this transformation the American industry seized and maintained worldwide leadership in a field intimately related to the fine chemical and optical industries, both of which were at the time dominated internationally by German cartels. Changes in four key areas epitomized the transformation of the American industry and related to its success: (1) scale of enterprise; (2) character of product and market; (3) mode of production; and (4) business and technological strategy. An evaluation of the role of technology is central to an understanding of the stages in this transformation and of the forces underpinning it; yet, in the analysis presented here technology is seen not as an independent causal factor but as an interdependent element in the conceptualization, formulation, and implementation of those business strategies that ultimately shaped the character of the industry.

Central to this analysis is the framework of successive business-technological mind-sets. Periods of relative business stability and of rapid changes are understood in terms of certain business and technological conceptions that dominated practice during stable periods and that underwent substantial alteration during periods of rapid change. Certain key technological elements defined for a given period the ultimate boundaries of further technological and business innovation, although business conceptualization, personal values, and social and political restraints sharply defined practice *within* the technological boundaries. Precedent-shattering technological change, generally with its revolutionary implications unperceived and often originating from outside the mainstream of the industry, triggered each time a redefinition of the broad technological boundaries within which practice could occur. Hence, technological change disrupted the equilibrium of the stable period and lowered the barriers to entry for the agents of redefinition of the business-technological mind-set. These agents, which included new enterprises, technicians, and businessmen, entered with different conceptions, strategies, and values, many of which reflected new business and technical attitudes developed outside the photographic industry. Therefore, the fundamental change in the mind-set came as a consequence of the influx of new business ideas and practices and as a result of their interaction and redefinition in coordination with the new technological bound

Historically, the redefinitions of the mind-sets graphic industry corresponded to four "triggering changes. The change in the 1850s from the direct po otype to the negative-positive collodion processe first major redefinition. The alterations in the interrelated techni- cal parameters of product, production, and marketing stimulated the reversal in rank of the two leading firms as the production-oriented metallurgical firm of Scovill lost its preeminence to the more market-oriented Anthony company. Although the Anthony firm produced a broad spectrum of photographic supplies for professional photographers, it emphasized manufacture of chemicals and treated paper, new products of importance with the advent of collodion processes. Certain other new areas of production, for example, the tintype plate, emerged from the efforts of individuals who were initially independent of the leadership of the industry. The tintype sector only slowly became integrated into the sphere of the new marketing leadership of the industry.

The change in the early 1880s from wet collodion to dry gelatin processes for negatives stimulated the second redefinition. The introduction of gelatin, which served as carrier for the photosensitive salts and as a preserver of their photosensitivity, allowed for the first time factory mass production of the basic negative material. This not only simplified the professional photographers' task but heightened the importance of centralized production in the industry. While dry plate production in the United States developed from outside the traditional leadership of the industry, most of the new producers restricted their production to a single product line and depended upon the traditional marketing channels for distribution. Yet, a few of the new firms, most notably those of Eastman and Blair, entered the industry with quite untraditional conceptions, goals, and strategies. Reflecting the new business climate, which was characterized by large-scale corporations and growing respect for the role of technology and science in business, these new firms pioneered in technical and business innovation in an industry which, with the sudden enlargement of the producers' consumable market, provided remarkable opportunity for growth. In this setting the first forms of business integration and of pursuit of systematized technological strategies occurred. The new scale of operation and new conceptions of business-technological strategies also led to the employment of persons who were academically trained in science.

The third and in many respects the most revolutionary changes came in the late 1880s with the introduction and successful exploitation of the roll film system. With its adoption the amateur photographer supplanted the professional in the dominance of the practice of photography and, consequently, transformed the entire character of the industry. The enormous expansion of the market for both durable and consumable products fostered the emergence of large-scale corporate enterprise which paralleled a pattern seen in other sectors of American industry. Within this new economic framework the leadership switched from the Anthony and Scovill companies, dominated by a marketing conception of the industry, to the Eastman Kodak Company, committed to a broader and more sophisticated conception of the integrated relationship between production and marketing. The enlarged scale of enterprise encouraged such new directions in technological-business strategy as the systematic acquisition and enforcement of patents, continuous product innovation, and employment of academically trained engineers to design and supervise new forms of production technology. Moreover, this enlarged scale supported new directions in business strategy, principally, horizontal and vertical integration and rapid expansion of international operations. By the early twentieth century these strategies had transformed Eastman Kodak into a multinational firm.

The fourth change, the introduction of cinematography in the late 1890s, substantially expanded the boundaries of the industry. Once again innovation came from outside the traditional leadership of the photographic industry; nevertheless, the leadership in this new sector had to depend substantially upon the traditional leader, Eastman Kodak, for a reliable supply of quality photosensitive film. Eastman Kodak carefully and narrowly defined its role in order to avoid interference with the interests of this new market for its film. Accordingly, it eschewed entry into commercial production of film plays and commercial motion picture apparatus but, somewhat later, inaugurated and fostered a small but growing amateur movie business. Many of the new firms in the cinematographic sector, entering with business conceptions and attitudes prevelant in business generally in the early twentieth century, immediately pursued various methods of business integration and product innovation. The companies that acted most vigorously upon those perceptions arrived at positions of leadership in the cinematographic sector in the nineteen twenties. Meanwhile, Eastman Kodak conservatively pursued its traditionally defined markets in still photography while continuing to innovate technically within the traditional framework and to innovate organizationally with backward vertical integration. Moreover, its creation and successful exploitation of an industrial research laboratory climaxed the institutionalization of a long standing, informally executed function. Eastman Kodak continued to expand its sales and to maintain its world leadership and, as a part of its vertical integration strategy, to exercise new leadership in the critically important production of fine chemicals for a previously German-dependent market at home during World War I.

In this framework of successive business-technological mind-sets, exogenous forces provided in large measure the crucial dynamic elements, i.e., the triggering technological developments and the new business conceptions. The initial triggering elements interacted with one another and with the traditions of the photographic industry to shape unique structures and forms. Hence, the sources of technological change played a critical role in the character and vitality of the American industry. During the middle decades of the nineteenth century, the industry in the United States remained subordinate to that in Europe because of Europe's initial position as innovator in commercial photography, her original and continued superior position in the photographically related chemical and optical industries, and the intellectual and institutional support of photography-related research by Europe's relatively large scientific community. Although American investigators such as Draper studied photochemistry, and American technical people introduced several important developments—the tintype and the electroplating of daguerreotypes—most technical improvements originated and obtained their earliest development in Europe. Most notable among these were increased photosensitization, gold toning, hypo fixing, and specially designed portrait and landscape lenses, in addition to the three earliest "triggering innovations": wet collodion processes, dry gelatin emulsion, and the roll film system.

After 1880 the American industry continued to rely to some extent upon European technology and science, but the earlier relationship of dependence diminished as certain new entrants, most notably Eastman and Blair, sought aggressively to alter the predominant style in production methods, products, and market. George Eastman brought to the industry not only a fresh technological perspective but also a business point of view gained from contemporary business circles. With determination to achieve sustained business growth, Eastman employed mass production, technological innovation, and patent control of invention as cornerstones of his business strategy. The shy, publicly austere entrepreneur elicited from his senior and extroverted partner, Henry Strong, a combination of capital, business advice, and warm personal support. Yet, Eastman personally combined a keen facility for technology, marketing, and finance, with regard to both their fine details and their larger interrelationships.

During the 1890s Eastman and Strong overcame the vicissitudes of finance and technology. Despite European superiority in such critical areas as photosensitivity of emulsions, photographic chemicals, photographic optics, raw paper, and raw gelatin, the Eastman company succeeded in creating a mass amateur market both at home and abroad. Depending upon its initial lead and innovator's experience, upon the American position of leadership generally in all areas of celluloid production, and upon its innovative production engineering, Eastman Kodak mass-produced the critical celluloid roll film base, achieving excellent quality at relatively low cost. Accordingly, at the turn of the century the American industry wrested international leadership in production of photographic materials and apparatus from Western European firms that were rooted in traditions of relatively labor-intensive modes of producing high-quality and high-value products for a limited market. Hence, the key element in American leadership in the early twentieth century was not superiority in theoretical science but the advantage of an initial lead in the mass production of celluloid film and amateur roll film cameras. The American industry maintained that lead later in the twentieth century through a strong commitment to the preservation of trade secrets, support of a comprehensive patent policy, pursuit of continuous innovation, employment of academically trained mechanical, optical, and chemical engineers to insure the use of the latest methods of mass production, and establishment of laboratories devoted to both basic and applied chemical and physical research.

Critical in this pattern of development from 1839 to 1925 was the symbiotic relationship between technological creativity and scale of enterprise. During the first forty years of the American photographic industry, the dominating companies were relatively small. They sought novelty of design but rather than risk sustained effort to conceive and develop new products within their firms, they sought instead to initiate and exploit the creations of others, most notably those of Europeans. After 1880 certain new firms employing new attitudes and strategies entered an industry with a substantially enlarged producer consumables market. These small firms assumed the risk of systematic pursuit of product innovation without the assurance of previous market testing. As the scale of enterprise in the industry increased, the commitment to product and production innovation increased accordingly.

The scale and organization of enterprise existing at the turn of the century reflected not only the influence of the new technological boundaries created by the introduction of dry gelatin emulsions and the amateur roll film system but also the new business conceptions of the time. Among the most critical business influences were (1) European vertical integration strategies, most notably those of the Dresden-based photographic paper industry; (2) the ambiguities of American judicial position with regard to business organizaton and marketing practice; and (3) the unique opportunities provided by the raw materials and market characteristics of the American photographic industry. The photographic industry essentially imitated certain American and European business innovations that reflected the changing scale of the market and the changing character of technology, but it proved innovative in the specific way it implemented these new strategies and forms of enterprise in a new setting.

Early in the twentieth century many critics regarded the change of scale and organization of enterprise that was occurring generally in American business as threatening to the integrity of the American marketplace. The analysis presented here, while admitting that the new scale of enterprise may have had important sociopolitical implications, indicates that within the American photographic industry the new market structure involving large-scale enterprise did not differ substantially from the structures that had prevailed during the earlier years of the industry. Within each business-technological mind-set from 1839 to 1925, the market pattern passed from imperfect toward perfect and back to imperfect competition once again. That the market structure should tend toward imperfect competition is not surprising, since enterprise with profit maximization goals naturally sought means of legally avoiding perfectly competitive conditions. During the late nineteenth century patented product innovation and continuous innovations accompanied horizontal and vertical integration as strategies directed to this end. As the costs, complexities, and uncertainties of patent litigation increased and the courts began to define more narrowly the limits of corporate organization and marketing practice, large-scale enterprise in the photographic industry relied increasingly upon the cost savings derived from backward vertical integration and upon continuous product innovation. The industrial research and development laboratories became the foundation of corporate security protecting the leadership from unanticipated external innovations and providing a steady stream of new products that served as effective barriers to entry into the industry.

With markets located in societies that valued the new and the novel, the American industry obtained financial rewards in the marketplace by appealing to the desires of people at home and abroad. Moreover, the leadership of the American industry carefully fostered the quality and reliability of its photosensitive materials and the simplicity and dependability of operation of its apparatus. To the novice photographers who constituted the mass market, these features were critical to the pursuit of visual reminders of those unique, fleeting, and highly personal events to which so many of their photographic endeavors were directed. Despite the vicissitudes of American judicial and social opinion, the perils of technological change, and the vigorous international competition of the German fine chemical and optical cartels, the American photographic industry created a socially and economically viable organizational, marketing, and technical strategy, and upon that foundation built a position of international leadership, a position it has maintained to the present day.

Appendix / Technological Diffusion

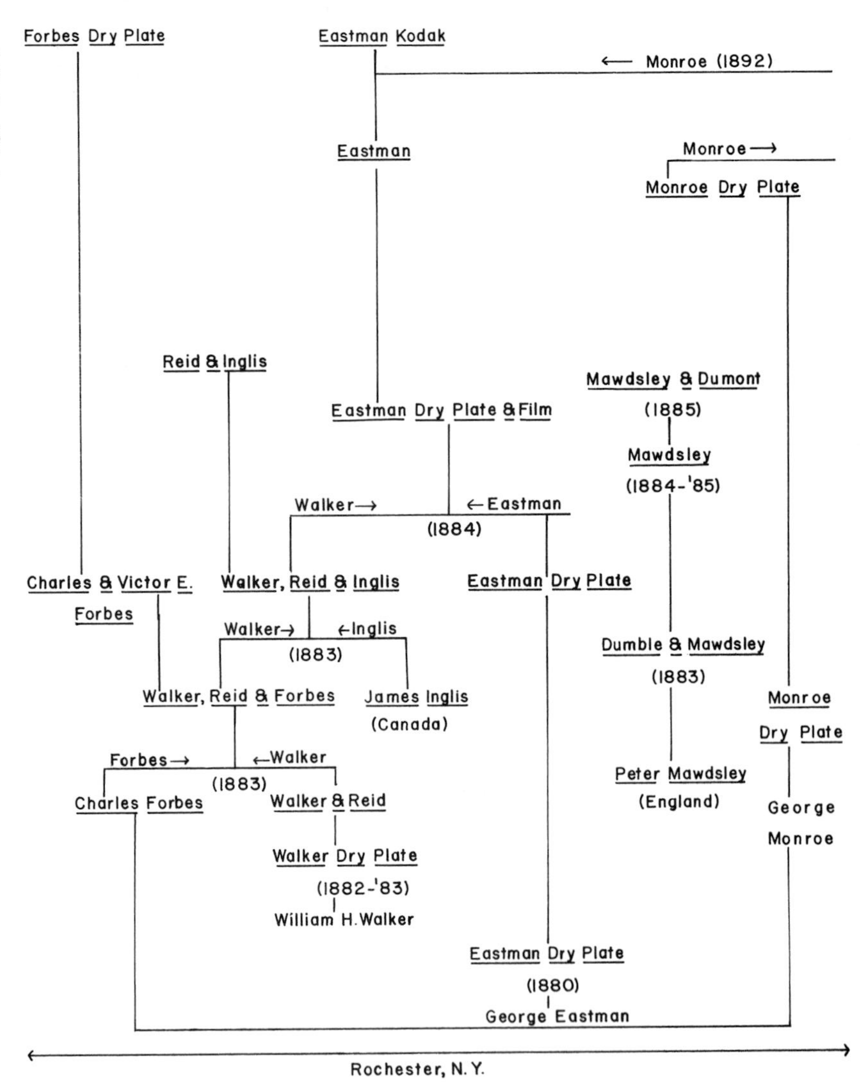

A.
Diffusion of emulsion knowledge (1879–94): Rochester, N.Y.; St. Louis, Mo.; Iowa City, Iowa; and Rockford, Ill.

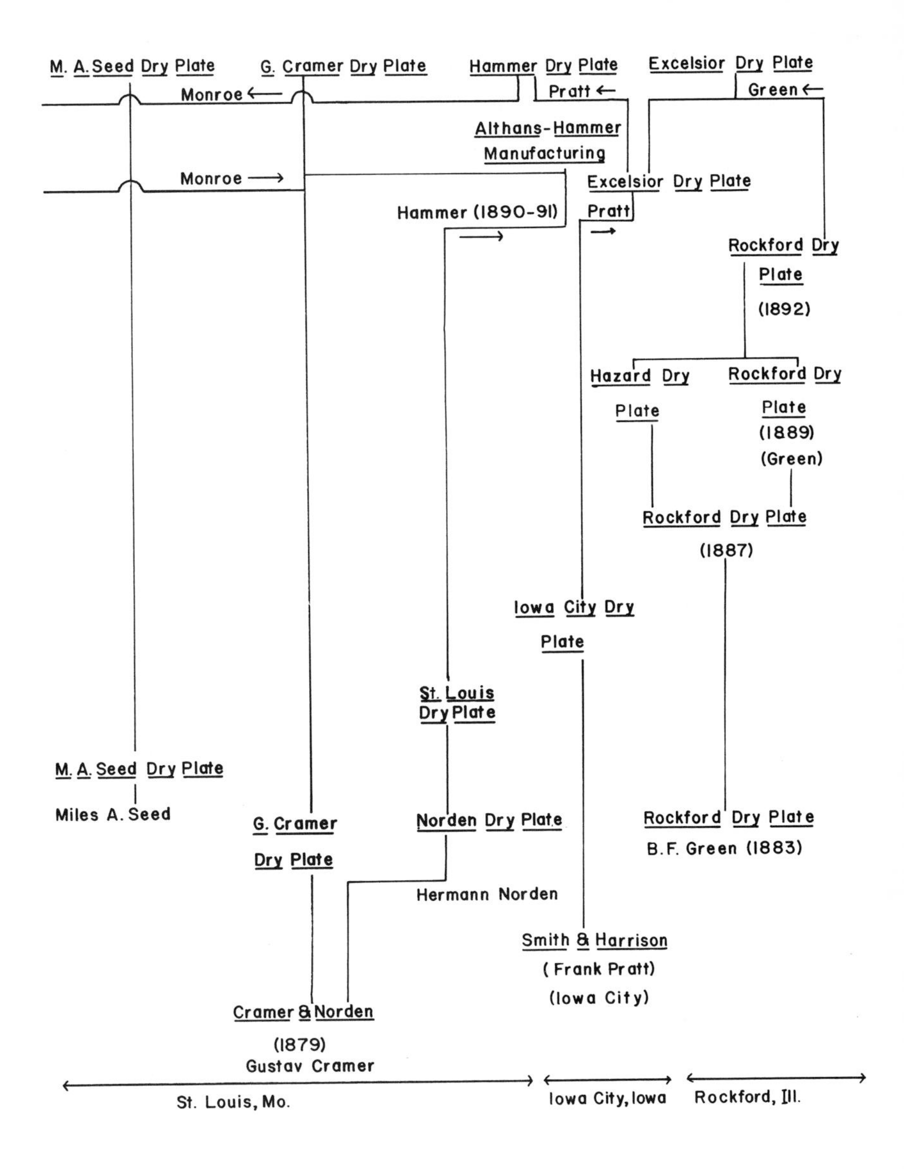

M. A. Seed Dry Plate
G. Cramer Dry Plate
Hammer Dry Plate
Excelsior Dry Plate
Monroe
Pratt
Green
Althans-Hammer Manufacturing
Monroe
Excelsior Dry Plate
Hammer (1890-91)
Pratt
Rockford Dry Plate
(1892)
Hazard Dry Plate
Rockford Dry Plate
(1889)
(Green)
Rockford Dry Plate
(1887)
Iowa City Dry Plate
St. Louis Dry Plate
M. A. Seed Dry Plate
Miles A. Seed
G. Cramer Dry Plate
Norden Dry Plate
Rockford Dry Plate
B.F. Green (1883)
Hermann Norden
Smith & Harrison
(Frank Pratt)
(Iowa City)
Cramer & Norden
(1879)
Gustav Cramer
St. Louis, Mo.
Iowa City, Iowa
Rockford, Ill.

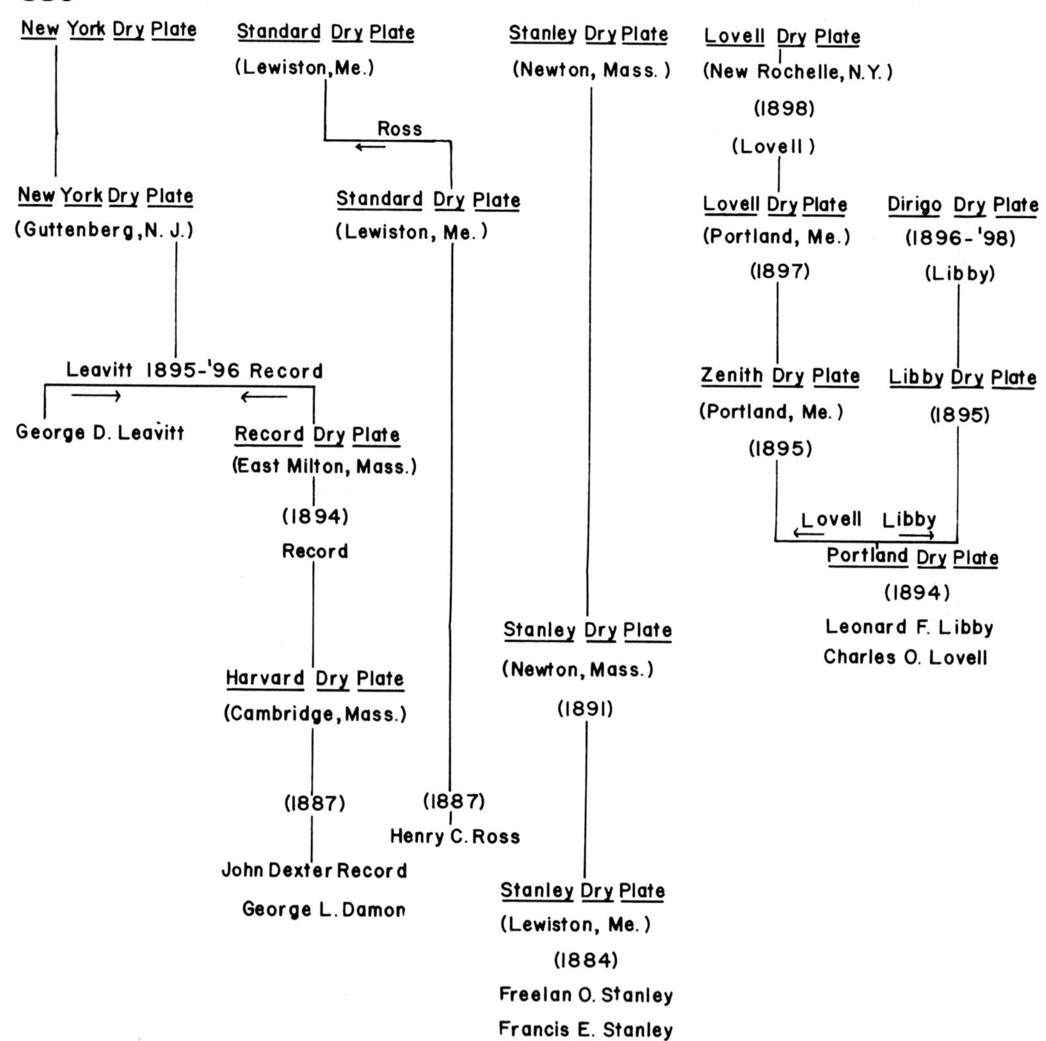

B.
Emulsion knowledge:
eastern seaboard companies.

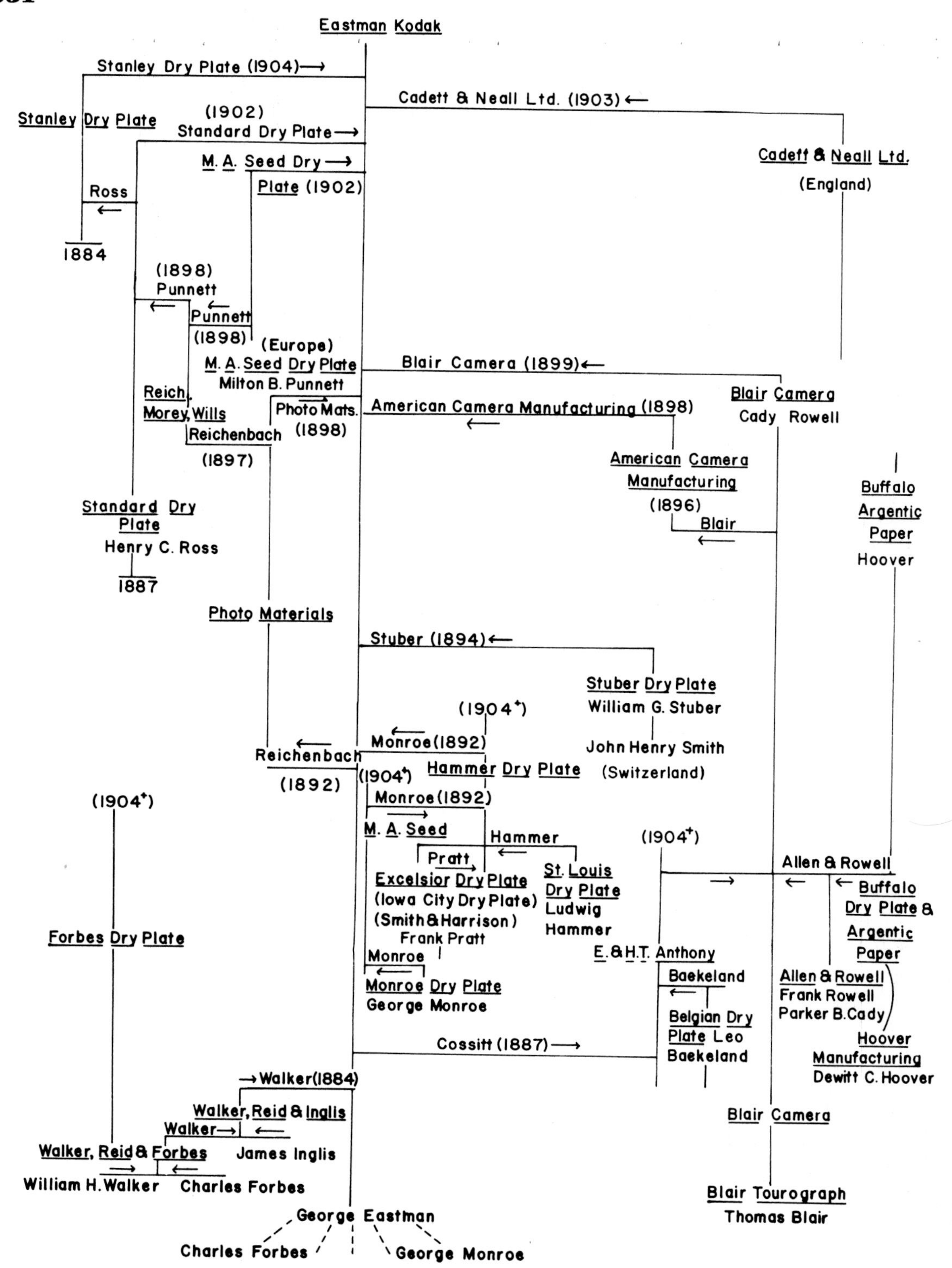

*C.
Origins of Eastman Kodak
Company emulsion
knowledge.*

D.
Key technical men at Eastman Kodak Company, originally with acquired camera companies.

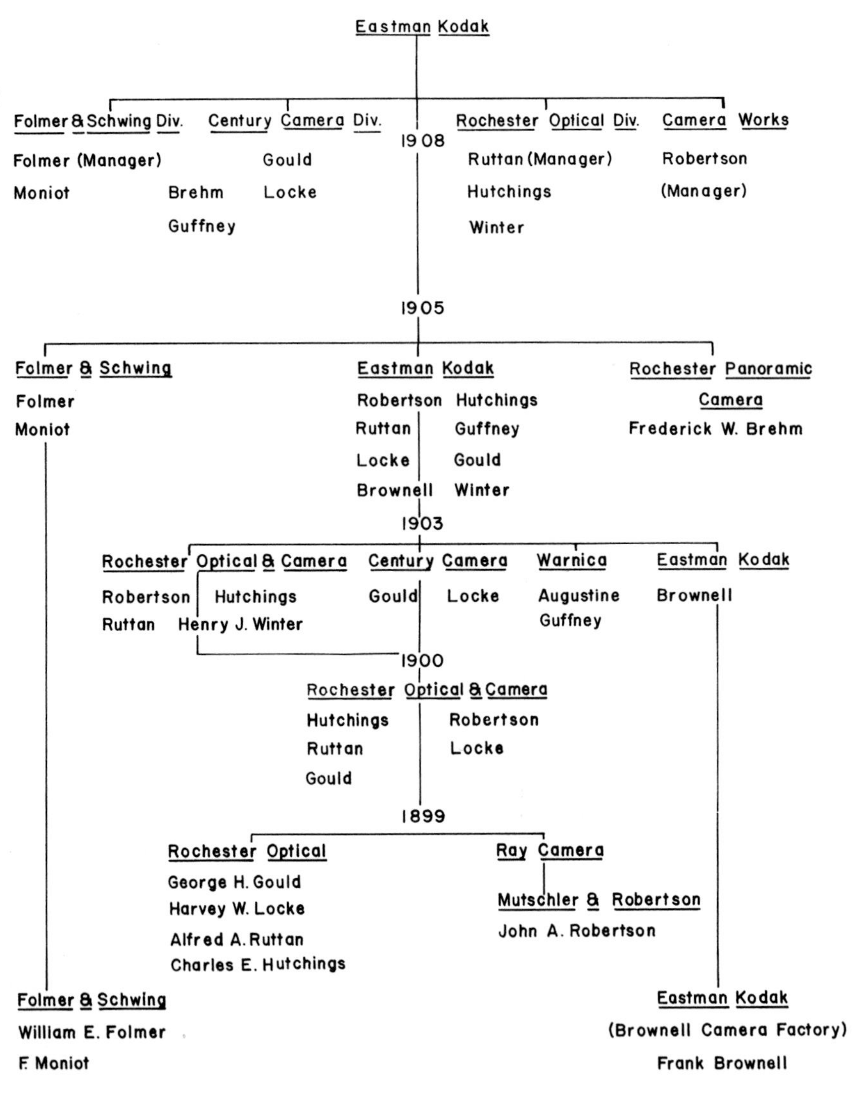

Primary Sources of Information

This study has drawn upon information from a large variety of primary sources, including manuscripts and business records, public documents, correspondence, and interviews.

Manuscripts and Business Records

1. Scovill Manufacturing Company Records, Baker Library, Harvard University, Cambridge, Mass.
   Documents, letterpress books, correspondence, journals, ledgers, inventory books from 1811 to the twentieth century. Materials relevant to photography cover the periods 1839–42 and 1850–82 and may be found among (1) Scovill Manufacturing Company, (2) New York Store, and (3) S. Peck and Company materials.
2. Dun & Bradstreet Manuscript Records, Baker Library, Harvard University, Cambridge, Mass.
   These records, constituting credit rating information for small and moderate-sized businesses for the entire country, provide valuable data on companies in the photographic industry in the period from 1850 to 1890. This is one of the principal primary sources for E. & H. T. Anthony & Company for the period 1850 to 1880.
3. Ansco Company and Goodwin Film & Camera Company Business Records, Department of Photography, Museum of History and Technology, Smithsonian Institution, Washington, D.C.
   Ansco Company, *Journal*, vol. 1, October 1907–December 1910.

Goodwin Film & Camera Company, *Cash & Ledger,* 26 March 1914–31 December 1915.

4. Charles F. Chandler Collection, Butler Library, Columbia University, New York, N.Y.
   This is an extensive collection of materials of the former technical counsel to the Anthony company and former editor of *Anthony's Photographic Bulletin,* containing correspondence, clipping files, notebooks, and legal records. Numerous sections of the collection are relevant to the photographic industry, particularly the boxes marked "Correspondence" and "Clippings Box" (Celluloid Company, 1881–91; Anthony Case; and Recovery of Waste).
5. Blair Materials, Northborough Historical Society, Northborough, Mass.
   Scattered correspondence, newspaper clippings, and mention in diaries of local residents, e.g., Rev. J. C. Kent. Also records from the Town of Northborough and the Northborough Schools Register.
6. Gustav Cramer Collection, Missouri Historical Society, St. Louis, Mo.
   Laboratory notebooks, clippings, correspondence, and legal documents.
7. George Eastman Correspondence and Other Historical Materials, Eastman Kodak Company, Rochester, N.Y. (at the Office of Corporate Information unless otherwise noted).
   A. George Eastman Correspondence. Outgoing, Letter Press Books.
      "Original Business Letter Book" (OB), 1879–85.
      "Personal Letter Books" (GECP), 28 vols., 1879–1932.
      "Business Letter Books" (GEC), 12 vols., 1885–1926.
   B. George Eastman Correspondence. Incoming (GEC).
      Filed chronologically in boxes, 1890–1932.
   C. Henry A. Strong Correspondence. Strong to Eastman, file 1884–98.
   D. Eastman Dry Plate Company, Comptroller's Office.
      Ledger, January 1881–May 1885.
      Journal, January 1881–May 1885.
      (Other account books of Eastman and acquired companies are in the custody of the Comptroller's Office.)
   E. "George Eastman Notebook," Patent Museum.
      Background material on acquired paper, plate, and camera companies, apparently prepared for briefing George Eastman prior to his testimony in company litigation. Materials were drawn from correspondence with officers of acquired companies, company records, and account books.
   F. Wyatt Brummitt, "Brummitt Manuscript," unpublished.
      This is a history of George Eastman and of the Eastman Kodak Company (1879–1955) prepared in the late 1950s and early 1960s by the director of corporate information, who had access to correspondence, company records, and

superannuated employees. Unfortunately, the manuscript was not documented.

8. George Eastman House Materials. International Museum of Photography, Rochester, N.Y.
   A. Early correspondence and materials of George Eastman. Early correspondence with his mother and other relatives, diaries, memoranda books, manuscript copy of first *Kodak Manual*, and George Eastman artifacts.
   B. C. E. Kenneth Mees correspondence and biographical materials. Correspondence from 1899–1906 and 1950–60.
   C. Miscellaneous correspondence of photographers from the 1840s to the 1900s.
9. Walter S. Hubbell Collection, Archives, University of Rochester, Rochester, N.Y. Four volumes of clippings, correspondence, and memorabilia, including letters and cards from and to George Eastman and Henry A. Strong covering the period 1871 to 1919. Hubbell was an Eastman Kodak attorney and member of the Board of Directors.
10. Miscellaneous Materials, Rochester Historical Society and Local History Division of Rochester Public Library, Rochester, N.Y. A large part of the holdings of RHS is at RPL. Files include manuscripts and data on Charles Forbes, financial and legal data on a number of local photographic companies, and an index to the Rochester newspapers for the nineteenth century.
11. Peter Neff Materials.
    A. Kenyon College Archives. Kenyon, Ohio. Neff autobiographical manuscript.
    B. Western Reserve Historical Society. Cleveland, Ohio. Miscellaneous Neff clippings and publications.

Public Documents and Records

1. *Eastman Company* v. *Blair Camera Company* (*Eastman* v. *Blair*), U.S. Circuit Court, District of Massachusetts, 1 June 1894, no. 2883; U.S. Circuit Court of Appeals, First Circuit, 31 October 1894, no. 105, *Transcript of Record,* 2 vols. (c. 550 pages); available from the Clerk, First Circuit Court of Appeals, Boston.
2. *Eastman Kodak Company* v. *J. Edward Blackmore* (*EKCo.* v. *Blackmore*), U.S. Circuit Court of Appeals, Second Circuit, 14 December 1921, no. 65, *Transcript of Record,* 2 vols. (c. 3600 pages); available from the Clerk, Second Circuit Court of Appeals, New York.
3. *Goodwin Film & Camera Company* v. *Eastman Kodak Company* (*Goodwin* v. *EKCo.*), U.S. Circuit Court, Western District of New York, 9 September 1913; U.S. Circuit Court of Appeals, Second Circuit, 10 March 1914, no. 194, *Transcript of*

Record, 6 vols. (c. 4000 pages), and *Briefs,* 3 vols. (c. 300 pages); available from the Clerk, Second Circuit Court of Appeals, New York.

4. *United States* v. *Eastman Kodak Company* (*U.S.* v. *EKCo.*), U.S. Circuit Court, Western District of New York, 24 August 1915, no. A–51; appeal to U.S. Supreme Court, petition: 8 March 1916; withdrawn: February 1921, *Abridged Transcript of Record,* 10 vols. (c. 4200 pages)—available in Appellate Case File No. 25,293, Legislative, Judicial and Diplomatic Records Division, National Archives, Washington, D.C.—and *Supplemental Brief and Argument for Defendants* (c. 80 pages) —available at the Rochester Public Library, Rochester, N.Y.
5. Bureau of Corporation Records, National Archives, Washington, D.C.
6. Corporation Records, Commonwealth of Massachusetts, Boston.
7. Corporation Records, State of Maine, Augusta.
8. Judicial decisions published in the *Federal Reporter, Cases Argued and Determined in the Circuit Courts of Appeal and Circuit and District Courts of the United States* (*Fed. Rep.*), the *United States Reports, Cases Adjudged in the Supreme Court* (*U.S.*), and state courts. The standard form of citation to these legal reference works available in most law school libraries is, for example, 64 *Fed. Rep.* 381, meaning vol. 64 of the *Federal Reporter,* p. 381.
   A. *Blair Camera Company* v. *Barker,* 53 *Fed. Rep.* 483 (1893).
   B. *Blair Camera Company* v. *Eastman Company,* 64 *Fed. Rep.* 491 (1894).
   C. *Celluloid Manufacturing Company* v. *Eastman Dry Plate & Film Company,* 42 *Fed. Rep.* 159 (1890).
   D. *E. & H. T. Anthony & Company* v. *Gennert,* 99 *Fed. Rep.* 95 (1900) and 108 *Fed. Rep.* 396 (1901).
   E. *E. & H. T. Anthony & Company* v. *George Murphy,* 64 *Fed. Rep.* 381 (1894).
   F. *E. & H. T. Anthony & Company* v. *United States,* 90 *Fed. Rep.* 802 (1898).
   G. *Eastman Company* v. *Blair Camera Company,* 62 *Fed. Rep.* 400 (1894).
   H. *Eastman Company* v. *Getz,* 77 *Fed. Rep.* 412 (1896) and 84 *Fed. Rep.* 458 (1898).
   I. *Eastman Company* v. *Reichenbach,* 20 *N.Y.S.* 110 (1892).
   J. *Eastman Kodak Company* v. *Anthony and Scovill Company,* 139 *Fed. Rep* 36 (1905) and 145 *Fed. Rep.* 833 (1906).
   K. *Eastman Kodak Company* v. *J. Edward Blackmore,* 277 *Fed. Rep.* 694 (1921).
   L. *Eastman Kodak Company* v. *Southern Photo Materials Company,* 234 *Fed. Rep.* 955 (1915); 295 *Fed. Rep.* 98 (1923); and 273 *U.S.* 359 (1926).
   M. *Goodwin Film & Camera Company* v. *Eastman Kodak Company,* 207 Fed. Rep. 351 (1913) and 213 *Fed. Rep.* 231 (1914).

N. *Motion Picture Patents Company* v. *Carl Laemmle and the Independent Moving Pictures Company of America*, 178 *Fed. Rep.* 104 (1910).
O. *Motion Picture Patents Company* v. *Champion Film Company*, 183 *Fed. Rep.* 986 (1910).
P. *Motion Picture Patents Company* v. *New York Motion Picture Company*, 174 *Fed. Rep.* 51 (1909).
Q. *Motion Picture Patents Company* v. *Yankee Film Company*, 183 *Fed. Rep.* 989 (1911).
R. *United States* v. *Eastman Kodak Company*, 226 *Fed. Rep.* 62 (1915).

Interviews and Correspondence

1. Blair Camera Company Materials.
   Interview with Albert W. Mentzer, Northborough, Mass., 1 July 1970.
2. George Eastman and Eastman Kodak Company Materials.
   A. Interview with Marion B. Folsom, Folsom home, Rochester, N.Y., 28 July 1970.
   B. Interview with Albert K. Chapman, Eastman Kodak Co., Rochester, N.Y., 27 July 1970.
   C. Interview with Charles F. Ames (by Jean Ennis), Rochester, N.Y., 28 September 1950, and a second interview about 1954. On file at Office of Corporate Information, Eastman Kodak Company, Rochester.
   D. Interview with Charles F. Hutchison (by Jean Ennis), Kodak Park, Rochester, N.Y., 17 October 1950. On file at Office of Corporate Information, Eastman Kodak Company, Rochester.
3. Forbes Materials.
   A. "Charles Forbes, M.D., 1844–1917," manuscript by Florence Forbes Cornwall, in private possession of Mrs. Cornwall of Pultneyville, N.Y.
   B. "Charles Forbes," manuscript by Mary Forbes (1947) at Rochester Public Library, Local History Division.
   C. Interview with James Forbes at University of Rochester, 7 June 1969.
   D. Interview with Florence Forbes Cornwall, Pultneyville, N.Y., June 1970.
4. Phenix Plate Company, Worcester, Mass.
   Charles A. Hill to Reese V. Jenkins, 24 August 1971.
5. St. Louis Dry Plate Companies.
   Rudolph Schiller to Reese V. Jenkins, 1969–71.
6. Stanley Materials.
   A. Interview with Raymond Stanley, York Harbor, Maine, 18 August 1970.
   B. Newton Public Library, Newton, Mass.
      Miscellaneous addresses to clubs in Newton, 1890–1920.

Index

Pages listed in **bold** contain illustrations.

A Abbott, Charles S., **197**, 319; with American Aristotype, 87, 94–95, 196–201; as entrepreneur, 94–95, 196–99, 225–26, 239, 240–41, 243
Abney, Captain W. De W., 87
Acworth (English firm), 241
Adamantean plates, 52
Adams, Washington Irving, 55–56, 162–64
Adams, Washington Irving Lincoln, 162–64
Addyston Pipe & Steel Company, 175
Advertising: Ansco, 331; Eastman Kodak, 116, **117**; Independent Moving Pictures, 286–87; New York Dry Plate, **222**; Stanley Dry Plate, 221; Universal Film, 291; William H. Walker, **99**
Agfa (German firm), **168**, 168, 278, 289, 295, 305, 337
Agfa Ansco, 337. *See also* Ansco Company
Albumen, 81. *See also* Paper, photographic
Albums, 50
Allen & Rowell Company, 83, 86, 89, 104, 124, 138
Allix, Colonel J. N., 117 n
Alter, David, 26
Althans. *See* Hammer-Althans Company
Ambryotype, 40, **41**, 41
American Albumenizing Company, 60, 88
American Aristotype Company: Anthony as sales agent for, 165, 248; early history of, 89, 91–95; integration of, 199–201; offices and factory of, **201**, **202**; sales and profit position of, 192–93, 200, 205–6. *See also* Abbott, Charles S.; General Aristo Company
American Camera Company, 181, 183. *See also* American Camera Manufacturing Company
American Camera Manufacturing Company, 145, 182, 190, 191, 195, 248. *See also* American Camera Company; Blair, Thomas Henry; Blair Camera Company; Eastman Kodak Company of New York
American Daguerre Association, 19–20
American Dry Plate Company, 227
American film, 115. *See also* Film, stripping
American Film Manufacturing Company, 288
American Mutascope and Biograph Company, 277, 283, 285
American Mutoscope Company, 270–71
American Optical Company, 56, **57**, 62, 135, 162–63, 209, 251
American Photographic Paper Company, 202
American Photo-Relief Printing Company, 70
American Self-Toning Paper Company, 198
Ames, Charles F., **145**, 145, 216

Andrews & Thompson, 26
Anschütz, Ottomar, 264
Ansco Company: dependence on Goodwin patent of, 331–34; Eastman's attitude toward, 319, 321; financial status of, 255–57, 331, 332, 334–37; formation of, **255,** 255–57; leadership and organization of, 255–57, 334, 336–37; plant of, 336–37, **337**; products of, 204, 205, 206, 208, **252, 253,** 331; research laboratory at, 335–36, **336**; sales of, 256, 331–32, 338. *See also* Anthony and Scovill Company; Goodwin Film & Camera Company
Ansco Limited, 331
Ansco Photoproducts Company, 337
Anthony, Edward (person), **21,** 21, 22, 32–33, 86, 165
Anthony, Edward (company): approaches technological change, 32–33, 48–52, 165; early history of, 20–28; 32–33; as manufacturer, 21–23, 25–28, 49, 50, **51**; production facilities of, **22, 23, 24, 27, 29, 30**; pursues integration, 21–23, 25, 46; as supply house and jobber, 20–23, 25. *See also* Anthony & Company, E. & H. T.; Daguerreotype; Supply houses
Anthony, Frederick A., **164,** 164, 165, 248, 250–52, 254, 255
Anthony, Henry T., **22,** 22, 51, 61
Anthony, Richard A., 91, **164,** 164–65, 251–52, 254, 255
Anthony & Company, E. & H. T., 56, 59, 246–48, 250, 342; attitudes and policies of, 69, 76, 80, 160–61, 164–66, 246–48, 250–51; Blair merges with, 89, 138–39, 144–45; as camera producer, 58–61, 89, 138–39, **164,** 164–67, 211, 248; as jobber during collodion period, 52, 54, 58–61; as jobber during gelatin period, 69, 71, 76, 80, 87, 106, 124, 136, 190, 248; as paper producer, 82–86, 104, 106; as product innovator, 61–63, 69, 82–86, 124, **164,** 164–65, 167; relations of, with Blair, 89, 136, 138–39, 143–44, 190; relations of, with Eastman, 71, 85, 104, 106, 166; sale of plate camera division of, 211, 248; Scovill & Adams merges with, 204, 213, 248, 250–51. *See also* American Aristotype Company; Anthony, Edward; Anthony, Frederick A.; Anthony, Henry T.; Anthony, Richard A.; Anthony and Scovill Company; Goodwin Film & Camera Company
Anthony and Scovill Company, 248, 250–57, 318–19
Antitrust: investigations of Eastman Kodak, 318–22; legislation, 175, 253; legislation, influence of ambiguities in, 245, 328, 338
Archer, Frederick Scott, 37–38, 99 n
Argo paper, 208, **338**
Aristotypie, 86–87
Armat, Thomas, 273–74
Artura paper, 339
Artura Photo Paper Company, 204–8, 322, 339
Associations (price maintenance), 19–20, 175; Dry Plate Manufacturers Association, 74–76, 98; Photographic Merchant Board of Trade, 92–93, 198, 220–21, 225
Atlas Motion Picture Company, 288
Auburn Box Company, 226–27
Austin-Edwards (England), 278
Autochrome process, 303
Axtell, F. C., 145–46

B

Bacon, Francis, 308–9, **316**
Badeau, H., **50,** 58–59, **61**
Baekeland, Leo Hendrick, **65,** 124, 165, 186–87; at Nepera Chemical, 90, **193,** 193–94, 197, 223
Baker & Rouse (Australian firm), 241
Barker, Erastus B., 164
Barnett, John, 46 n
Barr, Captain H. J., 99 n
Barriers-to-entry: in celluloid film, 279, 288–89, 333; in cine film, 279, 285, 288–89, 299; continuous innovation as, 184–85, 333, 341, 346; in daguerreotype supplies, 16, 17, 18; defined, 4 n; in gelatin dry plates, 71–72, 74, 76–77, 162, 220–21; patents as, 118–20, 162, 167, 189, 285, 299, 333; in tintype production, 45
Bates, Joseph L., **58,** 58
Bausch, John J., 214
Bausch & Lomb Optical Company, 111, 114, 211, 214, 215, 244, 323–24
Bayer (German firm), 278, 289, 305–6
Bellsmith, H. W., 117 n
Benecke (European firm), 202
Bertsch (German firm), 82
Biograph, 269–72, **270, 271, 272**
Biograph Company, 277
"Biograph Girl," 292
Bishop, Fred, 314
Black Maria, 267–68
Blair, Thomas Henry, 91, 159, 169, 344; with Blair Camera, 137–40, 143–44; as early photographer and innovator, 134–36; relations of, with Eastman, 137–40, 143–44; relations of, with other companies, 145, **181,** 190. *See also* American Camera Company; American Camera Manufacturing Company; Anthony & Company, E. & H. T.; Blair Camera Company; Buffalo Dry Plate and Argentic Paper Company; Eastman, George; Eastman Company; Eastman Kodak

Company of New York; Litigation, patent
Blair Camera Company, **137**, 136–45, 156–59, 251; as division of Eastman Kodak, 181, 190–91, **195**, 195; finance and capital of, 137, 139, 144, 156, 159; horizontal integration efforts by, 91–92, 138–39, 143–44, 181, 190–91; imitation of Eastman by, 134, 136–43, 145; innovation by, 89, 138, 140–43, 145, 166, 183, 342; production of roll film and cameras by, 136–43, 166, 183, 267, 277; relations of, with Anthony, 137, 138–39, 165–66; relations of, with Eastman, 91–92, 140–44, 158, 166, 181, 190–91. *See also* Ames, Charles F.; Blair, Thomas Henry; Boston Camera Manufacturing Company
Blair International trademark, **145**
Blair Tourograph and Dry Plate Company, 135–36, **136**. *See also* Blair, Thomas Henry; Blair Camera Company
Blanchet Frères & Kleber of Rives, 49
Blanquart-Evrard, Louis-Désiré, 49
Bobbs-Merrill Company, 176, 239–40
Boissonnas, Edward V., 78, 223
Bornmann, Carl, 162, 331
Boston Camera Company, 137–38. *See also* Blair, Thomas Henry; Blair Camera Company; Boston Camera Manufacturing Company; Turner, Samuel N.
Boston Camera Manufacturing Company, 158–59, 189–90. *See also* Blair, Thomas Henry; Blair Camera Company; Boston Camera Company; Turner, Samuel N.
Bostwick and Harrison patent, 182
Bosworth, Incorporated, 293
Bradfisch & Hopkins, 87
Bradfisch & Pierce, 92
Bradfisch Aristotype Company, 87
Brady, Mathew, 16–17, 19
Brassart, August, 48
Brown, A. M., 89
Brownell, Frank A.: as assembler and manufacturer, 106, 111, 148 n, 183–85; as camera designer, 113, **142**, 142, 149–50, 158, 184, 301
Brownie camera, 238
Brulatour, Jules E., 287–88, **290**, 290–91, 334
Buffalo Dry Plate and Argentic Paper Company, 86, 89, 138–39, 158
Bullet camera, **158**, 158
Bulls-Eye camera, 145, **158**
Bureau of Corporations, 176

C Cadett & Neall Dry Plate, Limited, 229–30, **230**, 240–41
Cameras, daguerreotype and collodion, **12**, **31**, **57**, **58**, **100**, 134–36, **136**; producers of, 17–18, 46–48. *See also* Collodion photography; Daguerreotype; Lenses
Cameras, dry plate: popularity of, 208–9, 212; production of, 111–12, 162–63, 164, 178, 216–19
Cameras, dry plate (by brand): Anthony view, **164**; Blair tripod, **137**; Cyclone, **167**, 211; Eastman detective, **111**; Eastman Genesee, 111; Graflex reflex, 219; Henry Clay, 209; Irving-Clay, **209**; Lucidograph, 136; Poco, 168, 211; Premo, 211; Trochenette, 168; Waterbury detective, 163. *See also* Cameras daguerreotype and collodion
Cameras, Eastman Kodak roll film: daylight loading cartridges for, 149–50, 210, 237; effect on industry of, 166–68; introduction of, 112–16, **114**, **115**, **119**; new models of, 118, 148; production of, **120**, **148**; sales of, 116, 154, 209, 210. *See also* Eastman, George; Eastman Company; Eastman Dry Plate & Film Company; Eastman Kodak Company of New York
Cameras, Eastman Kodak roll film (by model): ABC, 210; Autographic, 316–17, **317**; Brownie, 210, 237, **238**; Bullet, **158**, 158–59; Eureka, 210; Falcon, 210; First, **114**, 114–15, **115**, **119**; Flexo, 210; Folding, 148; Folding Pocket, 210; Kodako, 210; Kodet, 210; No. 1, **118**, 118, **119**; No. 2, 118; Pocket, 158, 210
Cameras, magazine, **167**, 167
Cameras, motion or cine, **264**, 267, 269, **270**, **271**
Cameras, reflex, **218**, 218–19
Cameras, roll film (by company other than Eastman Kodak): Ansco, 256, 331; Anthony and Scoville, **253**; Blair, **141**, 141–42; Boston Camera, 137–38; Boston Camera Manufacturing, **158**, 158, 210; Houston (non-commercial), **108**; Warnerke, **100**
Cameras, roll film holders attached to, **100**, 100–103, **103**, 137–38, 163
Camera obscura, 11, **13**
Capital and finance, 118, 139, 148, 159, 233, 235, 294–95. *See also* Eastman, George; Eastman Kodak Company of New York; *other companies by name*
Capstaff, John George, 311, 314, 316
Carbutt, John, 74, 164, 232, 277; background of, 69–70, **70**; innovations of, 69–70, 78–80, 124. *See also* Keystone Dry Plate Company
Carbutt Dry Plate and Film Corporation, 232
Carlton, Harvey B., 168, 211, 212

Carlton, William F., 166, 211, 212
Carlton Motion Picture Company, 288
Cartes de visite photographs, 49
Cases, production of, 46. *See also* Daguerreotype, cases
Casler, Herman, 269–72
Celluloid Company, 181–82; litigation involving, 146, 153, 156, 334; relations of, with Eastman, 145–46, 153, 156, 182; sales of unsensitized roll film by, 138, 140–41, 153, 190, 252, 278–79, 288
Celluloid Manufacturing Company, 123–25, 127, 138
Celluloid Varnish Company, 124, 127
Celluloid-Zapon Company, 146, 154
Centeur Film Company, 288
Centralization of production, 69, 86–90
Century Camera Company, 212–13, **213**, 216, 331
Century Camera Company (Division of Eastman Kodak), 216, 217, 322
Champion Film Company, 288
Chandler, Charles F., 146, 165, 335
Chapman, Albert, 313–14, 329
Chapman, George, 47 n
Chapman, Levi: as manufacturer, 23, 25, 26, 27, 28, 32–33, 45–46; as supply house and jobber, 20, 23, 25, 32–33, 45–46
Chapman & Wilcox, 59
Chemical production, photographic, 168, 242–44, 313, 324, 343; during daguerreotype and collodion periods, 16, 26, 48, 59. *See also* Collodion photography; Film, celluloid roll; Sodium thiosulfate
Chevalier, Charles, 17
Chevalier, Vincent, 17
Christensen, Carl, 87, 89, 94, 203
Christofle, Charles et Ce, 22, 25, **26**
Cine-Kodak, 316
Cinematography, 258–79, 282–99, 316, 343
Clark (relative of George H. and Brackett H.), 328
Clark, Brackett H., 106 n
Clark, George H., 106 n
Clarke, Joseph Thatcher, 107, 303, 306
Clayton Act, 176, 291
Climax plates, 164–65
Cochrane, R. H., 286–87
Colas (German firm), 82
Collodion photography, 35–63, 262–63, 344; dry process of, 52, 100; wet process of, 37–40, **38–40**
Color photography, 303–4, 317–18
Columbian Dry Plate Company, 224
Columbian Photographic Paper Company, 251
Competition, 82–95, 345–46; approaching perfect, 5, 6, 16, 45, 50–51, 54–55; duopolistic, 44–45, 54–55; imperfect, 5, 16, 20; oligopolistic, 5, 52, 220–23, 225, 233, 282, 295, 298; perfect, 279, 282. *See also* Antitrust
Conley Camera Company, 212
Cooper & Company, Charles, 124–25, 130, 146, 242
Cooper, David, 85, 112
Corbett, E. M., 47 n
Corporate enterprise, large-scale, 66–68, 172–77, 342
Cossitt, Franklin Millard, 85, 89, **111**, 111, 112, 166
Cotton States Exposition, 273
Coxe, Judge Alfred Conkling, 332–33
Crabtree, John I., 311
Cramer, F. Ernest, 226
Cramer, Gustav, **70**, 70–71, **71**, 72, 74, 225–26, **232**, 295; technical information of, 153, 223, 224. *See also* Cramer Dry Plate Company
Cramer & Norden Company, 70–72
Cramer Dry Plate Company: marketing position of, 154, 221–22, 227–30, 232; relations of, with Eastman, 230–31, 233; technical capacity of, 78–79, 154, 223, 224. *See also* Cramer, Gustav
Crane, Frederick, 127
Critchlow & Company, A. P., 46
Crockett, James L., 125 n
Cutting, James A., 63
Cyko paper, **204**, 256

D Daedaleum, **261**, 261. *See also* Zoetrope
Daguerre, Louis Jacques Mandé, 10, 48, 63
Daguerreotype, 10–33, 262; camera boxes and lenses, **12**, 17–18, 28–32; cases, **16**, 16, 26–28, 31–32; equipment, miscellaneous, **12**, **15**, **17**, **18**; galleries, 3, **14**, 18–19, **19**, **31**; plates, **15**, 22, 25–26, **26**, 266; process, 10, **17**, 22, 31, 344; supply houses, 20–25, 32–33
Daguerreotypy, science and innovation involving, 32–33
Dallmeyer, John H., 57–58
David (French photographer), 124
Davis, Daniel, 22
Davison, George, 117 n, 240
Daylight loading cartridge, **142**, 142–43, **143**. *See also* Brownell, Frank A.; Eastman Company; Eastman Kodak Company of New York; Turner, Samuel N.
Dean, John, 52, 55
Dean & Company, John, 52–55
Dean, Emerson & Company, 52
Defender Photo Supply Company, 295; marketing of, 256, 338–39; paper of, 204, 205, 206, 208, 322, 338, 339; relations of, with other firms, 232–33, 256, 319, 322, 330, 338–39

Defiance plate, 69
DeForest, Lee, 297
DeForest Brothers, 46 n
de Lancey, Darragh, 153, **180**, 180–83; as production organizer and innovator, 80, 149, 180–83, 235, 330, 333. *See also* Eastman Kodak Company of New York
Delaunay (French firm), 82
Developing and printing service, 84–85, 138
Dickman, George, 150–51
Dickson, William Kennedy Laurie, 265–72, **266**, 274, 296. *See also* Edison, Thomas A.
Diffusion, technological, 32, 73, 77, 79, 89, 150, 186, 215, 223–27, 233
Di Nunzio Company, Joseph, 204
Diversification, 137–40, 167–69; as pursued by Eastman, 94, 110, 235, 324
Donnelly Act, 253, 318–19
Donnelly, Thomas, 253
Draper, John W., 12, 343
Driffield, Vero Charles, 312, 335
Dry Plate Manufacturers Association, 74, 76, 98
Du Pont de Nemours & Company, E. I., 176, 305, 328, 339

E

Eagle Dry Plate trademark, **76**
Eastman, George: attitude toward Ansco, 252, 255, 321; attitude toward innovation, 118–20, 133, 156, 179–87, 184–85; background of, 69, 71–72, 96–97, 152–53; as corporate officer, 105–6; creation of Kodak system by, 112–16, 148–49; as entrepreneur, 94–95, 128–29, 148–50, 159, 179–87, 329–30, 344; evaluation of business skills of, 96–97, 151, 169, 233–35, 300–318, 324–25, 328–30; as financier, 118, 133, 151–52, 201; goals of, 96–97, 112, 171; as integrator (backward vertical), 202, 242–45, 324; as integrator (forward vertical), 283–85, 290–93; as integrator (horizontal), 196–201, 211, 216, 225–26, 229–31, 254, 338–39; as nontechnical innovator in technology, 186, 302, 303–4, 307, 316–18, 329–30; as organizational innovator, 133, 201, 234–35; patent litigation involving, 141–42, 250, 253; patent or patent license purchases of, 119 n, 186, 188–91; attitude toward patents of, 97, 130, 184, 249; philanthropy of, 307, 328–29; photographs of, **71**, **187**, **197**, **234**, **329**, **330**; quality as a concern of, 151, 278, 281; relations of, with Walker, 150–51; research laboratory fostered by, 305–8, 310; respect for technical people held by, 149, 186–87; sources of technical information of, 72, 100, 107, 119, 150, 224; staff created by, 98–99, 180, 326; as technical innovator, **72**, 72, 82–86, **83**, 96–105, **103**, 127–33, 329–30; as technologist, 89–90, **111**, 111, 148–50, 182–83; trade secrets of, 186; withdrawal from direct technical matters by, 110, 300–318. *See also* Eastman Company; Eastman Dry Plate & Film Company; Eastman Dry Plate Company; Eastman Kodak Company of New York; Strong, Henry A.; Walker, William H.
Eastman Company (24 Dec. 1889–23 May 1892), **148**, 150, 224, 267; financial data of, 157; Kodak Park developed by, **147**, 147–48; organization of, 117–18; photographic paper of, 86, 90–92; plate innovations of, 79–80; production of celluloid form by, 127–33, **131**, 137; production difficulties of, 138, 152–55; relations of, with Blair, 137–38, 143–44, 158–59; relations of, with Celluloid, 146, 153; relations of, with Goodwin, 146–47; relations of, with Photo Materials, 152–53. *See also* Eastman, George; Eastman Dry Plate & Film Company; Eastman Kodak Company; Reichenbach, Henry M.; Strong, Henry A.; Walker, William H.
Eastman Dry Plate & Film Company (1 Oct. 1884–24 Dec. 1889): advertising of, 116, **117**, development of Kodak camera system by, 109–10, 112–13, **114**, **115**, **118–20**; domestic marketing of, 106, 110–11, 157; international marketing of, 107, 117, 157; introduction of paper by, 82–88; organization and finance of, 105–6, **106**, 107, 115, 117–18, 157; printing and enlarging service of, 109–13; relations of, with Anthony, 85, 106; relations of, with Scovil, 85, 163. *See also* Eastman, George; Eastman Company; Eastman Dry Plate Company; Strong, Henry A.; Walker, William H.
Eastman Dry Plate Company (1 Jan. 1881–1 Oct. 1884), 71, **73**, 73, 76, 98 n, 105
Eastman Kodak Company of New Jersey (holding company), 235
Eastman Kodak Company of New York (operating company, 23 May 1892–): antitrust issues at, 318–24; chemical production of, 242–43, **243**; emulsion technology at, 154–55, 223; Experimental and Testing Department of, 184–86, 301–5; film production of, 153–56, 277–79, 287–91; financial operations of, 148, 151–52, 157, 200; horizontal integration in dry plates by, 229–30; horizontal integration

in paper by, 194–95, 198–204, 206–8; horizontal integration in plate cameras by, 216, 218, 219; horizontal integration in roll film by, 158–59, 189–91, **195**; international operations of, 203, 207–8, 229–30, 241–42, 290–91; international sales of, 178–79, 290–91; organization and personnel of, 152, 226, 233–35, 324–26, 328, **329**; paper sales and production of, 93–95, 192–95, 205–6; product and production innovations of, 158–59, 179–87, 203–4, 219, 238, 248, 288–90; production of dry plates and plate cameras by, 215–19, 227, 228; relations of, with Ansco and Goodwin, 252–54, 332–34; relations of, with Bausch & Lomb, 214–15; relations of, with Defender, 338–39; relations of, with Motion Picture Patents, 284–86, 300; relations of, with Rochester Optical and Camera, 211, 212; research and development at, 300–318; response to World War I by, 321; sales by, 157, 177–78, 205–6, 210, 279, 291, 295, 322 n; sales policies of, 225, 231–33, 237–40, 248, 251, 319–20; vertical integration of, 236–45, 322–24. *See also* Eastman, George; Eastman Company; Eastman Photo Materials Company, Limited; Strong, Henry A.

Eastman Kodak Research Laboratory: contributions of, 312, 316, 323, 343, establishment of, 305–11, **315**; personnel and their early activities at, 311–14. *See also* Eastman, George; Eastman Kodak Company of New York; Mees, C. E. Kenneth

Eastman Photo Materials Company, Limited, 117, **151**, 155–56

Eclectic Film Company, 334

Eder, Josef M., 72, 223

Edison, Thomas A., **265**, 265–72; attitude of, toward projection, 274–75; early cinematographic conceptions of, 265–68, **266**, 273, 296; role of, in cine industry, 274, 277. *See also* Edison Manufacturing Company

Edison Manufacturing Company, 282–85; 305. *See also* Edison, Thomas A.

Edwards, Austin, 223, 287, 289

Edwards, B. J., 78, 223

Elliott, Arthur H., 165 n

Ellis, Edgar E., 127

Emerson, Samuel P., 52

Ensign film, 287, 289

Entrepreneurship: of Eastman and Blair, 159; of Edison, 265–66

Essanay Film Manufacturing Company, 277, 284

Eureka paper, 85

Europe, 10, 337, 343–45; early production facilities of, **73**; as source of technological stimulus, 12, 13, 15, 48, 50, 61, 107, **145**, 154–55, 223–24, 303–6. *See also* Eastman Kodak Company of New York; Eastman Kodak Research Laboratory; Eastman Photo Materials Company, Limited; Kodak Limited

European Blair Camera Company, Limited, **145**

Excelsior Dry Plate Company, **224**, 224

Exhibition of Manufactures of the American Institute of New York (1843), **15**

Experimental and Testing Department, Eastman Kodak Company, 185–86

Exports of photographic goods, 178–79

F

Famous Players Company, 292

Famous Players-Lasky Company, 294–95

FBO Productions, 298

Fenn, Bert, 252, 325

Ferrier (Paris firm), **50**

Film, celluloid roll, 333, 344; introduction of, 122–33, **132**; introduction of noncurling, 148, **186**, 186; introduction of noninflammable (safety), 288–90, 295–96, 302; production of, **131–32**, 131–32. *See also* Anthony and Scovill Company; Eastman, George; Eastman Company; Goodwin, Hannibal; Goodwin Film & Camera Company; Litigation; Reichenbach, Henry M.

Film, cinematographic, **273**; with paper base, 265–66; production of, 256, 277–79, 296

Film, paper, **109**, 109, 265–66

Film, stripping, 84–85, 100–102, **101**, **102**, 108–9

Fireproof Film Company, 295

First National Exhibitor's Circuit, 294–97

Fitts, Edmund, 314

Fitz, Henry, 17, 29

Flammang Camera Company, 168

Flimco, General, 287

Folmer, William E., 301. *See also* Folmer & Schwing Manufacturing Company

Folmer & Schwing Manufacturing Company, 168, 217, 219

Folsom, Marion B., 326, **329**

Forbes, Charles, 224

Fox, William, 276, 292–93, 295–98

Franklin Institute (Philadelphia), 50, 273

French & Company, Benjamin, 239

Friese-Greene, William, 264–65, 267

Frohman, Daniel, 292

G Gaisman, H. J., 316
Galleries, **3**, **14**, 18–19, **19**, **31**
Gallup, Harriet, 155, 185, 301
Gammon, Frank, 268
Garrigues & Magee, 26
Gaudin, Alexis, 26
Gaumont Company, 288, 296–97
Gelatin revolution, 66–95, 274–75, 323, 344–45. *See also* Film, celluloid roll; Film, cinematographic; Plates, gelatin dry
General Aristo Company, 200–202, 211, 226. *See also* Abbott, Charles S.; Eastman, George; Eastman Kodak Company of New York
General Film Company, 287
General Motors Corporation, 328
General Paper Company of Brussels, Belgium, 196–99, 202, 206–8, 243. *See also* General Aristo Company
General Sales Company, 293
Gennert, Gustav, **61**, **71**
Germany, 185, 215, 323, 333; chemical industry in, 186, 244; paper industry in, 195–96
Gevaert (Belgian firm), 278
Gifford, William S., **325**, 325–26
Gilbert, Frederick B., 335
Gillespie, Edward and James, 26
Globe lens, **47**, 47
Goerz American Optical Company, C. P., 209, 214, 218
Goff, Darius L., 136, 137, 139, 143–45, 191
Goodwin, Hannibal: patent litigation involving, 125–27, 129–30, 146–47, 154, 156–58, 248–51, 332, 333; personal background of, **125**, 125–27, 249, 332; role of, in development of celluloid roll film, 125–27, 333. *See also* Anthony and Scovill Company; Goodwin Film & Camera Company
Goodwin Film & Camera Company, 248–57, 331–34. *See also* Anthony and Scovill Company; Goodwin, Hannibal
Gordon, E., 47 n
Gouraud, François, 12–13, 17
Graf (optician), 324
Graflex, Incorporated, 322
Green, B. F., 224
Green, Newton, 314
Greenpoint Optical Company, 139
Griffin, Colonel J. T., 117 n
Griswold, Victor Moreau, 35, **43**, 43–44, 54
Gundlach, Ernest, 209
Gundlach Optical Company, 167, 212, 214
Gurney, Jeremiah, 19

H Hagotype Company, 125
Hahn, Albert G., 90, 223
Halation, **78**, 78–79
Hall, Ogden, 27
Haloid Company, 205, 206
Halverson, Halver, 28
Hammer, Ludwig F., **75**, 75–77
Hammer-Althans Company, 224
Hammer Dry Plate Company: early history of, 75–76, **76**, 165; relations of, with Eastman, 225–26, 230–31; sales and marketing by, 221–22, 225, 227, 228, 229, 231, 233, 251, 254
Harrison, Charles C., 28–30, 46, **47**, 47, 48, 56
Harrison Optical Company. *See* Harrison, Charles C.
Harrow, England: Eastman facilities at, 132, **151**
Harvard Dry Plate Company, **76**, 76, 222
Haste, James: at Kodak Park, 235, **242**, 242, 325, 333; as part of management, 325, 328, **329**
Havens, James S., 326, 328, **329**
Hazel, Judge John R., 320, 332
Hedden, Horace M., 52, 54, 55
Hendricks, Gordon, 267
Hetherington & Hibben Company, 167
Hill, Charles A., 55
Hill, Levi L., 33
Hodkinson, William W., 293–94
Holding company, 175–77, 225–26, 235
Holmes, Booth & Hayden, 28 n, 47–48
Holmes, Oliver Wendell, **58**, 58
Holmes, Samuel, 20–21, 46, 55–56
Holst, Lodewyk J. R., 218
Hoover and Getz. *See* Buffalo Dry Plate and Argentic Paper Company
Hopkins, C. E. (company), 91, 92
Horgan, Robey & Company, 239
Horner, William George, **261**, 261
Houston, David H., **108**, 108, 119, 141, 158, 190, 191
Hovey, Charles F., 168
Hovey, Douglas, 60, 88, 168, 199
Hub Dry Plate Company, 74 n
Hubbell, Walter S., 150, 319, 325, 328, 330
Hudson, J. Harold, 314
Hurter, Ferdinand, 312, 335
Hurter-Driffield, 335
Hutchison, Charles Force, 223, 302
Huygens, Christian, 259
Hyatt, Isaiah, 123–24
Hyatt, John, 123–24
Hypo fixing, 334. *See also* Sodium thiosulfate

I I. G. Farbenindustrie, A.G., 337
Ilford, Limited, 240–41
IMP, 286–88
Imperial Camera Company, 212–13
Imperial Dry Plate Company, 79, 240–41
Independent Moving Pictures Company of America, 286–88
Inglis, James, 224

Innovation, technical: attitudes during daguerreotype and collodion periods toward, 33, 61; attitudes during gelatin period toward, 87, 145, 159, 247–48, 335–36; continuous, at Eastman, 184–85, 342–43; Eastman institutionalization of, 133, 149–50, 179–87; Eastman's philosophy and strategy of, 148, 151, 235, 329–30, 344–45; related to scale of enterprise, 345–46; as strategy of entry, 6, 167–69, 203–4. *See also* Research and development
Integration, backward vertical, 172 n; at Ansco, 334; of cinematography, 288, 293–95; at Eastman, 207, 242–45, 322–24
Integration, forward vertical, 172 n; at Anthony, 21–23, 25; at Anthony and Scovill, 246–57; at Blair, 137; at Eastman, 109–10, 236–40; at Motion Picture Patents, 287; at Scovill, 20–21, 25; at Warner Brothers, 297–98
Integration, horizontal, 66, 172 n, 188–235; of cameras, 56, 89, 91–92, 138–39, 143–44, 158–59, 163, 165, 204, 208–19, 246–57; of dry plates, 225–31, 233, 338–39; of motion pictures, 287, 293–95; of papers, 94, 197, 338–39; of tintypes, 55
Irving-Clay camera, **209**

J Janssen, Pierre Jules C., 262
Japanning of tintypes, 43
Jenkins, Charles F., 273–74
Jennings, W. N., 127
Jobbers, 88, 92–93, 161–62, 164, 192, 220–21, 246. *See also* Supply houses
Jones, Lewis Bunnell, 150, 328, **329**
Jones, Loyd A., 311, 313, 315
Journals, photographic: influence of, 72–73, 79, 100, 108, 124, 150, 223–24, 264, 286–87; location of, 43, 56, 68, 77, 86, 87

K Kalem Company, 277, 284, 292–93
Kamaret camera, **141**, 141–42
Karsak paper, 93–94
Keith-Albee-Orpheum, 298
Kelvin, Lady, **187**, 187
Kelvin, Lord, **187**, 187
Kent, John H., **106**, 106
Keystone Dry Plate Company, 70–73, 338. *See also* Carbutt, John
Kilborn (paper), 205–6
Kilburn Brothers, 60
Kinetoscope; 267–68, **268–69**, 272, 274–75. *See also* Edison, Thomas A.
Kinetoscope Company, 268
Kinora, 273, 275
Kirkland Lithium Paper Company, 90, 193, 200, 201. *See also* General Aristo Company
Kleine, George, 277, 285
Kloro paper, 93–94, 192–93
KMCD, 269–72
Kodacolor, 318
Kodak camera. *See* Cameras, Eastman Kodak roll film; Eastman, George; Eastman Company; Eastman Kodak Company of New York
Kodak Limited (England), 199–200, 226, 235, 240–42
Kodak Park. *See* Eastman Company; Eastman Kodak Company of New York
Koopman, Elias, 269–72
Krystalline Company, 124
Kuhn, Henry, 295
Kuhn Crystallograph Company, 87

L Laboratories, research. *See* Research and development
Laemmle, Carl, 276, 286–88, 292–93
Lamoutte, Alexander C., 162, 251–52
Langenheim brothers (Philadelphia firm), 50
Lasky, Jesse L., 292–93
Lauste, Eugene, 268–69, 272
Lawrence, Martin, 19
Lasky Feature Play, 293–94
Latham family, 269, 272
Leavitt, Fred L., 227
Lenses, 17–18, **31**, 57–58, 323–24, 344; Globe, **47**, 47; Petzval, 28, 31; Steinheil, **58**
LePrince, Augustin, 265
Lerebours, N. P., 17
Lewis, Henry J., 47 n
Lewis, William and William H. (company), 17–18, 28, 59, 62, 164
Libby, Leonard F., 224
Liesegang (Düsseldorf firm), 87
Litigation, antitrust: Eastman Kodak, 318–22, 330, 339; Motion Picture Patents, 292
Litigation, patent: Anthony, 85; Blair, 136; Edison Mfg., 283–84; Independent Motion Pictures, 287; Motion Picture Patents, 291–92; Scovill, 85, 136
Litigation, patent, involving Eastman companies: Anthony, 85, 255; Biograph, 277; Blair, 141–42, 158; Boston Camera Manufacturing, 158–59; Buffalo Dry Plate and Argentic Paper, 89–90, 158; Celluloid, 153, 156; Edison, 277; Goodwin Film & Camera, 129–30, 156, 248–49, 252–57, 331–34; Vitagraph, 277
Littlefield, Parson & Company, 28 n, 46
Loew, Marcus, 276, 295, 296, 298

Lomb, Henry, **187**, 214
London Stereoscopic Company, 50
Lovejoy, Frank: as chemist and technical supervisor, 98, 180, **182**, 182–83, 185, 235, 301, 333; as member of management, 235, 314, 325–26, 328, **329**, 330
Lovell, Charles O., 224
Lovell Dry Plate Company, 222
Lubin Manufacturing Company, 277, 284
Lucidograph, 136
Lumière brothers (Auguste and Louis), 168, 273, 303
Lumière, Société, 334; as producer of film, 229, **273**, 273, 275, 278, 279, 287–89; Eastman negotiations with, 240–42, 290

M McDaniel, A. S., 311
McReynolds, James Clark, 320
Maddox, Richard L., 68
Magazine camera, **167**, 167
Magee & Company, J. F., 59
Magic lantern, 259–60, **260**
Mallinckrodt Chemical Works, 59, 168
Mallory, George, 20, 33, 46
Manhatten Optical Company, 167–68, 214
Marey, E. J., 262, 263, **264**, 264, 265, 267
Marion & Company, 60 n
Market, amateur: 76, 116–17, 149–50, 166, 191, 346; Ansco, 208; Anthony, 161, 164; Blair, 140; Defender, 339; Eastman, 67–68, 84–85, 134–40, 215, 236, 344; Scovill & Adams, 161–63; Walker, 98. *See also* Marketing conceptions and strategies
Market, foreign, 178–79, 203, 240–41
Market, mass, 174, 177–79, 342
Market, serious amateur, 168
Marketing conceptions and strategies (by company): American Aristotype, 87; Ansco, 331–32; Anthony, 87, 161–62, 165–66, 247–48; Cramer Dry Plate, 232; Defender, 338–39; Eastman, 84, 106–110, 112–13, 131, 215–16, 231, 232, 236–38; Scovill & Adams, 161–62, 247–48; Scovill Manufacturing, 56. *See also* Market, amateur
Marketing conceptions and strategies (by sector): cinematography, 292–93, 294; collodion process, 62–63; daguerreotypy, 32–33; gelatin dry plate, 66–68, 76–77, 225; sensitized paper, 88, 90–95, 192; tintype, 43, 44–45. *See also* Competition
Martin, Adolphe Alexandre, 42
Marvin, Harry, 269–72
Massachusetts Institute of Technology, Eastman employees from: 149, 180, 245; de Lancey, 180, 235; Gallup, 155; Haste, 235, 242, 325; Lovejoy, 182, 235; Packard, 303
Masury & Silsby, 19
Mawson and Swan, 83
Meade, Charles, **19**, 19
Meade, Henry, **19**, 19
Mechanized production, 72, 86, 96–97, 103–5, 108, 109, 342, 344
Mees, C. E. Kenneth, **306**, 307–18, 328, **329**, 330
Melhuish, Arthur, 99 n
Melies Manufacturing Company, George, 277, 284
Miethe, Adolf, 79
Milburn, Gustave D., 152–53
Miles, Harry J., 276
Mind-set, business-technological, 4–6, 264–65, 274–75, 298–99, 340–43, 345
Monarch Paper Company, 251–52
Monroe, George, 77, 153, 224
Monroe Camera Company, 209–11
Monroe Dry Plate Company, 74 n
Morgan (John Dean & Company), 52, 55
Morosco Photoplay Company, Oliver, 293
Morse, Samuel F. B., 12, 13–14, 21
Motion Picture Distributing and Sales Company, 287
Motion Picture Patents Company, 284–87, 291–93
Motion pictures: home, 316; sound, 296–98. *See also* Cinematography
Multiscope & Film Company, 209, 216, 218
Mutograph camera, **270**
Mutopticon, 271
Mutoscope, 269–72, **270**, 274–75
Muybridge, Eadweard James, 262–64, 267

N National Bureau of Standards, 311
Neff, Peter, Jr., **42**, 42–43, 44, 54
Nepera Chemical Company, 90, 193–95, **194**, 199–202, 223. *See also* Baekeland, Leo Hendrick
Nestor Film Company, 287, 288
Neue Photographische Gesellschaft, 303
New Jersey Aristotype Company, 192–93, 197, 198, 200, 201. *See also* General Aristo Company
New Jersey incorporation laws, 175
New York Aristotype Company, 87, 89, 92, 93
New York Dry Plate Company, **222**, 222
New York Motion Picture Company, 288
New York State Daguerreian Association, 20

Nickelodeon, 276
Niepce, Joseph Nicéphore, 10
Norden, Hermann, 70–71, 74, 78
North American Lumière, 278. *See also* Lumière, Société
Northern Securities Company, 175–76
Nutting, Dr. Perley G., 311, 312, 313, 315

O Oberlin College, 314
Optics. *See* Cameras; Lenses

P Packard, Emerson, 303
Paget, Leonard, 146
Palmer & Croughton, 193, 198, 199
Panoptikon, 272
Paper, photographic: coating techniques used in producing, **83**, **104**, 196 n; listing of manufacturers of, 60, 92, 193, **205**
Paper, photographic (by brand name): Angelo, 204, 205; Argo, 208; Artura, 204–8, 339; Azo, 194; Cyko, 204–6, 208, 256, 331; Dekko, 194–95; Kloro, 93–94, 192–93, 194; Permanent Bromide, 109–10; Solio, 89, 93–94, 192–94; Velour Black, 339; Velox, 193–95
Paper, photographic (by type), **88**; albumen, 48–50; aristotype, 192–93; collodion, 38, 86–87; DOP, 81–82, 86, 88; gaslight, 193–94; gelatin bromide, 81–82, 109–10; platinum, 60, 204; POP, 87, 193–94, 204–8
Paper, raw (photographic), 195–96, 198–99, 202–3, 206–7, 243
Paramount Pictures, 293–96, 298
Parkes, Alexander, 122
Passavant, S. Carl, 90, 149, 152–53, 185, 301
Patentees: Armat, 273–74, 282; Barker, 164; Blair, 135, 181–82, 190; Bornmann, 331; Bostwick, 182; Brownell, 142; Church, 186; Cutting, 32; Daguerre, 32; Eastman, 72, 73, 83–84, 102, 103, 111, 115, **118**, 128–30; Gaisman, 316–17; Good, 141; Goodwin, 125–27, 146–47, 248–51, 334; Gray, 141; Griswold, 43–45; Harrison, 182; Hedden, 54, 55; Houston, 119, 137, 141, 190; Hyatt, 123; Jenkins, 273–74, 282; Langenheim, 32; Lathan, 287; Lewis, 164; Reichenbach, 146–47, 333; Robertson, 212; Roche, 85; Smith, 42–45; Stammers, 141; Stanley, **77**, 77; Talbot, 32; Turner, 142, 252–53; Walker, 83–84, 98, 137; Waterman, 181–82, 190; Whipple, 32; Whitney, 141. *See also* Litigation, patent
Patents: number of photographic, **62**; role of, 6, 65, 159, 162, 169, 342; in strategies of Blair, 136, 159; in strategies of Eastman, 90–91, 97–98, 105–8, 118–20, 131, 143, 150–51, 158–59, 184–85, 188–91, 218, 316, 329–30, 344, 345. *See also* Litigation, patent
Patents (by area): camera, 98, 115, **118**, 119 n, 135, 137, 139, 141, 164–65, 190, 209, 247–48, 316–17, 331; case, 46; cine, 267, 273–75, 282, 284–85, 287; collodion, 32, 43, 61–63; daguerreotype, 32; film, 102–5, 125–30, 146–47, 181–82, 190, 212, 248–51, 267, 333; film cartridge, 142, 252–53; paper, 83–86, 192; plate, 72, 73, 77; plate holder, 103; roll holder, 103; tintype, 42–45, 54, 55. *See also* Litigation, patent
Pathé, Charles, 283–84, 288
Pathé Exchange, Incorporated, 344
Pathé Frères, 277, 279, 285, 289–90, 295, 334
Peck and Company, Samuel, 28, 46
Persistence of vision, 259–61
Petzval, Joseph M., 31
Phantascope, 260–61, **261**, 273
Phantasmascope. *See* Phantascope
Phenix Plate Company, 52–55
Philipp, M. B., 253, 330
Photographers: amateur, 112–14, 342; field, **40**; professional, 112, 155. *See also* Market
Photographic Merchant Board of Trade, 92–93, 192, 198, 220–21, 248
Photo Materials Company, 90, 93–95, 153, 167–68, 192–95, **195**, 198, 199
Pittman, W. Davies, 226
Plateau, Joseph Antoine, 260–61, **261**
Plates, gelatin dry, 96, 164, 225–31; cine use of, 263, 264; introduction of, 69–80, **73**, **77**, **80**; spectral sensitivity of, 77–78, 315, 338. *See also* Integration, horizontal
Platinotype Company, 204
Plumbe, John, 18
Potter, Rowland S., 232–33, 338
Powers & Weightman, 59, 168
Powers Company, 288
Powrie, J. H., 303
Pratt, Frank, 77, 224
Price, Waterhouse & Company, 326, 328
Price-cutting, 85, 90–95, 163, 191, 192, 240. *See also* Competition
Price maintenance, 56, 91–93, 175, 176, 192, 198, 213, 220–21, 239–40, 284–85. *See also* Associations; Competition; Litigation, antitrust
Pringle, Andrew, 117 n
Projector (cine), 258–61, **271–72**, 273–74, **274**
Punnett, Milton B., 80, 223, 226–27, **227**, 314

R Radio Corporation of America, 298
Radio-Keith-Orpheum, 298
Raff, Norman C., 268
Ramsay, William, 306
Raw Film & Supply Company, 334
Ray Camera Company, 209, 211, 212
Record Dry Plate Company, 76, 227
Rector, Enoch J., 272
Reese & Company, 18–19
Reflex Camera Company, 209, 218–19
Reich, John C., 214
Reichenbach, Henry M., 194; employed by Eastman, 80, 90, 98, 110, 128–33, 148–49, 152–53, 180, 185, 301, 307, 330; engaged on celluloid film, 128–33, 146–49, 248–49, 333
Reid, David E., **289**, 289, 302, 322
Research and development, 343, 346; at Ansco, 335–36, **336**; at Eastman, 186, 233–34, 300–18, **304**, 328, 343; key personnel in, 80, 90, 223; laboratories in, 6–7, 305, 344. *See also* Eastman, George; Goodwin, Hannibal; Innovation, technical; Reichenbach, Henry M.
Rex Motion Picture Manufacturing Company, 287–88
Réynaud, Emile, 265
Rintoul, William, 310, 311
Rives paper, 101
RKO, 298
Roach, John, 17
Robertson, John A., 212, 216, 301
Rochester, N.Y., as photographic center, 71, 73–74, 215, 219
Rochester Camera and Supply Company, 168, 211
Rochester Optical and Camera Company, 166, 211–13, 216
Rockford Dry Plate Company, 224, 227
Roll film: cine use of, 259, 274–75, 288; production of, by Anthony, 250–52, 256; sale of, by Defender, 259, 274–75, 288. *See also* Film, celluloid roll
Roll film system: effect of, 67–68, 163, 177–78, 212, 345; introduction of, 96–122, 153–56, 342, 344; modifications of, 134–59, 164–65, 188–91. *See also* Cameras, Eastman Kodak roll film; Film, celluloid
Roll holder, **100**, 100–103, 125, 137–38, 163
Rosengarten & Denis, 16
Rosengarten & Sons, 59, 168
Ross & Company (English firm), 57, 167–68
Rudge, John, 264

S Sage, Edwin O., 106
St. Louis, Missouri, as center of dry plate industry, 70–71, 73–74, 221–22, 231
Salzgeber (Hammer Dry Plate Company), 226
Sandell, Thomas, 79, 223
Sarony and Johnson, 82–83
Scale brand daguerreotype plates, **26**, 26
Schering (German firm), 59, 168
Schnitzer, I., 47
Schönbein, Christian F., 37
Schroeder, Ludwig Hugo, 167–68, 214
Scovill, James Mitchell L., 13, **14**
Scovill, William H., 13, **14**
Scovill & Adams Company: as manufacturer, 137, 167, 168, **209**, 209, 212, 331; organization of, **162**, 162, 204; relations of, with Anthony, 204, 213, 248, 250–51; relations of, with Eastman, 137, 166; strategies and policies of, 80, 160–64, 246–48, 162–64
Scovill Company, J. M. L. and W. H., 13–16, 20–22, 25. *See also* Scovill Manufacturing Company
Scovill Manufacturing Company: attitudes and strategies of, 32–33, 55–56, 61–63, 69; as jobber, 30, 54, 69, 71, 135; as manufacturer, 26–28, 46–47, 53–54; organization of, 21, 25, 46; relations of, with Anthony, 25, 46, 56, 85, 341–42; relations of, with Blair, 135, 136; relations of, with Eastman, 107, 166. *See also* Scovill & Adams Company
Sears & Roebuck Company. *See* Conley Camera Company
Seed, Miles A., **75**, 75
Seed Dry Plate Company, M. A., 75, 79, **231**; diversification at, 80, 212; emulsion of, 77, 79, 153, 154, 221–24, 226–27; production at, 80, 124; relations of, with Eastman, 225–26, 229–30, 322, 339; sales and marketing at, 74 n, 76, **225**, 225, 227, 228
Seidel, Ludwig P., 57
Selden, George, 97, 108, 130, 150, 330
Selig Polyscope Company, 283, 284
Sellner, Albert, 80, 223
Seneca Camera Company, 212, 331
Sensitometry, 312–15
Sheppard, Samuel E., 306–7, **311**, 311, 314–16, 323, 335–36
Sherman Act, 175, 321
Shutters, camera, 160–61, 214–15, 244
Simpson, George Wharton, 86
Smith, Hamilton L., 42–43
Smith, H. M., 240
Smith, James O., 43 n
Smith, J. H., 154–55, 223
Société d'Encouragement à l'Industrie Nationale, 273
Societies, photographic, 32, 71, 115–16
Sodium thiosulfate, 10, 26, 38, 39
Solio paper, 89, 93–94

Sound motion pictures, **297**, 297–98
Southworth & Hawes, 19
Späth, Carl, 303–4
Spencer, Joseph, 99 n
Spill, Daniel, 122
Stampfer, Simon Ritter, 260, 261
Standard Dry Plate Company, 76, 226–28, 231, 322, 339
Standard Oil Company of New Jersey, 173, 175, 176, 319
Stanford, Leland, 262
Stanley, Francis Edgar, 75, **77**, 230. *See also* Stanley Dry Plate Company
Stanley, Freelan Oscar, 75, **77**, 230. *See also* Stanley Dry Plate Company
Stanley Dry Plate Company, **75**, 75–77, 154, 192, 220–22, 225–30, 322, 339
Star daguerreotype plate, 26
Star system in motion pictures, 292–94
Steinbach of Malmedy, 49
Steinheil & Son, 57, **58**
Stereoscopic viewer and cards, 41, 49–52, **50–51**, **58**, 58, 60
Stillman, W. J., 113
Stock & Company, John, 47 n
Stolze, Franz, 79
Stroboscope, 261, 263
Strong, Henry A., 71, 116, 325; business philosophy and contributions of, 96, 98, 118, 234, 325, 330, 344; as officer in Eastman enterprises, 71, 105–6, 117 n
Structure, industrial, 4, 10, 26, 80, 220–33, 345
Stuber, William G., 154–55, **155**, 180, 223, **328**, 328, 330
Student Camera Company, 167
Sunart, 127
Sun plate, 53
Supply houses: collodion period, 45–52; daguerreotype period, 13–16, 20–25, 27, 32–33; gelatin period, 69, 139, 236, 239, 240, 338, 339. *See also* Jobbers
Sweet & Wallach Company, 239

T Taggers iron, 54 n
Talbot, William Henry Fox, 63
Taylor, Edward G., 46 n
Tennessee Eastman Company, 324
Tents, dark, **40**
Terms of sale, 237–40, 251, 256. *See also* Competition; Marketing conceptions and strategies (by company)
Thanhouser Company, 288
Thatcher & Whittemore, 244
Thomas, William, 187
Thompson, Warren, 22
Thomson, William. *See* Kelvin, Lord
Thornton, J. E., 212
Tintype, 40–45, **41**, 52–55, **53**, 344
Tompkins, Kilbourne, 115
Tourograph system, 134–36, **136**
Trademarks, 162. *See also names of specific companies*
Trade secrets, 162, 221, 344; methods of retaining, 158–59, 186, 322; modes of diffusing, 152–53, 226–27, 229, 230, 233
Trapp & Munch (German firm), 60 n
Trivelli, A. P. M., 314
Turner, Henry H., 214
Turner, Samuel N., 137, 254; as designer of Bulls-Eye camera, 145, 158–59, 166–67; as designer of daylight loading cartridge, 142–43, **143**, 150, 158–59, 189–90

U Uchatius, Franz, 261
United States Aristotype Company, 90, 192–93, 205
United States Federal Courts or Supreme Court. *See* Litigation
Universal Film Manufacturing Company, 287, 291, 293, 296

V Van Voorhis, Arthur, 224
Velox paper, 193–95
Vereinigte Fabriken Photographischer Papier, 203, 229, 241
Vereinigte Kunstseide Fabriken, 303
Victoria Manufacturing Company, 139
Villard, Henry, 268
Vitagraph camera, 277, 284
Vitagraph Company of America, 277
Vitascope, 273–74, **274**
Vogel, Hermann Wilhelm, 223
Voigtländer & Son, 17, 28, 31, 58

W Walker, William H.: as administrator and officer, 106, 107, 110, 117 n, 118, 127, 132–33, 190; relations of, with Eastman, 110, 118, 150–51; as technical innovator, **83**, 83–86, 96–105, **99**, **103**, 111, 133, 224, 307, 330
Walker-Reid-Inglis, 74
Warner Brothers, 296–99
Warnerke, Leon, 99–100, **100**, 102–3, 108, 264
Warnica & Company, 216
Watson (Eastman Kodak), 328
Wentzel, Fritz, 335–37
Western Camera Manufacturing Company, 209, 211
Western Collodion Paper Company, 87, 94, 199. *See also* Western Camera Manufacturing Company

Westinghouse Electric and Manufacturing Company, 282, 311
Wheatstone, Charles, 50, 261
Whipple, John, 19
Whitney, Alice K., 150, 326
Whitney, William R., 310, 311
Wilcox, Vincent M., 58, **59**, 164, 165, 182–83
Willis & Clements, 60, 204, 205, 206
Wilmot, Frank, 338, 339
Wilsey, Rex B., 313
Wolcott, Alexander, **31**, 31
Wollensak, Andrew, 214–15, 244
Woodward, David A., 47 n
World War I, 295, 313, 321, 323, 324, 338–39
Wratten & Wainwright, 306–8, 311
Wright, Nelson, 46–47
Wuestner, Edward, 79, 223
Wuestner Dry Plate Company, **76**, 76, 222

Y Yankee Films Company. *See* Independent Moving Pictures Company of America
Yawman & Erbe Company, 111

Z Zoetrope, **261**, 261, 263, 265
Zukor, Adolph, 276, 292, 294–95, 299
Zukor's Famous Players, 293–94